New Technologies
for Supercolliders

ETTORE MAJORANA
INTERNATIONAL SCIENCE SERIES
Series Editor:
Antonino Zichichi
European Physical Society
Geneva, Switzerland

(PHYSICAL SCIENCES)

Recent volumes in the series:

A Continuation Order Plan is available for this series. A continuation order will bring delivery of each new volume immediately upon publication. Volumes are billed only upon actual shipment. For further information please contact the publisher.

New Technologies for Supercolliders

Edited by

Luisa Cifarelli

University of Naples
Naples, Italy

and

Thomas Ypsilantis

Collège de France
Paris, France

Plenum Press • New York and London

Library of Congress Cataloging-in-Publication Data

New technologies for supercolliders / edited by Luisa Cifarelli and
  Thomas Ypsilantis.
       p.   cm. -- (Ettore Majorana international science series.
  Physical sciences ; v. 57)
    Includes bibliographical references and index.,
    ISBN 978-1-4684-1362-5
    1. Supercolliders--Congresses.   I. Cifarelli, L. (Luisa)
  II. Ypsilantis, Thomas.   III. Series.
  QC787.P7N49 1992
  539.7'9--dc20
                                                              91-39834
                                                                   CIP

Proceedings of the Twelfth Workshop of the INFN Eloisatron Project
on New Technologies for Supercolliders, held September 15–20, 1990,
in Erice, Sicily, Italy

A Division of Plenum Publishing Corporation
233 Spring Street, New York, N.Y. 10013

ISBN-13: 978-1-4684-1362-5   e-ISBN-13: 978-1-4684-1360-1
DOI: 10.1007/978-1-4684-1360-1

PREFACE

The present volume is based on the proceedings of the 12th Workshop of the INFN ELOISATRON Project, held at the "Ettore Majorana" Centre for Scientific Culture (EMCSC), Erice (Trapani), Sicily, Italy, in the period September 15-20, 1990.

The proceedings deal with the presentation of "New Technologies for Supercolliders". Three new energy frontiers (16, 40 and 200 TeV) are now opened up for the future of Subnuclear Physics. Basic problems above the Fermi-energy are crowding up: but no one knows the energy levels needed for their solution. This is why the technology for experiments with the new generation of Supercolliders needs to be pursued having in mind the problems which are of common interest in the three energy frontiers. The primary purpose of the Workshop was to contribute towards the highest energy limit in the search for new instruments and new technologies. Furthermore, the present status and performances of various detector technologies were reviewed. The possible options for a powerful apparatus whose goal would be the discovery of the top, Higgs and SUSY particles in a very high energy, high rate environment, were finally analysed.

The Workshop was sponsored by the Italian National Institute for Nuclear Physics (INFN), the Italian Ministry of Education, the Italian Ministry of Scientific and Technological Research and the Sicilian Regional Government. We are thankful to the staff of EMCSC for their efficient and warm support.

Luisa Cifarelli, University of Naples, INFN-Bologna, Italy,
and CERN/LAA, Geneva, Switzerland
Thomas Ypsilantis, Collège de France, Paris,
and CERN/LAA, Geneva, Switzerland
Editors

Geneva, May 25, 1991

CONTENTS

INTRODUCTION

CALORIMETRY

SUBNUCLEAR MULTICHANNEL INTEGRATED DETECTOR TECHNOLOGIES

LARGE AREA DEVICES

WHY 200 TeV
Opening Lecture

A. Zichichi

CERN, Geneva

Switzerland

This introductory note is intended to illustrate the motivations for an increased effort, both theoretical and experimental, towards the design of an hadronic collider of the highest possible energy and luminosity: ELOISATRON[1].

At present, four big colliders, whose main characteristics are given in Table 1, are either operating, or under construction, or under study: LEP, HERA, LHC, and SSC.

Table 1. Present colliders, operating, under construction, or under study.

Collider	Colliding particles	Centre-of-mass energy	Status
LEP, CERN, Geneva, CH	e^+e^-	$100 \div 200\,\mathrm{GeV}$	operating
HERA, DESY, Hamburg, FRG	ep	$314\,\mathrm{GeV}$	construction
LHC, CERN, Geneva, CH	pp	$16\,\mathrm{TeV}$	study
SSC, Ellis County, Texas, USA	pp	$40\,\mathrm{TeV}$	study

It is of interest to think about what can be the output, in terms of physics discoveries, of these four colliders.

Two scenarios are possible.

New Technologies for Supercolliders, Edited by L. Cifarelli and
T. Ypsilantis, Plenum Press, New York, 1991

1. PESSIMISTIC

LEP will be a great check of the "Standard Model".

HERA will allow superb measurements of the structure functions, with the "leading effect" correctly accounted for (i.e., measured).

LHC will produce a formidable proof that the Standard Model is OK.

SSC will confirm, with a superproof, that the Standard Model is OK.

These four colliders represent an enormous investment, both in money — about 20 billion US dollars in a period of 15 years (1985 ÷ 2000) — and in personnel. Nearly all the High Energy Physics community will eventually be engaged with one of them.

Should it happen, the result of this scenario will be that **Subnuclear Physics will not survive**. Therefore, we need, now, to start thinking about the **highest possible energy** and the **outmost luminosity** reachable.

2. OPTIMISTIC

Great discoveries will come from LEP, HERA, LHC and SSC. In this case, the energy jump towards the highest possible energy would be even more compulsory. But the implementation of a $(100 + 100)$ TeV supercollider needs at least 10 years. Therefore, the conclusion is the same as for scenario 1.

WHAT NEEDS TO BE DONE

Both scenarios point to the need to start now thinking about a collider with the

$$\text{highest possible energy} \rightarrow 100 + 100 \text{ TeV}$$

and the

$$\text{outmost luminosity} \rightarrow 10^{34} \div 10^{35} \div 10^{36} cm^{-2}s^{-1} \ .$$

This collider can not be an e^+e^- machine, because, even if the largest e^+e^- collider could be built in the near future $((1+1)$ TeV), this would correspond, in terms of the costituent centre-of-mass energy, to a pp collider of $(6+6)$ TeV only.

Let me synthetically report what is being done to implement the crucial steps for Subnuclear Physics towards the above targets, in Energy and Luminosity.

The conceptual design has been done[1]. As quoted above, the energy is $(100 + 100)$ TeV, and the Luminosity is, at present, $\sim 10^{-33} cm^{-2}s^{-1}$ (see Table 2). Theoretical studies are going on to work out the highest level of luminosity.

With a circumference of 300 km, the bending field of the collider will have to be about 10 T. The total bending length will, in that case, be 209.6 km/ring. The rest of the circumference will be used for focusing quadrupoles, correcting magnets and other auxiliary magnets, acceleration cavities, beam instrumentation, injection and extraction systems, and, most important of all, the insertion of the interaction

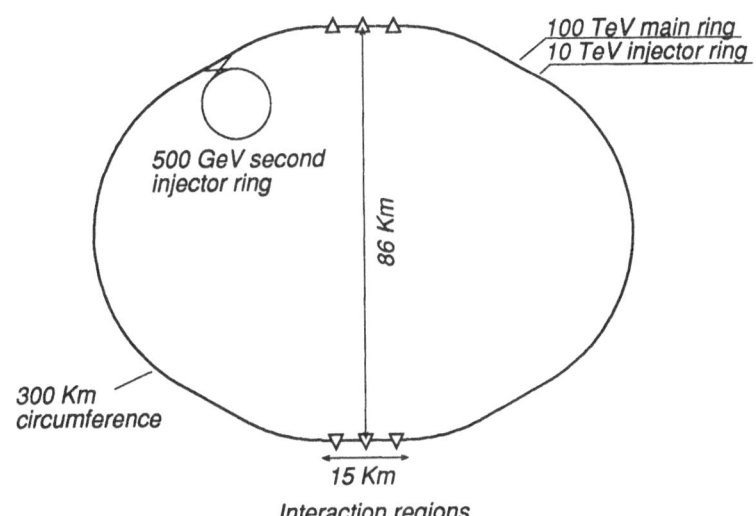

Fig. 1. Schematic layout of the *ELOISATRON* accelerator system.

Table 2. Basic parameters.

Energy per beam	100 TeV
Number of bunches	39600 per beam
β-value at interaction point	$1.25 \div 0.6$ m
Normalized emittance	$0.75\pi \times 10^{-6}$ m
r.m.s. beam radius at interaction point	$1.25 \div 0.9 \times 10^{-6}$ m
Circulating current	16.43 mA
Particles per bunch	2.56×10^9
Beam-beam tune shift per crossing	
(with 6 active crossings)	1.67×10^{-3}
Bunch spacing	25×10^{-9} s
Stored beam energy	1.623×10^9 J
Luminosity $(\text{cm}^{-2}\text{s}^{-1})$	$0.9 \div 1.8 \times 10^{33}$
Energy loss per turn due to synchrotron radiation	23.34 MeV
Radiated power (per beam)	385 KW
Power per unit length of one beam	1.89 W/m
Transverse em. damping time	1.2 h

Table 3. Lattice parameters.

Length of period	200 m
Phase advance per period	$\pi/3$
Betatron wavelength	1200 m
Bending angle per normal period	4.7 mrad
Number of quads per period	2
Effective length of each quad	13.6 m
N° of quadrupoles (without insertion)	2664
Maximum dipole field	10 Tesla
Bending radius	33356 m
Number of dipoles per normal period	12
Effective dipole length	13.1 m
N° of dipoles	15984

Table 4. Steps towards ELOISATRON.

◇ 1979 - CS Project (Superconducting Cyclotron) resumed.

◇ 1981 - Start of the CS construction (Ansaldo - LMI - Zanon). The HERA Project is presented to the INFN Council.

◇ 1982 - First Meeting of the HERA designers with Ansaldo - LMI - Zanon.

◇ 1985 - The construction in Italy of the superconducting magnets (prototypes) for HERA begins.

◇ 1986 - R&D for Detectors - the LAA Project is approved and financed.

◇ 1987 - R&D for LSCM (Long SuperConducting Magnet) prototypes – 10% ELOISATRON Model – starts at Ansaldo.

◇ 1989 - The construction of the HERA magnets is completed.

◇ 1990 - Magnet construction for the 10% ELOISATRON Model is expected to start.

◇ ☐ - The Full Scale ELOISATRON PROJECT:

  • 1st Conceptual Design.
  • 2nd R&D in fundamental technologies.
  • 3rd Construction of the Superconducting Long Magnets.

Table 5. The 200 TeVClub.

S. Brodsky		G.'t Hooft	
M. Chanowitz		V. Koze	
L. Cifarelli	$+12+N_C$	T.D. Lee	$+2+N_L$
Y. Dotshitzer	$+4+N_D$	G. Marchesini	
M. Duff		R. Peccei	$+2+N_P$
J. Ellis	$+3+N_E$	A. Ringwald	
G. Farrar		V. Rubakov	$+N_R$
J. Gunion		M. Voloshin	
L. Hall		C. Williams	
P. Higgs		A. Zichichi	$+N_Z$

Total: $43+N_{C,D,E,L,P,R,Z}$

regions. The main rings will be fed from a cascade of synchrotrons (most likely three in succession) which in turn will be fed from a linear accelerator. This is shown schematically in Fig. 1.

For other details see Tables 2 and 3.

The project has started in 1979, and its basic steps are shown in Table 4. So far the time schedule has been followed as expected.

At present we are fully engaged in the R&D (LAA) project for new Detectors to be used at the MultiTeVSupercollider such as the 10% ELOISATRON model. At the same time an increasing group of physicists (named "the 200 TeV Club" — see Table 5) is studying some selected topics of the 200 TeV physics.

The next step is to start the construction of the 10% model: having in mind the highest energy and luminosity as the real targets of the R&D. The conceptual design of the 200 TeV collider having been accomplished, intensive R&D in fundamental technologies is going to be an integral part of the next step. This is where we are.

A closing remark. The ELOISATRON Project[2] has been presented in 33 different Meetings and Conferences. In particular:

- VERSAILLES Summit Meeting (France)   June 1982

- Washington, DC (USA)   3 October 1983
  National Science Foundation
  Working Group Meeting on High Energy Physics

- Livermore, California (USA)   9 April 1985
  Lawrence Livermore National Laboratory

- Boulder, Colorado (USA)   11 April 1985
  University of Colorado - Conference on World Affairs

- Geneva (Switzerland)   11 November 1985
  CERN - Senior Staff Consultative Committee

- STRASBOURG (France)   20 November 1985
  Council of Europe - Parliamentary Assembly
  Committee on Science and Technology

- Moscow (USSR)   29 January 1986
  Presidium of the USSR Academy of Sciences   15 January 1987

- Beijing (China)   18 October 1986
  Academia Sinica and University of Beijing

- Rome (Italy)   10 June 1988
  European Particle Accelerator Conference

- Rome (Italy)   19 December 1988
  National Conference on
  Scientific and Technological Research

- Lecce (Italy)   17 March 1990
  INFN Symposium

References

1) A. Zichichi, "The ELOISATRON Project", August 1989.
   A. Zichichi, "The ELOISATRON Project: Eurasiatic Long Intersecting Storage Accelerator", in *New Aspects of High-Energy Proton-Proton Collisions*, Plenum Press, 1988 (ed. A. Ali), page 1.
2) The authors are 570 physicists and engineers from the world over (see Ref. 1).

A TOTALLY ACTIVE LIQUID XENON OR KRYPTON ELECTROMAGNETIC CALORIMETER FOR HIGH LUMINOSITY HADRON COLLIDERS *

J. Seguinot and T. Ypsilantis.[a]
M. Bosteels, A. Gougas, G. Passardi and J. Tischhauser.[b]
Y. Giomataris.[c]

[a] Collège de France, Paris, France
[b] CERN, Geneva, Switzerland
[c] World Laboratory, Lausanne, Switzerland

Abstract

A totally active liquid Xenon or Krypton electromagnetic calorimeter is proposed for a dedicated Higgs search experiment at LHC, SSC or ELOISATRON. Detection of fast scintillation permits excellent energy resolution ($\sigma_E \leq 1\% \sqrt{E}$) while detection of ionization (by drift) gives precise determination of the direction (≈ 1 mr) and vertex origin (≈ 1 mm) of high energy ($E \geq 25$ GeV) photons or electrons.

Slow (≈ 750 ns) drift at high luminosity ($2 \cdot 10^{34}/cm^2 s$) leads to a pileup background of transverse energy (per tower) $E_t \leq 1$ GeV at polar angles $\theta \geq 20^0$.

The liquid Xenon scintillation is found to be fast ($\tau \leq 20$ ns) and intense ($3 \cdot 10^4$ photons/MeV) and the liquid is transparent to its own radiation. Large surface area in-situ photodiodes have been developed which efficiently detect the fast scintillation signal.

To transport 99% of the ionization electrons 15 mm or scintillation photons 1250 mm requires an impurity level of about 13 ppb. In pure Xenon a large electric field (≈ 100 kV/cm) is needed to collect 99% of the ionization charge however, doping with 3% methane increases the electron drift velocity from 3 to 20 mm/μs and so reduces the field required to 15 kV/cm.

The energy needed to produce a free electron is measured to be W=42 eV with (15 to 67 keV) electron excitation whereas, the canonical value W=15.6 eV obtains for higher energy (1 MeV) electron excitation.

A dedicated Higgs experiment will require 60 tons of Xenon or 110 tons of Krypton. The availability and cost of this materiel is addressed.

1. INTRODUCTION

Liquid Xenon is attractive as the showering and detection medium of a totally active electromagnetic (em) calorimeter because it is a fast ($\tau \leq 20$ ns) and efficient scintillator ($3 \cdot 10^4$ photons/MeV) and an efficient ionizer ($6.8 \cdot 10^4$ electrons/MeV). Drift of ionization electrons can provide highly sensitive positional information. Its relatively short radiation length (X_0) of 2.8 cm allows construction of a totally absorbing calorimeter ($28X_0$) in a length of 78.4 cm.

Observation of either ionization or scintillation gives an excellent measure of energy ($\approx 1\%$) but simultaneous observation of both can give even better resolution because the anticorrelation of these quantities allows suppression of Landau-like fluctuations.

Accurate position determination ($\sigma_x \approx 0.1$ mm) can be obtained by drift in short cells (d=15 mm) and directional sensitivity ($\sigma_\theta \approx 1$ mr) by depth sampling. Direction and position determinations are made without

---

* Presented by J. Seguinot and T. Ypsilantis at the 12th Workshop " New Technologies for Supercolliders" Erice, Sicily: 15-20 September 1990.

referance to the vertex origin of the photons (and electrons) hence, these quantities may be used to determine this point along the z axis of the collider.

The slow drift velocity (v=3 mm/μs) of electrons in pure liquid Xenon poses a problem at high luminosity LHC because 333 beam crossings occur during a driftout time (t=v/d) of 5 ms. Yoshino, Sowada and Schmidt [1] have shown that doping with 3% methane increases v to 20 mm/μs hence reduces t to 750 ns (50 crossings). To further increase v, studies of other dopants are planned. However, even with the still long integration time of 750 ns, double hits in the same tower (but different crossings) are resolved by the fast tower scintillation signal which provides an amplitude tag for the position sensitive (but slow) ionization signal. At the highest luminosity ($2 \cdot 10^{34}$/cm^{2}s) the slow drift gives rise to a pileup of transverse energy $E_t \leq 3$ GeV (per tower) at polar angles $\theta \geq 14^0$.

Since cost and availability of Xenon are important factors in such a project, we shall evaluate the performance of a liquid Krypton calorimeter through laboratory measurements similar to those we are carrying out for Xenon. Doping of Krypton (with 1 to 5% Xenon) will probably be required to supress the slow (85 ns) scintillation component. Krypton has a radiation length of 4.6 cm hence a totally active 28X$_0$ calorimeter will be 128.8 cm long. This length is acceptable in a dedicated LHC, SSC or ELOISATRON Higgs search experiment via photon and electron decay modes, in forward collider spectrometers for B physics (see SPSC proposal P238+Add. 1, 2, 3 re B$_s$ mixing) or in fixed target experiments (see NA 31' re CP violation).

In the following text, the prototype calorimeter will be described, the work done so far will be presented and the future work outlined.

2. THE CALORIMETER CONCEPT

A preliminary design of the prototype calorimeter was presented earlier [2], and now an updated version is described. It is made with rectangular rather than trapezoidal towers to keep the volume within the available 100 liters, however an LHC, SSC or ELOISATRON design would indeed be with trapezoidal towers and pointing geometry. Resolution in energy is obtained from the fast scintillation signal while position and direction are from depth sampled drift cells. Updates to this design will be made as new data from the test cell (described in section 3) become available.

The unit cell and a single tower are shown in fig.'s 2.1 a and b. The tower is defined by the transverse walls, floor and roof, all 200 μm thick. The walls are of carbon fiber while the floor and roof are non-conducting. The collection plane, which runs down the middle of the tower, has 100 mm φ metalized quartz wires (pitch 1 mm) strung through the 3 mm φ ceramic posts (pitch 56 mm). The tower is oriented so that the axis of the electromagnetic shower is parallel to the wire array. Each drift cell has a depth of two radiation lengths (Δz=56 mm) with transverse size (Δx=Δy=30 mm).

The transverse walls, which are parallel to the wire array in the prototype calorimeter, are coated with an aluminum film and overcoated with MgF$_2$ (see fig. 2.1b). These walls serve as equipotential surfaces (15 KV) and also as reflecting mirrors. The non conducting floor and roof are coated with metallic strips (95% coverage) which serve as field shaping traces as well as reflectors. A uniform electric field of 10 KV/cm, between the transverse wall and the earthed wire array, causes ionization electrons to drift up to and through the Frisch decoupling grid to the collection wire array. The Frisch grid is made of a reflecting electroformed metal mesh (90% transparent) located 1.5 mm from the wire plane. Each quartz wire (which runs the full length of the calorimeter) is metalized in 54 mm sections with 2 mm bare gaps, matching the cell pitch of 56 mm. The unmetallized gap allows each cell to be readout independantly. As the wire enters a readout post, it is soldered to a resistive carbon fiber wound helically around the post. Flash ADCs located at each end of the resistive fiber (i.e. at the floor and roof of each cell) measure the deposited charge along the wire array direction (y) by charge division and along the drift direction (x) by timing. The readout therefore gives the position (x_i, y_i) of deposited charge (q_i) in the cell at depth (z_i). The left-right ambiguity of the drift system can be resolved by detecting an above threshold digital signal on the corresponding left or right Frisch grid. Another method to resolve the left-right ambiguity is considered below.

A fast tower scintillation signal is detected at the tower end by a photosensor (Silicon or CsI photodiode), shown as the hatched region of fig. 2.1b.

Fourteen cells, in depth, constitute the tower and 10x10 towers make up the prototype calorimeter $(10.7X_0)^2 \times (28X_0)$, as shown in fig. 2.2.

Active liquid Xenon makes up about 99% of the calorimeter mass, the remainder being structural material (i.e. carbon fiber or ceramic). Monte Carlo simulations (GEANT 3.14, EGS 4) of em showers show that the inactive material contributes only a small term to the energy resolution ($\sigma_E \approx 0.2\%$ $\sqrt{E}$). The Xenon sensitive volume is 70 liters and its total volume is 100 liters weighing 308 kg.

The scintillation signal ($\tau \leq 20$ ns) and photosensor are fast hence this signal can be used as part of a fast transverse energy ($E_t=E\sin\theta$) trigger as well as to identify the beam crossing. Slow drift in the d=15 mm (=750 ns) cells can be tolerated because ionization and scintillation amplitudes are correlated hence the ionization charge, produced in the same beam crossing as the trigger, is amplitude tagged by the large scintillation signal (trigger).An example of this logic is shown in fig's. 2.3.

10

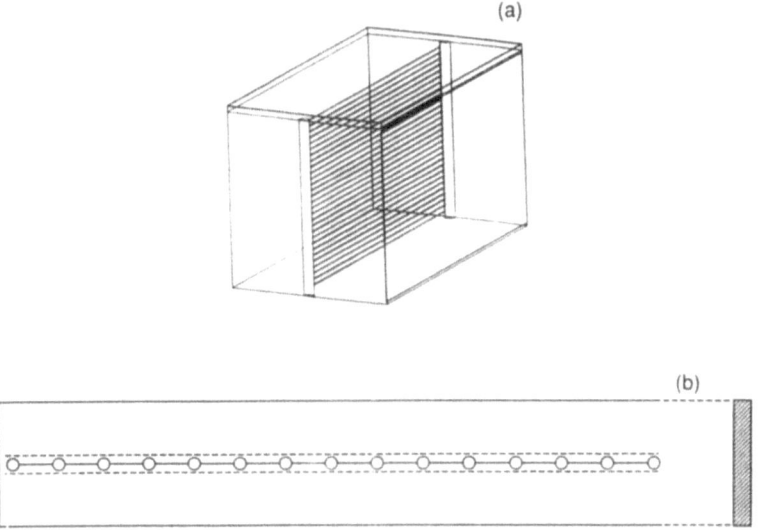

Fig. 2.1. a) A perspective view of the calorimeter cell. Its length along the wire direction is 56 mm and its transverse size is 30 mm. b) A top view of the tower with 14 drift cells, wires, readout posts, Frisch grids and photosensor (hatched).

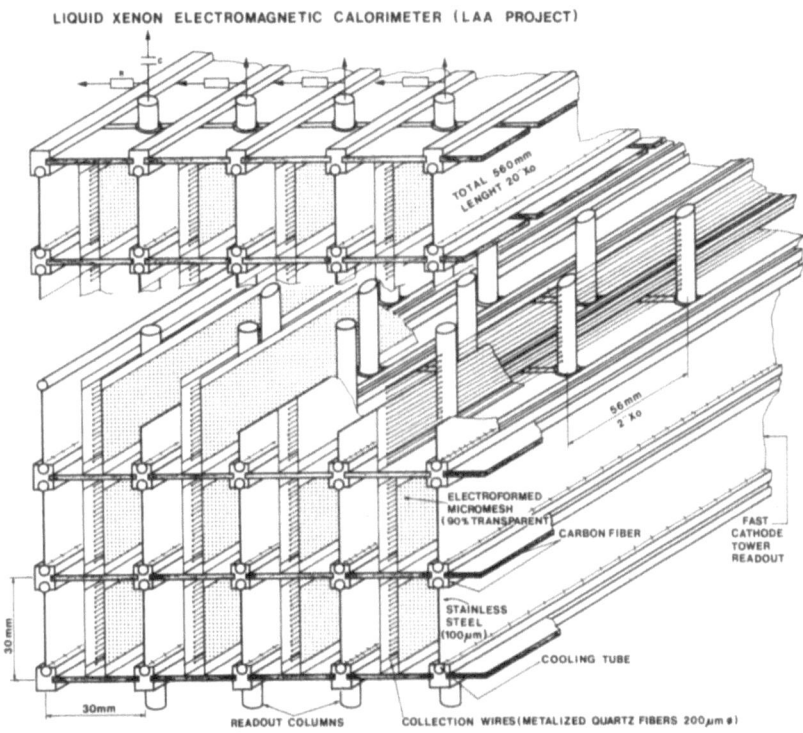

Fig. 2.2. A perspective view of the prototype calorimeter with its towers, readout columns, Frisch grids and the collection wire planes. The cooling tube structure was for cryopumping TMAE but is now suppressed.

11

Assume that the (t=0) trigger is initiated (in part) by a 25 GeV photon which enters a particular tower (say at $\sin\theta=0.4$ hence $E_t=10$ GeV) at a position (say) 10 mm to the left of the wire array. Its ionization signal I_1 therefore arrives at t=500 ns, as shown in fig. 2.3a. The shower ionization charge is collected isochronously because the shower direction is parallel to the wire plane and because the shower is columner, especially in the early cells (see fig.'s 2.6a, b). This signal also retains the intrinsic shower width since longitudinal and transverse diffusion are small (i.e. $\sigma_l=\sigma_t=7$ ns=140 μm for 15 mm drift). A second (say t=550 ns) trigger initiated (in part) by a 50 GeV photon which hits the same tower ($E_t=20$ GeV) at a position (say) 3 mm to the right of the wires produces an ambiguity because its ionization signal I_2 arrives at t=700 ns, before the first event is entirely collected. The scintillation signals (S_1 and S_2 in fig. 2.3b) are coincident with (and part of) the triggers (Tr$_1$ and Tr$_2$) and they amplitude tag the ionization signals. The ionization and scintillation signals are not generally of equal amplitude but, they are proportional hence the argument remains valid. Of course, these signals (I_1 and I_2) sit on a background of drift electrons arriving from particles produced in both earlier and later beam crossings. This background level is estimated below.

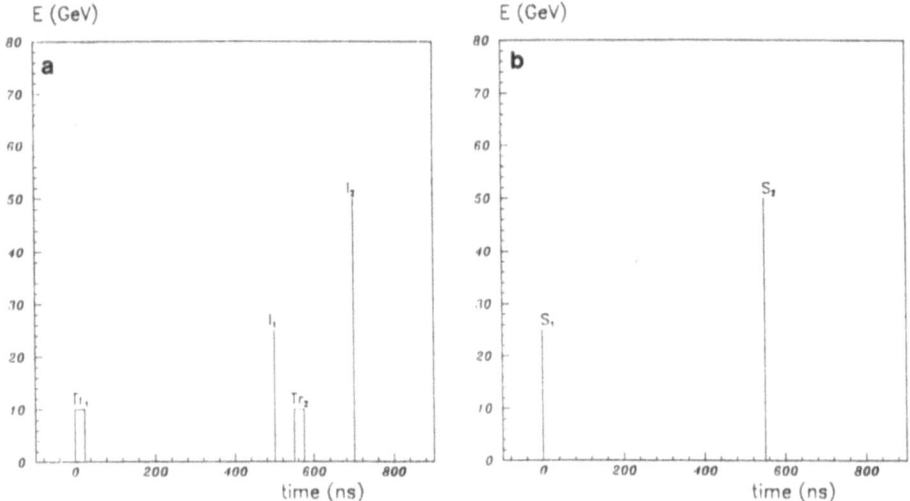

Fig. 2.3 a) The timing of the ionization and trigger signals produced in a single tower by a 25 GeV photon at t=0 followed by a 50 GeV photon at t=550 ns. b) The corresponding scintillation signals (which are part of the triggers) amplitude tag the time origin of the ionization signals, which are otherwise ambiguous.

A minimum ionizing (non interacting) hadron loses 21.4 MeV in traversing a cell and 300 MeV per tower. A fast scintillation tranverse energy threshold E_t of (say) 20 GeV could be triggered either by an em shower of the requisite energy (50 GeV at $\sin\theta=0.4$) or by the improbable event that 170 hadrons traverse the same tower. More important is the pileup background from minimum bias γs (from π^0's) which deposit all of their energy in the calorimeter. This source is eliminated by an E_t cut however, pileup imposes a lower limit on this quantity.

A Monte Carlo simulation of the (e or γ) and hadron pileup energy was performed for a spherically symmetric calorimeter geometry with a r=1000 mm inner radius and radial normals. The tower dimension at the inner surface is 30x30 mm^2 hence, at a fixed polar angle θ the number of φ towers is $N_\phi=2\pi r\sin\theta/30\approx209\sin\theta$. The tower widths are therefore $\Delta\theta=30$ mr and $\Delta\phi=(30/\sin\theta)$ mr. A total of 1000 minimum bias events (20 events/crossing x 50 crossings) were generated (at $\sqrt{s}=16$ TeV) and each particle of each event was tracked through the calorimeter.A charged hadron was assumed to deposit only 300 MeV (no interactions) whereas an (e or γ) deposited all its energy in its tower of incidence. This assumes that spill-out energy from tower equals spill-in energy from nearby towers. The φ summed pileup energy E_p is shown in fig. 2.4.

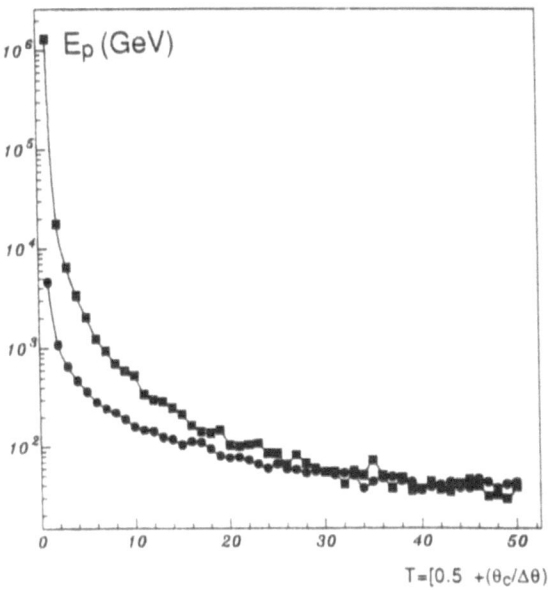

Fig. 2.4. The ϕ summed pileup energy E_p versus the tower index T. The solid circles show the hadron energy loss (no interactions) and the solid squares the (e, γ) total energy. The transverse energy per tower $E_t=E_p/209$.

The transverse energy per tower is therefore $E_t=E_p\sin\theta/N_\phi=E_p/209$. A pileup level of $E_t=10$ Gev per tower (i.e. $E_p=2090$ GeV) occurs only at T=5 (i.e. $\theta_c=135$ mr) which is exceedingly comfortable as this calorimeter is only intended to cover $\theta \geq 14^0=244$ mr where the pileup $E_t \leq 3$ GeV. It is also clear from fig. 2.4 that the hadronic part of the E_t background is smaller than the (e, γ), but not negligible. Hadronic interactions (in the 1.2λ calorimeter) will be included in the simulation but should not significantly alter these results.

In the angular interval $14^0 \leq \theta \leq 166^0$ (i.e. $\eta \leq 2.1$) the total (e, γ) energy deposit is 9.8 GeV and the hadronic energy loss is 6.0 GeV hence a total of only 16 GeV per interaction. In the simulated calorimeter ($14^0 \leq \theta \leq 166^0$) there are 13560 towers and 14 cells per tower hence a total of $1.9\cdot10^5$ cells. At high luminosity LHC there will be 20 interactions per beam crossing (15 ns) therefore the average pileup energy arriving per cell per 5ns time bin is 0.55 MeV. This arrival energy corresponds to a average cell current of 13200 e per 5ns time bin i.e. 0.4 μA, taking the presently measured value W=42 eV/e (see section 3.6.1). This background level is not excessive and seems to indicate that energy measurements from ionization (in addition to scintillation) are possible at high luminosity LHC. A total electron charge (contained in the cell at any time) is simply 150 times the arrival charge (i.e. $2\cdot10^6$ e) however, the positive ion charge is 10^3 times larger because of its lower mobility (i.e. $Q_p= 2\cdot10^9$ e). From Gauss' theorem, the electric field due to the positive ions is $E_d=Q_p/(\varepsilon_0\cdot S)=60$ V/cm (where S is the surface area of the cell).This field is negligible compared to the applied field of 10 kV/cm.

These simulations will continue, principally to calculate the shower spill-out and spill-in and the pileup time distribution. Fluctuations in this distribution will ultimately determine the resolution limits of the long drift technique at high luminosity hadron colliders.

2.1 THE EXPECTED PERFORMANCE

A 1 GeV electron shower is shown in fig. 2.5, projected onto the XZ plane of the prototype calorimeter. The practically full containment of the shower energy is apparent as is the strong collimation of the deposited energy in the first several cells.

An (x_i, y_i) perspective view of the z-integrated energy deposit in the first (0 to 56 mm) and second (56 to 112 mm) depth cells is shown in fig.'s 2.6a and b, respectively (from the Monte Carlo simulation program GEANT 3.14). Evidently, the energy centroid gives a precise determination of the shower position and many such depth measurements allow precise determination of the vector direction of the shower.

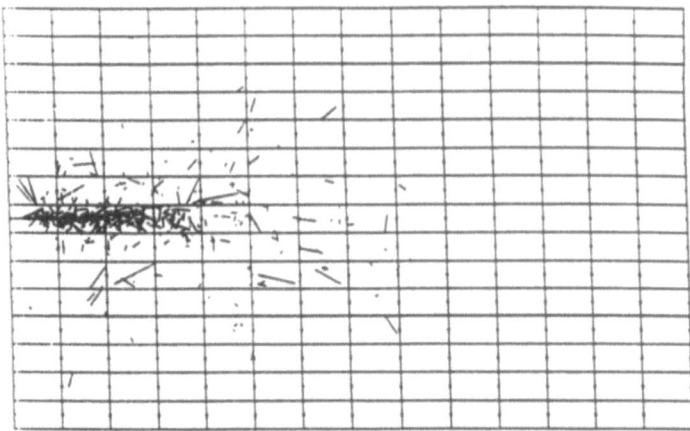

Fig. 2.5. A plane view of the prototype calorimeter cells (and towers) and the projected energy deposited by a 1 GeV electron incident from the left. The vertical lines are non physical (they show the drift cell limits) and do not impede light collection at the tower end (right edge).

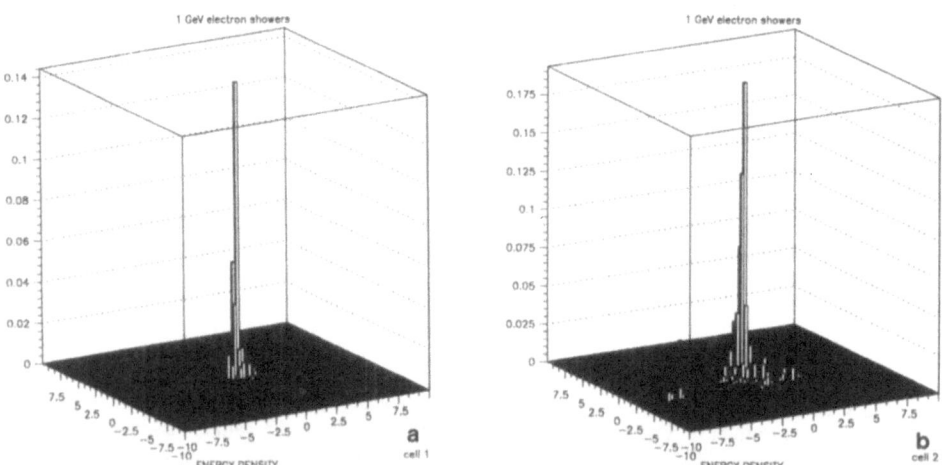

Fig. 2.6. A 1 GeV electron shower energy deposit profile versus the cell transverse coordinates (x, y) with $\Delta x = \Delta y = 0.5$ mm binning. a) in the 1st cell; b) in the 2nd cell.

A spherically symmetric calorimeter was simulated with a r=1000 mm inner radius and with radial normals. From EGS 4 simulated shower data, the directional resolution σ_θ for (25, 10, 1) GeV electrons was calculated from the energy weighted centroid points (x_i, y_i) at depth (z_i) with weights $(\sigma_{xi}, \sigma_{yi})$ obtained from the width of these distributions. To remove tails only the points with more than (10, 20 or 30)% of the peak energy were used. The directional errors σ_θ versus N (number of cells used in the fit) are shown in fig. 2.7a. The best directional errors $\sigma_\theta = (0.9, 1.4, 12.4)$ mr, respectively were obtained with the 30% peak cuts. The projected position coordinates ρ (at the interaction point) were found with errors $\sigma_\rho = (1.0, 1.5, 12.8)$ mm, respectively (again with the 30% peak cuts) as shown in fig. 2.7b.

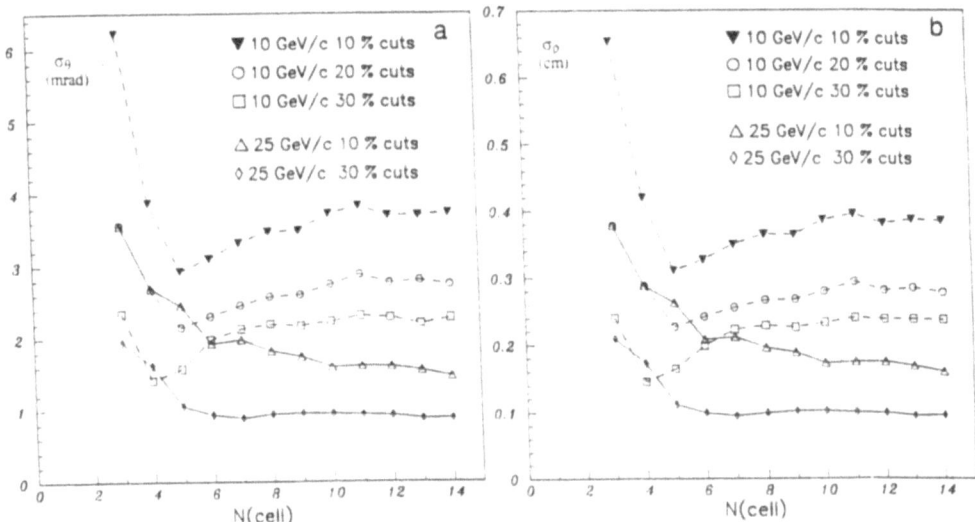

Fig. 2.7. a) The angular error σ_θ versus N (number of cells used in fit) for 10 and 25 GeV electrons with tail suppression cuts (10, 20, 30)% of peak energy. b) The vertex error σ_ρ versus N for 10 and 25 GeV electrons.

The vertex error along the beam axis $\sigma_z = \sigma_\rho / \sin\theta$ is good enough to resolve multiple interaction vertices if the photon energy $E \geq 25$ GeV and if the polar angle $\theta \geq 14^0$ (i.e. $\sigma_z \leq 4$ mm for $\eta \leq 2.1$). Since photons from Higgs decays are almost always greater than 25 GeV the only limitation will be in the rapidity range. The position and direction determination also eliminates the left-right ambiguity since only the true solution will intercept the beam axis. This vertex capability can reduce the 2γ combinatorial background by several orders of magnitude at high luminosity LHC ($2 \cdot 10^{34}$/cm^{2}s) with its 20 interactions per crossing in a 200 mm interaction region hence, with vertex spacing ≤ 10 mm.

Finally, the position errors σ_ρ' at the inner face of the calorimeter were found to be (50, 90, 640) μm, again for electrons of (25, 10, 1) GeV respectively. This calorimeter can function as its own pre-shower counter because of its extremely good position resolution. Off-line tagging of two showers with ≈ 1 mm separation seems possible due to the extremely fine shower profiles in the first several depth cells.

The energy resolution of this calorimeter will be of the form $\sigma_E/E = b + a/\sqrt{E}$ with a $\leq 1.0\%$, either from ionization or scintillation. The constant term is estimated as $b \approx (0.2-0.3)\%$, due to calorimeter leakage. A combined ionization and scintillation measurement may supress Landau-like fluctuations (due to δ rays) and further improve the resolution. The smallness of both (a, b) guarantees excellent resolution at both the highest and lowest energies of interest.

It is believed that the combination of unexcelled energy and angular resolution along with the asociated vertex finding capability of this calorimeter will be of decisive value for Higgs searches and other new physics at LHC, SSC or ELOISATRON. In addition, the 2γ shower separation capability of less than 1 mm will allow excellent off-line π^0 tagging.

3. EXPERIMENTAL RESULTS

3.1 SCINTILLATION DETECTORS

Liquid Xenon scintillation light is emitted at wavelength $\lambda = (175 \pm 5)$ nm, energy $E_{ph} = 7.1$ eV. Silicon photodiodes have been used as the photosensors for scintillating crystal calorimeters with tower geometry (e.g. L3 at LEP) in which the light traverses a large fraction of the crystal length to reach the photodiode. In liquid Xenon, the distance between shower maximum and the tower end is about 60 cm which will require excellent wall reflectivity and liquid transparency (see fig. 2b).

Two in-situ photosensors have been developed and/or tested. They are a reflective CsI photocathode and a Silicon strip photodiode. The photosensor for the prototype calorimeter has not yet been selected but these two candidates are shown to be viable.

3.1.1 THE CsI REFLECTIVE PHOTOCATHODE DIODE

A CsI reflective photocathode was made by vacuum deposition onto a stainless steel substrate [3]. The detected quantum efficiency Q (electrons per incident photon) was measured with electron extraction into 1 bar of methane gas, as shown in fig. 3.1. At 175 nm, the CsI reflectivity is calculated from its optical constants to be R=0.23=1-A and since Q=0.32 the "efficient quantum yield" Q_y=Q/A=0.42 (electrons per absorbed photon). This quantity is related to the usual definition of yield Y by the equation Y=Q_y/E_{ph}=.059 (electrons/eV) and W=1/Y=16.9 (eV/electron).

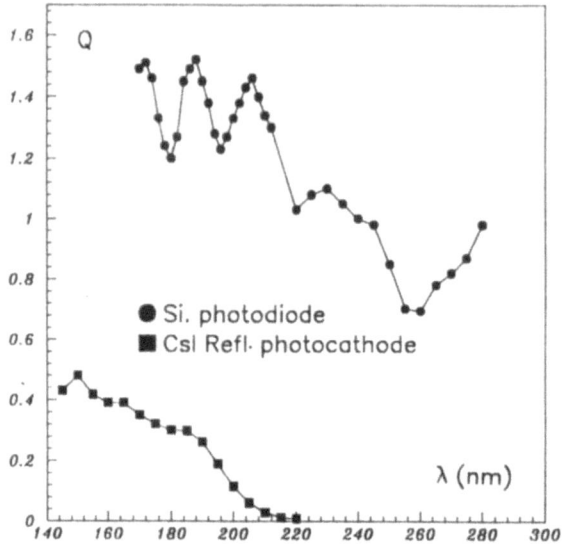

Fig. 3.1 The detected quantum efficiency Q (photoelectrons per incident photon) measured versus photon wavelength λ for a reflective CsI thin film photocathode and for the DELPHI Silicon strip photodiode.

A CsI photocathode could operate in direct contact with the liquid. The produced photoelectrons would be injected into the liquid and be drifted to a collector grid. This configuration is not very sensitive since the induced photocathode signal is proportional the (slow) drift velocity.

A gas filled diode with a reflective CsI photocathode has much higher sensitivity because all the produced photoelectrons are collected in about 10 ns. The diode is made with a 2 mm thick quartz entrance window followed by 1 mm of 1 bar methane gas and closed by a 2 mm thick slab of ceramic. The photocathode is deposited on the metalized inner surface of the ceramic and from which a photocurrent is detected. The true photoelectrons are collected on the (-100 V) metallized strips (10% coverage) deposited on the inner surface of the window. Such a photodiode is intrinsically fast because the electron drift time in methane is 10 ns/mm (at 1kV/cm) and its capacitance per unit area is only 885 fF/cm^2 hence a 9 cm^2 tower has a total capacitance of only 8 pF. This low capacitance guarantees fast electronic response (τ_{el} ≈ 0.16 ns) hence an overall time resolution determined only by the Xenon scintillation lifetime (τ ≤ 20 ns).

The inactive materiel of the photodiode will have a minimal effect on the measurement of shower energy since all showering electrons are absorbed before reaching the tower end (where the photodiodes are positioned). A hadron however, which traverses the photodiode loses 157 eV in the 1 mm of methane gas and produces a direct ionization signal D≈5 (e/tower) which may be compared to its scintillation signal S=9·10^6·ε_c·Q (e/tower), where ε_c is the photon collection efficiency. This signal ratio D/S=5.6·10^{-7}/(ε_c·Q) is extremely small and shows that hadrons cannot produce a significant effect in methane. A hadron traversing the 500 nm thick CsI film deposits 250 eV hence, assuming W=16.9 eV/e holds for ionization excitation, a direct signal 3 times greater than in methane, which is still negligible.

A CsI gas photodiode will be cheap compared to an equivalent area Silicon strip photodiode, but has about a factor four less detected quantum efficiency. This relative loss of quantum efficiency may be advantageous because the high CsI threshold (6.5 eV) guarantees that this cathode functions without intrinsic fluctuations i.e. the quantum yield Q_y is either 0 or 1 but never > 1. A related advantage is its high absorbtivity (A=0.77) compared to (A=0.38) for a Silicon photodiode (see next section).

3.1.2 THE SILICON STRIP UV PHOTODIODE

The detected quantum efficiency Q of a large area (18 cm^2) and high resistivity Silicon strip diode (DELPHI) was measured [4] versus photon wavelength λ, also shown in fig. 3.1. This efficiency is extremely high and is even greater than 100% in the UV region, including reflection and absorption losses. This indicates that the "efficient quantum yield" Q_y is much greater than one and also that electron collection is efficient in the strip configuration. This detector has 5 mm wide metallic strips with 50 mm pitch hence an opacity of 10%. It is covered with thin oxide layer (SiO$_2$) which must be crystaline because the diode is fully sensitive down to 140 nm, far beyond the transmission limit of amorphous quartz (160 nm). The measured Silicon reflectivity [5] is R=0.62 at λ=175 nm hence the total absorbtivity A=0.90x0.38=0.342 and since (from fig. 3.1) Q=1.35 hence Q_y= Q/A=3.95 electrons per absorbed photon. The yield therefore, is Y=Q_y/E_{ph}=0.56 e/eV and the energy to produce a conduction electron is W=1/Y=1.8 eV/e, whereas the usually quoted value in Silicon is W=3.6 eV/e [6]. Note that energy conservation only requires that W $\geq$ 1.12 eV/e (i.e. the conduction bandgap energy) and so the quantum yield $Q_y \leq$ 6.3 e/absorbed photon (7.1 eV). A possible problem then, with this diode, is that the energy resolution may be limited by fluctuations in the quantum yield ($Q_y \approx$ 4) unless a very small Fano factor intervenes.

The measured capacitance of the Silicon strip diode was 20 pF/cm^2. A strip detector covering the 9 cm^2 tower end however, has still only 180 pF capacitance which when coupled to a ($\Omega \approx$ 20 ohm) current amplifier gives the adequately fast response time of 3.6 ns. A trapezoidal tower with end area of 30 cm^2 will have a capacitance of 600 pF and so a response time of 12 ns. This may be too long for LHC and require development of a lower impedance current amplifier ($\Omega \approx$ 2 ohms) to obtain the requisite fast response.

Another possible problem with this diode is its extreme sensitivy to charged particles W_{Si}=(1.8 to 3.6) eV/e compared to Xenon scintillation W'=33 eV/photon (see section 3.4.2). In an em shower, all particles are absorbed before reaching the tower end however, a hadron traversal of the 300 μm Silicon diode produces a direct ionization signal D=3.3·10^4 (e/tower) compared to the scintillation signal S=9·10^6·ϵ_c·Q (e/tower). This signal ratio D/S=0.4%/(ϵ_c·Q) is small but not negligible and will require that the photon collection efficiency ϵ_c be kept high (i.e. $\epsilon_c \geq$ 0.5).

3.2 THE PULSED 100 KV ELECTRON ACCELERATOR

A pulsed 100 kV electron accelerator (fig. 3.2) was built in our laboratory and is used to inject excitation energy into the liquid. A light beam from a low pressure (200 torr) hydrogen flash lamp (FWHM=15ns) is focused onto a reflective CsI vacuum photocathode to produce the photoelectrons which are subsequently accelerated. This device can deliver as little as 1 MeV (400 electrons of 2.5 keV) or as much as 1 TeV (10^7 electrons of 100 keV) of excitation energy per pulse into the liquid, permitting the investigation of the scintillation and ionization over a wide range of excitation and total energies. Low energy measurements are important for high energy em calorimetry because minimum ionizing particles produce significant numbers of d rays and because most of the shower energy is eventually degraded into excitations of Xenon via low energy electrons and photons.

The charge entering the liquid Xenon cell is monitored by a Faraday-cage collimator which intercepts about 75% of the beam electrons. In an independant study, the Xenon cell was replaced by a solid copper Faraday-cage disc and the transmitted charge was measured. The correlation between the intercepted and transmitted charge showed that the transmitted charge q was monitored with fractional error σ_q/q $\approx$ 0.3%. The transmitted energy is obviously E=qV where V is the accelerator voltage, corrected for energy loss in the 12 μm thick mylar entrance window via a GEANT 3.14 simulation.

3.3. TEST CELL, CONDENSER AND FILTER SYSTEM

The test cell, condenser and filter system (fig. 3.3) was purposely built without high-vacuum specifications (i.e. a stainless steel vessel pumpable to high vacuum with bakeout at 300 ^{0}C) but with materiels which make construction of the calorimeter practical and with enough cleaning power to remove impurities by continuous closed gas circulation. The system contains ceramics, Boron Nitride and metals, all of which were cleaned and degassed before assembly. In addition, it contains valves with Viton O rings and a circulation pump with a neoprene diaphragm. The electron beam entrance window is made of a 12 μm thick metalized mylar foil glued over a 10 mm ϕ hole in the metal base of the test cell. The gas, which evaporates from the test cell due to the heat load, is forced by the circulation pump through an oxisorb cartridge and molecular sieves (13X and 4A at -80 ^{0}C) and then recondensed in the heat exchanger. The pressure in the liquid test cell is kept constant by feedback regulation of the liquid nitrogen flow in the heat exchanger, hence the temperature is constant ($\pm$0.1 ^{0}C) as is the Xenon level.

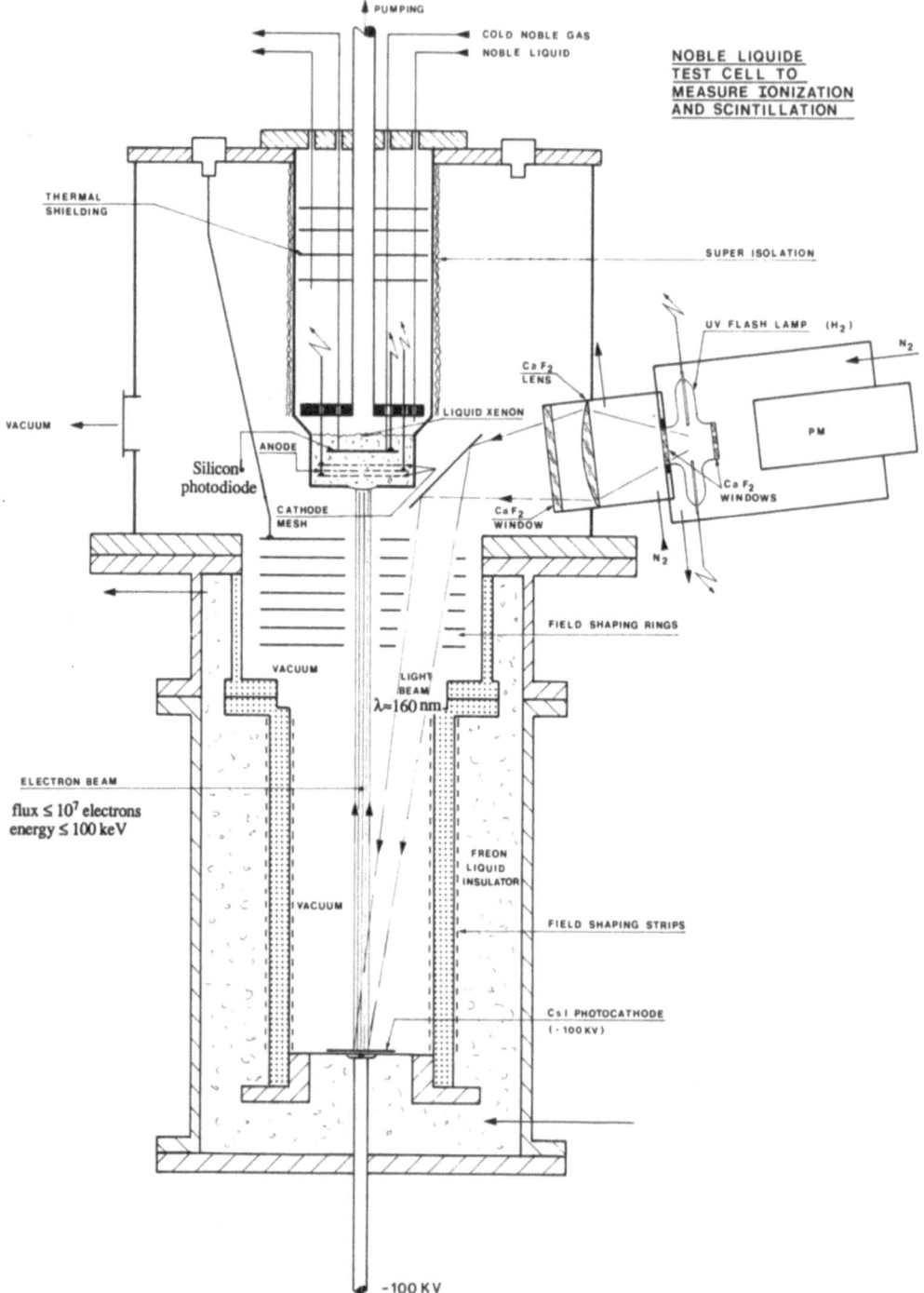

Fig. 3.2 A side view of the 100 kV electron accelerator with the UV flash lamp, accelerating structure and the liquid Xenon test cell.

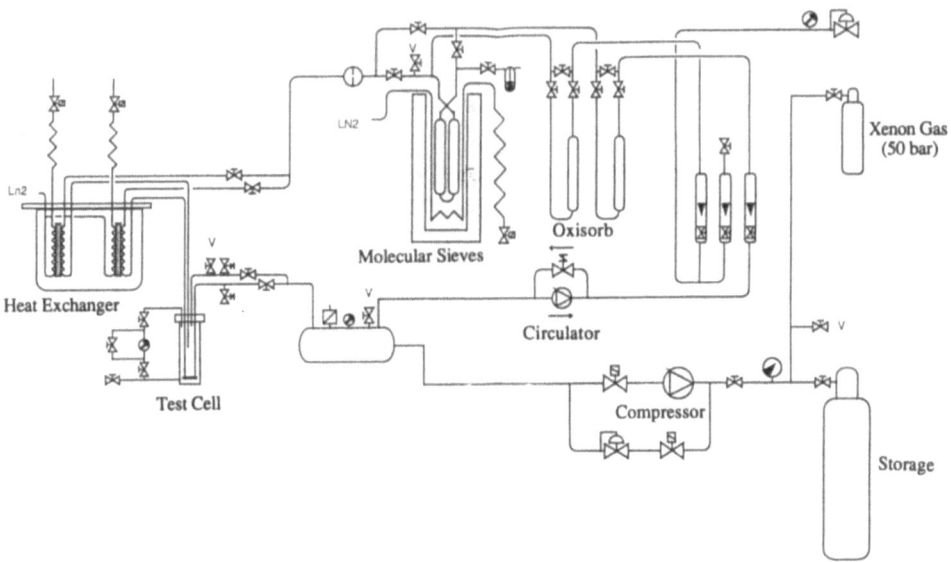

Fig. 3.3 The noble liquid cleaning system circuit with the circulation pump, filters, heat exchanger and test cell. The vacuum cryostat is not shown.

An impurity level of ≈13 ppb is needed to transport electrons 15 mm with 99% efficiency. The electron absorption length is then 1500 mm (corresponding to an electron lifetime τ=500 ms in pure Xenon). A photon absorption length of 1500 mm however, requires an impurity level of only 1 ppm hence a liquid sufficiently clean to drift electrons length d will allow efficient photon transport over length 77·d. This is almost exactly the ratio needed for a calorimeter designed to drift electrons 15 mm and transport photons about 1 m. Hence, an impurity level of ≈13 ppb suffices to transport electrons 15 mm or photons 1250 mm with 99% efficiency.

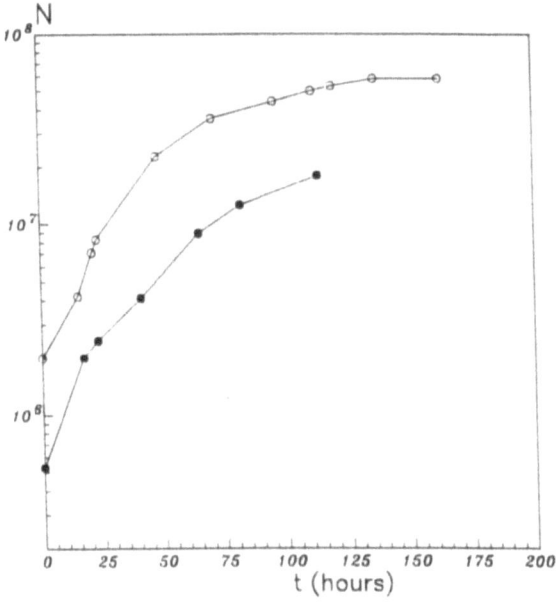

Fig. 3.4. The collected ionization charge N versus circulation time t through the filters. The closed circles are for the initial (badly contaminated) Xenon filling while the open circles are for the precleaned Xenon. The asymptotic lifetimes attained were 4 and 100 μs, respectively.

19

Medium purity Xenon gas (Carbagas N45) was initially injected into the system and liquid was condensed. Electron drift was immediately seen and the ionization signal was monitored as a function of time (the closed points of fig. 3.4). The initial lifetime of 100 ns was very small indicating strong contamination however, after 5 days of circulation through filters the lifetime increased to 4 μs. Even though the lifetime was still improving (albeit at lower rate) the run was terminated before reaching a plateau in order to test this cleaning technique with an initially purer gas.

High purity Argon (Carbagas N60) was therefore injected into the system and condensed. A long electron lifetime ($\tau \geq 70$ μs) was immediately observed without a long cleanup period. Measurements of both ionization and scintillation (and their anticorrelation) were made, with results given in section 3.5.2.

Improved Xenon lifetime was obtained by injecting precleaned gas into the system. The Xenon, of USSR origin, was cleaned with high vacuum technology to $\tau = 160$ μs by D.Schinzel and A.Gonidec of CERN (but with unknown contamination in transfer). As before, the ionization signal was monitored as a function of time (the open points of fig. 3.4). The initial lifetime was higher than previously and the rate of cleanup was somehat faster. After 5 days of circulation with cleaning, the electron lifetime increased to better than 100 μs. Further improvement in lifetime was probably attained but was difficult to measure.

These data show that a calorimeter can be constructed of ceramics, Boron Nitride, Viton O rings, rubber and metals without high vacuum technology. To obtain the required long lifetime it is preferable to start with a pure gas followed by continuous circulatation with purification through filters. The cleaning capacity should be sufficiently large to remove both initial and steady-state impurities in several days. Recirculation and recondensation capacity, consistent with this time scale, should also be installed.

3.4 XENON SCINTILLATION
3.4.1 LIFETIME

The first observation of the scintillation was made with a vacuum photomultiplier (PM) because it is a proven and reliable technique for photon detection. The test cell (fig. 3.2) was modified (fig. 3.5) to include a re-entrant CaF_2 window and a small PM with a MgF_2 window. Upon condensing Xenon, scintillation light was immediately observed with an amplitude that confirmed the expected large yield. It showed also that the scintillation is fast and that the liquid is transparent to its own radiation.

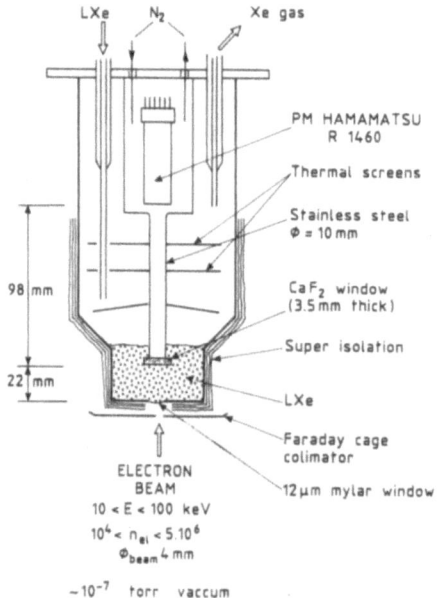

Fig. 3.5. The test cell geometry used to observe scintillation from liquid and gaseous Xenon. The scintillation signal was seen with a CsTe solar blind PM with a 25 mm ϕ MgF_2 window (ϕ = 6 mm aperature, $\Delta\Omega/4\pi = 1.7 \cdot 10^{-4}$). The liquid level height was varied to determine its transparency.

The liquid Xenon scintillation waveform was measured as shown in fig. 3.6a. The decay tail has a 20 ns time constant which (after the quadrature subtraction of the $\sigma_t = 6$ ns lamp pulse width) gives a lifetime $\tau \approx 19$ ns, in agreement with Kuroda et. al. [7] who report $\tau_3 = 22$ ns and $\tau_1 \approx 3$ ns for the triplet and singlet state lifetimes of the Xe_2 molecule. They also observed that the fast component comprised 77% of the scintillation light yield. We did not see this fast component however, at the time of these tests the Xenon was not yet highly purified. A faster lamp is being made and tests with highly pure Xenon are scheduled to clarify this issue. It is important to note, however, the absence of long lifetime recombination tails.

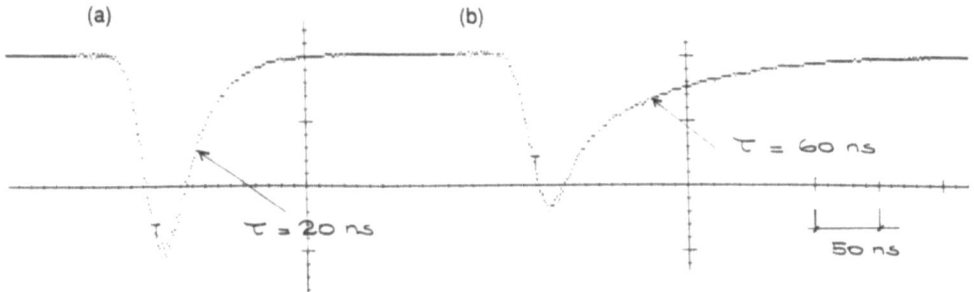

(a)   (b)

τ = 60 ns

τ = 20 ns

50 ns

Fig. 3.6. a) The liquid Xenon scintillation waveform.The decay tail corresponds to a lifetime $\tau \approx 19$ ns, after quadrature subtraction of the lamp signal. b) The 1 bar Xenon gas scintillation waveform.

The scintillation of 1 bar Xenon gas was also measured and its waveform is shown in fig. 3.6b where a 60 ns recombination tail is clearly visible.

3.4.2 YIELD AND LIQUID TRANSPARENCY

A PM scintillation data (shown in fig. 3.7) correspond to a yield $Y'=3 \cdot 10^4$ photons/MeV. The energy needed to create a photon $W'=1/Y'=33$ eV/photon is therefore about twice that needed to create an electron $W=15.6$ eV/e (see section 3.6.1).

To determine the liquid transparency the scintillation signal was monitored while the Xenon liquid level heigth h was varied. No significant absorption was seen in 20 mm of liquid path length (see also fig. 3.7) indicating that a photon absorption length $l_{ph} > 100$ mm. A longer test cell is needed to measure the actual absorption length.

The 1 bar Xenon gas scintillation yield was about 5 times less than the liquid i.e. $Y'(gas)=6 \cdot 10^3$ photons/MeV or $W'(gas)=167$ eV/photon.

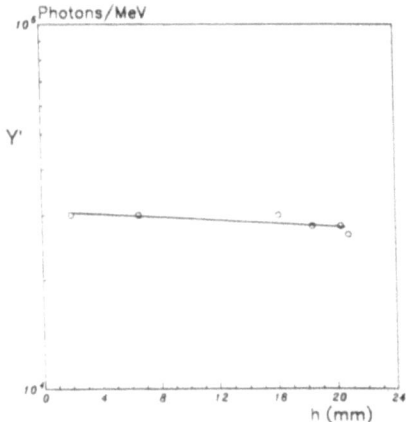

Fig. 3.7. The liquid Xenon scintillation vield Y' versus the liquid level height h.

3.4.3 ENERGY RESOLUTION

With the set-up of fig. 3.5 a linear PM response was observed. The number of photoelectrons N_{pe} versus total energy E is shown in fig. 3.8a. The energy resolution σ_E/E however, was limited by the photoelectron statistics $1/\sqrt{N_{pe}}$ as shown in fig. 3.8b. Even so, the resolution scaled correctly and a value $\sigma_E/E \approx 1.4\%$ was attained at the highest energy of 20 GeV. The PM solid angle was limited (by the $\phi = 6$ mm mask) to prevent signal saturation at high energy. This limitation will be removed in future Silicon or CsI photodiodes because they have larger surface areas and are in direct contact with the liquid.

21

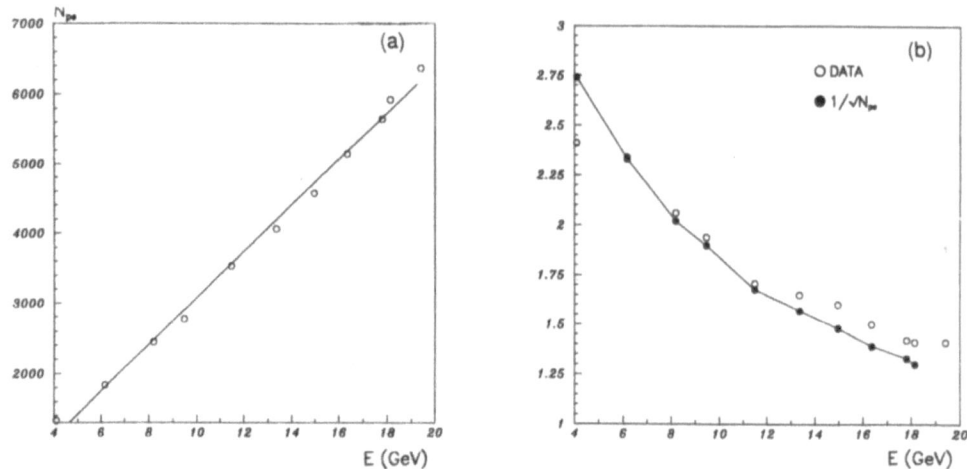

Fig.3.8. a) The liquid Xenon scintillation signal N_{pe} (photoelectrons) versus E, the total energy deposited in the test cell. **b)** The liquid Xenon scintillation energy resolution σ_E/E versus E. The open circles are data, the closed circles are the contribution to resolution from photoelectron statistics $1/\sqrt{N_{pe}}$.

3.5 IONIZATION AND SCINTILLATION
3.5.1 XENON LIQUID

In this set of measurements the PM of the test cell was replaced by a Silicon strip photodiode (glued to a ceramic backing plate) and a drift gap was installed to collect the ionization electrons (see fig. 3.9). The diode surface area is (22×22) mm^2 and it was placed 18 mm from the electron beam entrance window. Two 90% transparent electroformed metal meshes M_1 and M_2 were located 3 and 6 mm respectively from the entrance window. The mesh M_1, biased at positive potential, collected the ionization electrons through a decoupling capacitor into a low noise charge sensitive preamplifier (r.m.s. noise $\sigma_q = 300e$) while M_2 was biased to prevent collection of any Silicon photoelectrons. An ionization signal was observed but the liquid was still in the early stages of purification (fig. 3.4) and the electron lifetime was only 100 ns compared to the fastest collection time of 1 μs. This lifetime corresponds to an impurity level of 65 ppm (oxygen equivalent) and to a photon absorption length $l_{ph}=23$ mm hence to 55% light absorption in the l=18 mm path between the beam source and the Silicon photodiode.

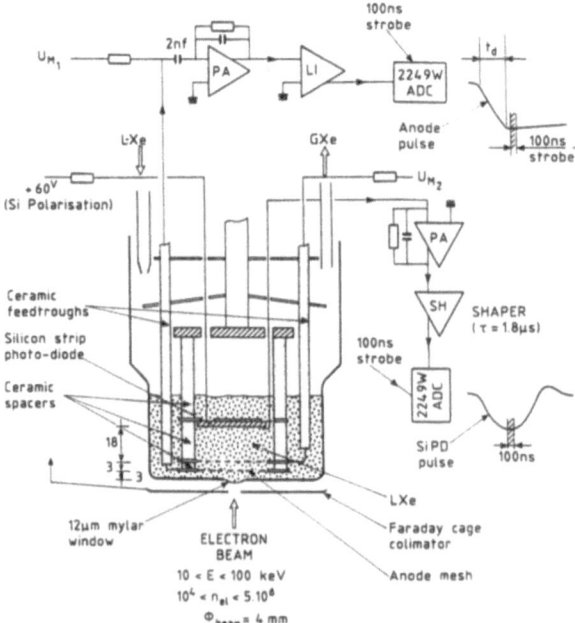

Fig. 3.9 The test cell geometry used for observation of ionization and scintillation. The ionization charge is collected on the anode mesh and scintillation light is detected by the Silicon strip photodiode. Their electronic readout chains are also shown. The electron beam collimator/Faraday-cage monitor is also indicated.

Notwithstanding the light absorption and the short electron lifetime, the ionization I and the scintillation S signals were measured as a function of the electric field E_d with the results shown in fig.s 3.10a, b (for fixed total energy E=9 GeV). Note that I rises and S decreases (slightly) with increasing E_d, as expected if an anticorrelation mechanism is operative. The scatter plot of I versus S (fig. 3.10c) shows, more succinctly, this anticorrelation. Improvement in energy resolution may be obtained by using that specific linear combination of the signals (I+αS) which projects the points onto an axis normal to this anticorrelation line. It should be emphasized that these tests were made before the liquid was highly purified, so this anticorrelation effect must be confirmed under better conditions (see section 3.5.2).

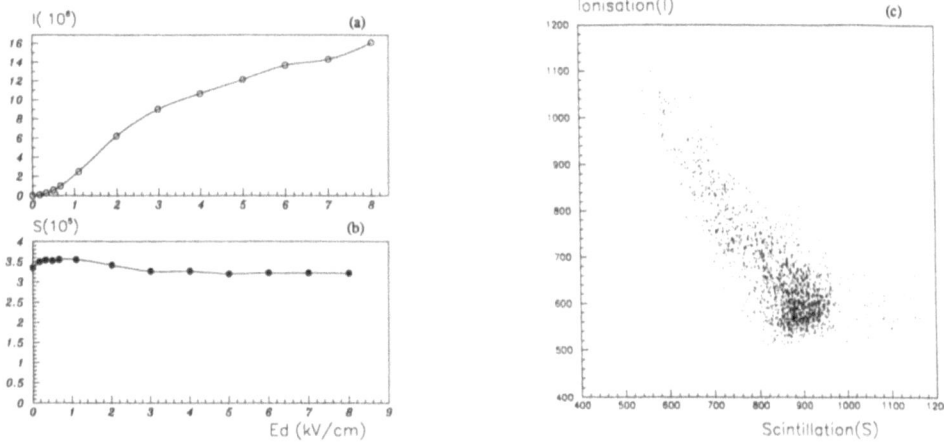

Fig. 3.10. a) The liquid Xenon ionization charge I versus the drift field E_d. b) The liquid Xenon scintillation signal S versus E_d. c) A scatter plot of I versus S for total energy E= 7.3 GeV at E_d=6.7 kV/cm.

The linearity of the individual I and S signals versus total energy E are shown in fig.'s 3.11 a, b. The energy resolutions (not shown) were reasonably good considering the short electron lifetime (in the ionization case) and a progressive deterioration (of unknown origin) of the Silicon photodiode signal with time (in the scintillation case).

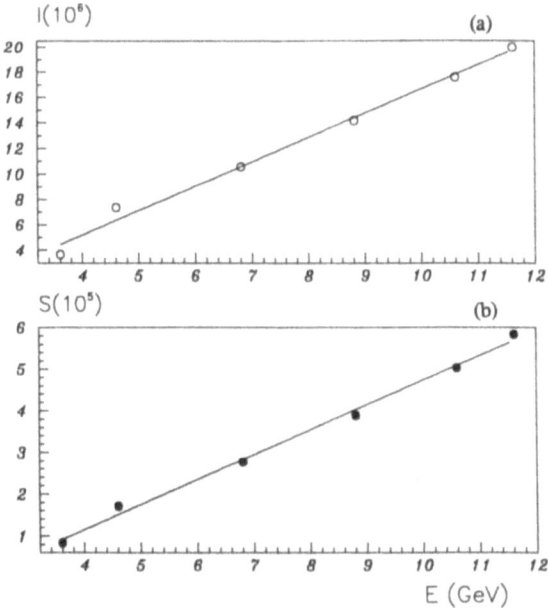

Fig. 3.11. Linearity of the liquid Xenon a) ionization signal I b) scintillation signal S, versus energy E.

23

3.5.2 ARGON LIQUID

High purity Argon gas (Carbagas N60) was condensed into the cell equipped with the PM (as in fig. 3.5) but with a drift cell added (as in fig. 3.9). Both ionization I and scintillation S (=N_{pe}=number of photoelectrons) signals were measured as a function of electric field E_d, as shown in fig.'s 3.12a and b. These data, obtained with a good electron lifetime ($\tau \geq 70$ μs), show that the anticorrelation mechanism is indeed operative because the ionization increases and scintillation decreases with electric field. The single charge-density model of Thomas and Imel [8] predicts the relation between collected ionization charge N and electric field E_d is

$$N=N_0\ln(1+\xi)/\xi \qquad (1)$$

where $\xi=K/v=(K/\mu)/E_d$ and N_0 is the asymptotic charge. The curve in fig. 3.12a shows a fit of this formula to the data. The fit quality is good and the fit parameters are $N_0=8.3\cdot10^7$e and K/m=5.2 kV/cm for E=2.25 GeV (i.e. $3.3\cdot10^4$ electrons of 67 keV). The energy needed to produce a free electron is therefore W=E/N_0=27.1 eV/e, which may be compared to the canonical value of 23.6 eV/e measured by Miyajima et. al. [9] for 1 MeV electron excitation. This 15% increase in W, due presumedly to the lower energy electron excitation, is in sharp contrast with the factor 2.7 increase observed in liquid Xenon (see section 3.6.1). The fit value of K/m=5.2 kV/cm however, is much higher than expected and must be further investigated.

Full anticorrelation requires I+αS=β=constant. The constants (α, β) were evaluated by using two pairs of measured points (I, S). The curve S=(β-I)/α, shown in fig. 3.12b, is obtained from the curve I of fig. 3.12a and (α, β). This transposed curve represents the data quite well, indicating full anticorrelation hence confirming the original observation of this effect by Kubota et. al. [10], for 1 MeV electron excitation of liquid Argon.

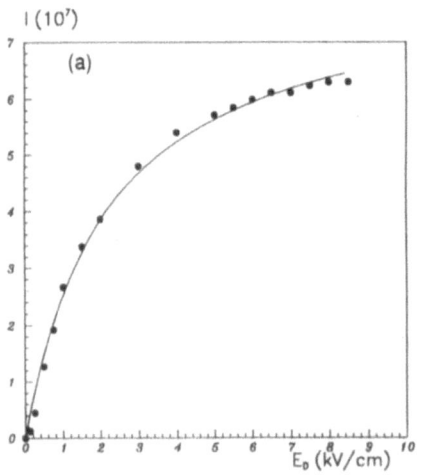 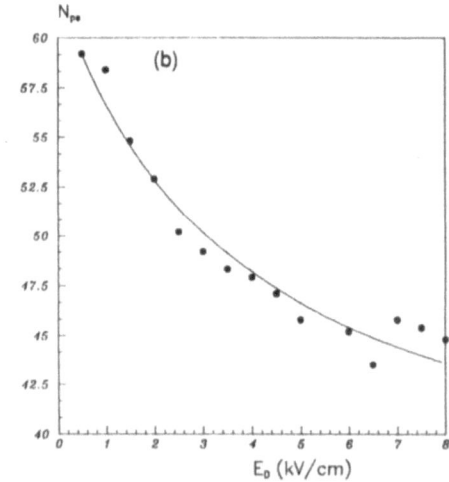

Fig. 3.12. a) The liquid Argon ionization charge I versus the drift field E_d. The fit curve is from the single charge-density model (eq. 1) of Thomas and Imel [8]. **b)** The scintillation signal S (=N_{pe}) versus the drift field E_d with the transcribed curve from the anticorrelation condition I+αS=β.

The energy resolution via the ionization signal is shown in fig. 3.13 and the fit $\sigma_E/E=(0.9+0.3/\sqrt{E})$% exhibits the excellent resolution which characterize totally active calorimeters. Of course, in a real em shower not all energy is deposited punctually and leakage must be added, but for sufficiently deep calorimeters (28X$_0$) this may be limited to 0.2%. The ultimate resolution may even be better than the fit because energy loss straggling (in the mylar entrance window) contributes non-negligibly to the 0.9 % constant term. Other (as yet) uncontrolled fluctuations in the electron beam flux (or energy) must be present since, in several runs a high energy resolution of 0.8% was attained.

The energy resolution via the scintillation signal is not shown because it was dominated by photoelectron statistics due to the small PM solid angle.

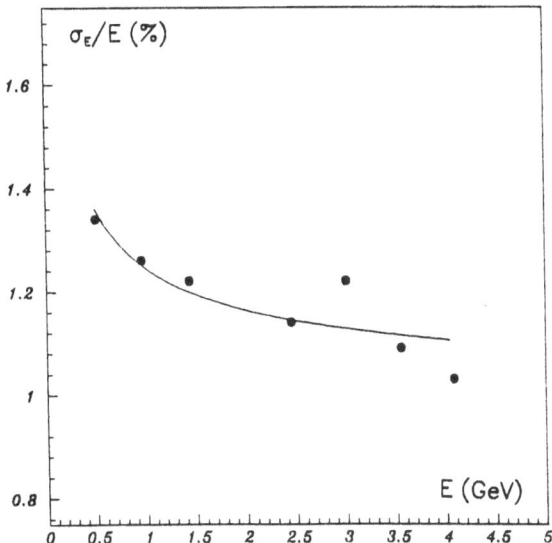

Fig. 3.13. The liquid Argon ionization energy resolution σ_E/E versus E, the total energy deposited. The fit curve $\sigma_E/E=(0.9+0.3/\sqrt{E})\%$ shows the excellent resolution which characterize totally active calorimeters.

3.6 XENON LIQUID IONIZATION
3.6.1 CHARGE COLLECTION EFFICIENCY AND FREE ELECTRON YIELD

The cell was as shown in fig. 3.9 except that the Silicon photodiode was removed to exclude extraneous sources of drift contamination. The electron lifetime during these measurements was, in fact, excellent (i.e. $\tau \geq$ 100 ms). To increase the drift sensitivity, the mesh M_1 was removed and charge was collected on mesh M_2 after d=6 mm drift.

The ionization signal amplitude was measured as a function of drift time t at a fixed electric field E_d. The ionization signal I(mV), was converted to collected charge Q by the preamplifier sensitivity 2.5 fC/mV, and was fit to the formula

$$Q=Q_0(\tau/t_d)(1-\tau/t_e)^{-1}(e^{-t/t_e}-e^{-t/\tau}) \tag{2}$$

where Q_0 is the asymptotic charge, $t_d=d/v$ is the maximum drift time, τ is the electron lifetime and $t_e=26.7\mu s$ is the preamplifier response time. An example of this signal is shown in fig. 3.14.

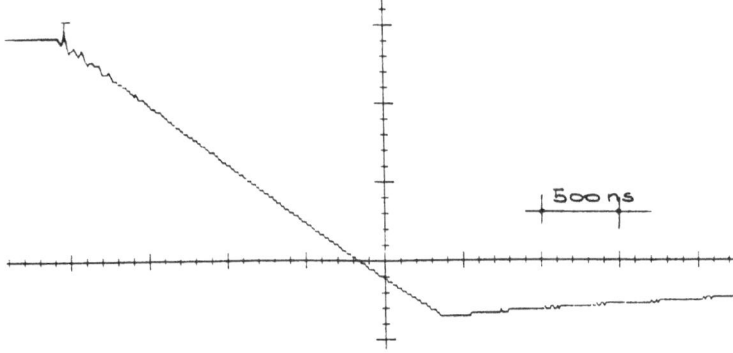

Fig. 3.14. The waveform of the collected ionization charge in liquid Xenon versus drift time. The electron lifetime here was excellent ($\tau \geq 100$ μs), the slight curvature is due to the electronic respose time ($t_{el}=26.7$ μs).

25

The charge increases almost linearly with time, the slight curvature is due to preamplifier response time. The quantities determined by the fit are Q_0, τ and the drift time t_d. At each electric field E_d the number of collected charges N ($=Q_0/e$) was plotted, as shown in fig. 3.15. The curves show the excellent fits of eq. 1 to the data, which were obtained with excitation energy of (15, 37, 67) keV and total energy E of (0.64, 1.97, 4.02) GeV, respectively. The fit parameters are K/μ=(3.8, 2.9, 2.1) kV/cm and N_0=(1.74, 4.51, 8.93)·10^7 hence the energy needed to produce a free electron $W=E/N_0$=(37, 44, 45) eV/e, respectively. These W values are much larger than the canonical value of 15.6 eV/e measured by Takahashi et. al. [10], for 1 MeV electron excitation. The corresponding observations in liquid Argon are only slightly different (i.e. 27.1 versus 23.6) eV/e. These W values are stable with variations of the incident flux however the K/μ values decrease with decreasing flux. The reason for this sensitivity is not yet understood but, perhaps implies that the fit values of K/μ should be considered as upper limits.

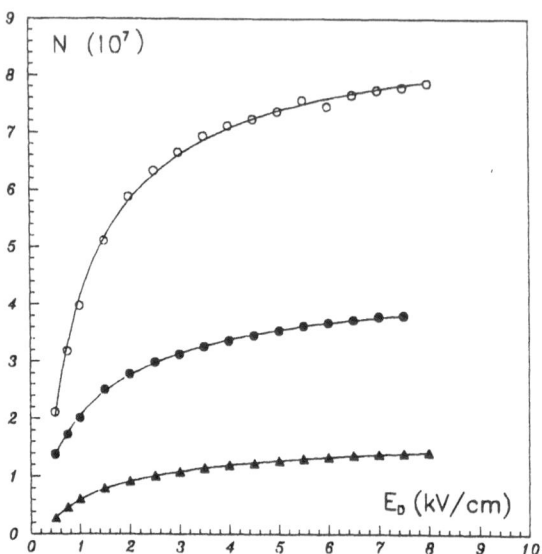

Fig. 3.15. The liquid Xenon collected ionization charge N versus drift field E_d for three total energies E=(0.64, 1.97, 4.02) GeV. The fit curves are to eq. 1, derived in the single-charge density model of Thomas and Imel [8].

Keeping in mind the above cautionary remark, the fit values of K/μ and eq. 1 imply that drift fields of (190, 145, 105) kV/cm respectively, are required to collect 99% of the charge. Improved charge collection should be obtained by increasing the drift velocity, since x (in eq. 1) depends only on $v=\mu E_d$. The addition of ≈3% methane is already planned to reduce pileup but it will have the added advantage of increasing the charge collection efficiency (equivalent to a 7 fold increase in field).

These measurements point out several basic problems in liquid Xenon ionization detectors related to δ ray energies (5 to 100 keV):

1) that the W value is 2.7 times larger than for minimum ionization.
2) that a faster drift velocity v is needed to collect the charge efficiently.

This increased W value may be characteristic of high density Xenon but perhaps not characteristic of lower density Krypton. Recall that an approximately normal value was seen in (low density) liquid Argon. A search for dopants to further increase the drift velocity will be valuable.

3.6.2 ENERGY RESOLUTION

During this same set of runs, energy resolution from the ionization signal was measured. The variation of the incident energy was made by changing the acceleration voltage from 40 kV (≈2.6 keV electrons transmitted through the mylar window) to 95 kV (84 keV electrons) and total energy E from 0.25 to 8.6 GeV. The measured energy resolution is shown in fig. 3.16 along with the fit curve $\sigma_E/E=(1.1+0.3/\sqrt{E})$%. The constant term is thought to be dominated by (as yet) uncontrolled fluctuations in the electron beam flux in addition to straggling in the window. This already excellent energy resolution was obtained only from ionization of (2.6 to 84) keV electrons which suffer most from recombination fluctuations and high W values. It will be indeed surprising if the correlated scintillation measurement does not improve markedly upon this result.

4. OUTLOOK

The long range objective of this project is to design and construct a full scale prototype liquid Xenon calorimeter. The planned calorimeter has transverse dimensions of 10.7 X_0 and length of 28 X_0 with a sensitive volume of 71 liters and total volume of 100 liters, weighing 308 kg. A refrigerator capable of condensing this quantity has been acquired and tested. The mechanical and cryogenic design of this calorimeter has been realized but construction awaits funding and the completion of the laboratory measurements. The goals are:

1) Laboratory demonstration of excellent (≤ 1%) energy resolution from scintillation or ionization and/or from their correlation.

2) Laboratory demonstration of high efficiency detection of liquid Xenon scintillation with CsI photodiodes.

3) Laboratory determination of the effect of dopants (methane, neopentane, butane etc.) on drift velocity, charge yield Y and scintillation yield Y'.

4) Test beam demonstration of the energy resolution attainable from scintillation (or ionization and/or from their correlation) in a totally active liquid Xenon calorimeter.

5) Test beam demonstration of the direction and position resolution from ionization (and drift) in a totally active liquid Xenon calorimeter.

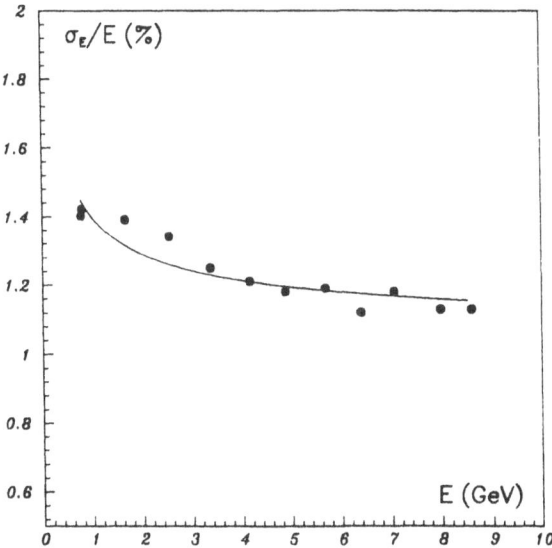

Fig. 3.16. The liquid Xenon ionization energy resolution σ_E/E versus E, the total energy deposited. The fit curve $\sigma_E/E=(1.1+0.3/\sqrt{E})\%$ shows the excellent resolution which characterize totally active calorimeters.

5. CONCLUSIONS

This work has attempted to show, by experiment and simulation, the attractive (and difficult) features of totally active liquid Xenon or Krypton calorimeters. The experimental work has demonstrated extremely good energy resolutions (by both scintillation and ionization) which may still be further improved. It appears possible to determine position and vector direction of (e, γ) however, this remains to be demonstrated. Pileup drift background appears manageable and only limits the tower E_t threshold to 1 GeV. Fast in-situ photosensors have been demonstrated which provide a fast triggering capability for the calorimeter.

A dedicated Higgs search experiment with a liquid Xenon totally active em calorimeter covering $14^0 \le \theta \le 166^0$ with an inner radius $r_i=1$ m and outer radius $r_0=1.8$ m will require a volume of 19 m^3, a mass of 59 tons and cost 59 MSF. The corresponding figures for a liquid Krypton calorimeter are $r_i=1$ m, $r_0=2.3$ m, 44 m^3, 108 tons and 11MSF. These quantities and costs are not excessive, especially if compared to other LHC, SSC or ELOISATRON calorimeters and detectors.

The usual objection made about Xenon concerns its availability and cost. This feature has recently been resolved by the independant proposals of Air Liquide, Matheson, Spectra Gases and Union Carbide to provide 45 tons (each) of new Xenon production by the year 1999 at a price of 2.5 M$/m^3 ≈ 1MSF/ton [12]. The price of Krypton is about ten times less, its natural abundance 20 times more and is now available in the required amounts.

A second usual objection is made relative to cryogenic calorimeters, specifically concerning the cryostat mass. In a dedicated experiment without a magnet or tracker the inner cryostat wall may be located at the beam pipe of the collider. This considerably reduces the surface area and therefore the cryostat mass moreover the beam pipe is itself at a much lower temperature than a Xenon or Krypton calorimeter. It will then simply suffice to thermally shield the beam pipe from the "warm" calorimeter.

If it is shown that a charged particle tracker is necessary for Higgs discovery, a gas tracker working at Xenon or Krypton temperatures can certainly be designed and produced. Alternatively, a small Silicon tracker would work admirably at these temperatures. Recall that (e, γ) tracking will be available in this totally active liquid calorimeter.

A hadron calorimeter or a muon detection system is not needed for Higgs discovery via (e, g) modes. Attempts to marry these devices with a totally active liquid calorimeter will compromise the best features of each system and multiply costs by large factors.

BIBLIOGRAPHY

[1] K. Yoshino, U. Sowada and W.F. Schmidt. Phys. Rev. A14(1976)438.

[2] The LAA Project, Liquid XENON Calorimetry, CERN-LAA/89-1, Pp 187.

[3] J. Seguinot, G. Charpak, Y. Giomataris, V. Peskov, J. Tischhauser and T.Ypsilantis. CERN-EP/90-88 and Nucl. Instrum. Methods A297(1990)133.

[4] J. Seguinot, Y. Giomataris, P. Weilhammer, and T. Ypsilantis. The UV Response of Silicon strip Photodiodes (in preparation).

[5] H.W. Verleur. Journ. Opt. Soc. Am. 58(1968)1356.

[6] L.R. Canfield, J. Kerner and R. Korde. Applied Optics 28(1989)3940.

[7] S. Kubota, M. Hishida and J. Raun. J. Phys. C. 11(1978)2645 & S. Kubota, M. Hishida, M. Suzuki and J. Raun. Nucl. Instrum. Methods 196(1982)101.

[8] J. Thomas and D.A. Imel. Phys. Rev. A36(1987)614.

[9] M. Miyajima, T. Takahashi, S. Konno and T. Hamada. Phys. Rev. A9(1974)9.

[10] S. Kubota, A. Nakamoto, T. Takahashi, T. Hamada, E. Shibamura, M. Miyajima, K. Masuda and T. Doke. Phys. Rev. B17(1978)2762.

[11] T. Takahashi, S. Konno, T. Hamada, M. Miyajima, S. Kubota, A. Nakamoto, A. Hitachi, E. Shibamura and T. Doke. Phys. Rev. A12(1975)1771.

[12] M. Chen et.al. LHC Collider Workshop Proceedings, Aachen, October, 1990.

PROGRESS ON BaF$_2$ CALORIMETERS

G. Charpak[1], V. Peskov[2] and D. Scigocki[1]

1) CERN, Geneva, Switzerland
2) World Lab., Lausanne, Switzerland

ABSTRACT

In this contribution we present our latest results concerning the development of BaF$_2$ calorimeters for experiments at the future large hadron colliders (LHC, SSC, and ELOISATRON). We are elaborating electromagnetic calorimeters made of BaF$_2$ crystals preceded by a BaF$_2$ preshower counter with a high granularity. The readout of this counter is done with parallel-plate avalanche chambers combined with photocathodes. For the BaF$_2$ calorimeter, we have tested new readouts using gaseous ionization chambers. Our latest results on new, stable photocathodes working at room temperature are described as well as recent measurements on the BaF$_2$ radiation hardness. We also present new photosensitive products that are unreactive with air, and a new family of inorganic scintillators which could probably replace BaF$_2$ advantageously in the future.

1. INTRODUCTION

It is commonly acknowledged that the main requirements for the electromagnetic calorimetry to be used at future colliders such as the Large Hadron Collider (LHC), the Superconducting Super Collider (SSC), and the ELOISATRON, are a high speed (15 ns bunch crossing), a high radiation resistance (up to 0.1 MGy), a good energy resolution, and the best possible hermeticity. Different calorimeters, which it was claimed would be equal to these problems, were discussed at recent conferences dedicated to instrumentation for the high-luminosity colliders [1]. Calorimeters made of dense inorganic scintillators were considered to be one of the best options, and a promising candidate is BaF$_2$. The question concerning BaF$_2$ calorimeters, which is still under intense discussion, is: What is the best choice for the readout? Three types of readout have been investigated. The first readout used in BaF$_2$ calorimetry consisted of low-pressure multiwire proportional chambers (MWPCs) filled with a photosensitive vapour [e.g. tetrakis(dimethylamine)ethylene (TMAE)] or using a liquid-TMAE photocathode [2–4]; the device was called a solid-scintillator proportional counter (SSPC). It was based on the overlap, in the spectral region between 170 and 230 nm, of the TMAE quantum efficiency and the BaF$_2$ fast emission. For such emission, the sensitivity of the readout by low-pressure MWPCs and TMAE vapour peaks at 193 nm. The TMAE vapour pressure is rather low (about 0.5 Torr at room temperature), and it is necessary to work at high temperatures, up to 40 °C, in order to increase the efficiency. Under these conditions, about 10 photoelectrons are produced in the detector per MeV deposited in the BaF_2. This figure determines the limit in energy resolution that can be achieved by a device such as an SSPC, i.e. about $2\%/\sqrt{E}$ (GeV) [5]. Experimentally, an energy resolution of about $3.9\%/\sqrt{E}$ (GeV) was measured with a good linearity, a position resolution close to 1 mm, and an e/π rejection factor of 10^{-3} at 5 GeV [5]. A time resolution of better than 1 ns was also obtained [5]. However, this readout is not attractive for BaF_2 calorimeters because of the rapid ageing effect caused by polymerization of the TMAE in MWPCs [6].

Another solution, developed by Lorenz et al. [7], is to read out BaF_2 crystals by means of solid (Si) photodiodes, which are simple and very compact. An energy resolution of about $2\%/\sqrt{E}$ (GeV) was measured between 2 and 40 GeV by recording the fast and slow components of the BaF_2 emission. Such photodiodes have no gain, which greatly simplifies the long-term stability problem, but their use is limited because of the high capacitance of the large devices needed in calorimetry. In addition, these standard photodiodes are sensitive to the BaF_2 slow component (600 ns decay time), which may induce an important pile-up in the high rates expected at the future high-luminosity hadron colliders.

The last readout candidate for BaF_2 calorimetry is a photomultiplier (PM) [8]. A PM is a fast device, but it occupies a lot of space. This does not permit high granularity, and PMs are very sensitive to magnetic fields.

We have developed a new approach to the readout of BaF_2 crystals, either by parallel-plate avalanche chambers (PPACs) combined with photocathodes that are sensitive to the BaF_2 fast emission, or by photosensitive gaseous ionization chambers.

2. NEW READOUT FOR BaF$_2$ CALORIMETERS: GASEOUS DETECTORS WITH PHOTOCATHODES

2.1 Hardness requirements

The bad ageing properties of TMAE vapour at atmospheric pressure is the main disadvantage of the BaF_2 readout by MWPCs [6]. The upper limit of the collected charge is about 10^{-3} C/cm of wire. After this limit, the polymerization of TMAE on the wires results in unstable operation and a continuous decrease of the gain. On the other hand, BaF_2 is extremely radiation-resistant. But until now, all the measurements with respect to BaF_2 radiation hardness were made at very high dose rates during a short time, and generally with small crystals [9, 10]. We have tested a BaF_2 crystal, of dimensions $2 \times 2 \times 5$ cm^3, submitting it to the γ-rays of a ^{60}Co source at a dose rate of 100 Gy/h, up to a total dose of 0.5 MGy. No effect on the scintillation was seen. The transmission before and after irradiation, measured between 190 and 350 nm, is presented in Fig. 1. The effect on the transmission is negligible below 230 nm, where the fast component is emitted. After these measurements, we placed the same crystal close to a CERN SPS beam for a short test in order to obtain the upper limit of the BaF_2 radiation hardness. The additional dose received by the crystal in three weeks was 1.7 MGy. The crystal became blue and lost half of its transmission below 230 nm (see Fig. 1). The gain of an SSPC used in a BaF_2 calorimeter is about 10^3–10^4, and the efficiency with TMAE vapour is about 10 photoelectrons per MeV deposited in the BaF_2 [5]. For a total charge of 10^5 per MeV, we estimate that 1 C

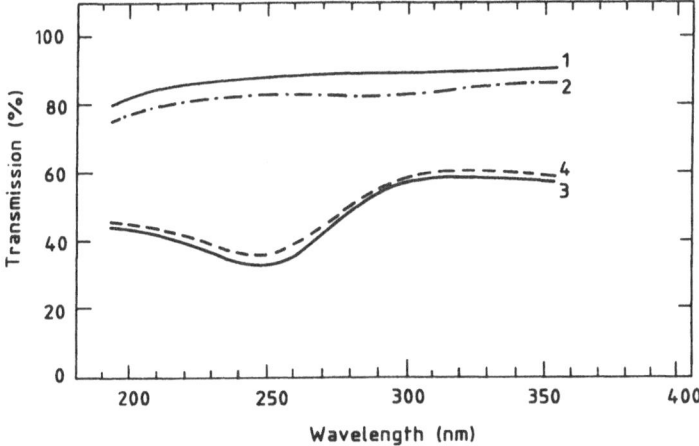

Fig. 1. Transmission of a BaF$_2$ crystal as a function of wavelengths: 1) before irradiation; 2) after 0.5 MGy; 3) after 1.7 MGy; 4) after three weeks recovery.

represents 2×10^3 Gy deposited in the BaF$_2$. For a wire chamber of 10 cm^2, with 2 mm wire spacing, the limit on the current is about 5×10^{-2} C, which represents about 10 Gy deposited in the BaF$_2$. It is clear that the radiation hardness of BaF$_2$ and that of MWPCs filled with TMAE vapour do not match. But until recently, TMAE was the only photosensitive compound that could be used with MWPCs to detect the fast component of the BaF$_2$ emission. We have therefore developed a new type of SSPC made of gaseous detectors, i.e. wire and parallel-plate avalanche chambers combined with new photocathodes as the photosensitive element, working at room temperature [11, 12]. In such detectors the photoelectrons are extracted from the photocathode only, and the gas does not play any role in the photoionization process. Therefore the detector can be filled with gases such as dimethyl ether, which have good ageing properties [13]. In this section we describe our latest measurements on photocathodes, as well as the results obtained with an SSPC made of a PPAC combined with photocathodes.

2.2 New photocathodes for gaseous detectors

During the last two years we have investigated different photocathodes [12, 14–17] for gaseous detectors. Some of them, such as solid TMAE + neopentane, have high efficiency in the wavelength region below 200 nm (more than 5%) [14]. Unfortunately, most of these organic photocathodes are based on TMAE, which interacts strongly with air and is very corrosive for many materials, and which also has bad ageing effects when used as a gas in wire chambers [18]. Thus a complicated technology is required in order for these substances to be deposited as photocathodes in SSPCs, because it is necessary to work under very clean conditions and to be able to cool the surface where the photocathode will be condensed in liquid or solid phases. For applications in calorimetry, where huge detectors will be built, it may be of interest to use a simpler technology; hence these requirements restrict the use of these organic photocathodes.

For these reasons, during the last year we concentrated our investigations on new photosensitive elements which do not interact with air and which can be used at room temperature, employing a simple technology to deposit them on surfaces as photocathodes. Good candidates are triethylamine (TEA) [19], some organometallic compounds [for example diethylferrocene and ethylferrocene (EF)] [14–17], and CsI

[17]. The experimental set-up used to study these photocathodes is described in Ref. [14]. We used a monochromator system to measure the quantum efficiency of these substances, as well as a small prototype SSPC made with BaF$_2$ crystals coupled to a wire chamber so as to be able to investigate these photocathodes in a real experimental situation.

2.2.1 TEA photocathodes

The technique for obtaining these photocathodes is simple. Vapours of TEA were introduced into the detector at room temperature. An adsorbed photosensitive layer is thus formed everywhere in the detector and can be used as a photocathode. The use of a TEA adsorbed layer as a photocathode was studied, and its quantum efficiency for different TEA vapour pressures is shown in Fig. 2. The TEA photocathode is sensitive to the fast emission of BaF$_2$, but has a poor quantum efficiency.

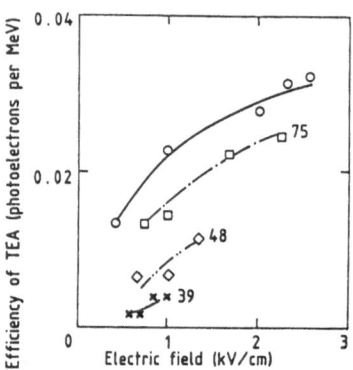

Fig. 2. Quantum efficiency in photoelectrons per MeV deposited in the BaF$_2$ crystal as a function of the electric field applied on the photocathode (in kV/cm), for TEA adsorbed layers at different TEA vapour pressures (75, 48, and 39 Torr) and for a TEA condensed layer deposited on the BaF$_2$ surface.

2.2.2 EF photocathodes

We have obtained better results (between 5 and 10 times higher) with organometallic photocathodes—especially with EF, which was finally chosen for use in further experiments. In the case of liquid EF, we obtained an efficiency of about 7% at 193 nm. In the case of an adsorbed layer, we have measured a quantum efficiency of about 1% at 193 mm. The same efficiency is obtained when the EF adsorbed layer is formed on a BaF$_2$ crystal covered by a mesh and working as a semitransparent photocathode. In this section we present the latest results of our investigation of these two types of photocathodes, which are called reflective and semitransparent, respectively.

a) Semitransparent liquid photocathodes

The installation required for the investigation of an EF liquid photocathode is presented in Fig. 3. It consists of a pulsed VUV source (15 ns duration), a window that is transparent to VUV, a photodiode for the monitoring of the light, and a wire chamber with a drift region closed by a grid over the window. The EF was deposited on the window in two ways. Thin layers (d $\lesssim$ 100 nm) were obtained by condensation when the warm EF vapour was introduced into the chamber at room temperature. Thick layers (0.1 to 1 mm) were obtained by trickling the EF directly through pipettes. In all cases, the wire chamber was operated with CH$_4$ at 0.1 atm. In the latter case,

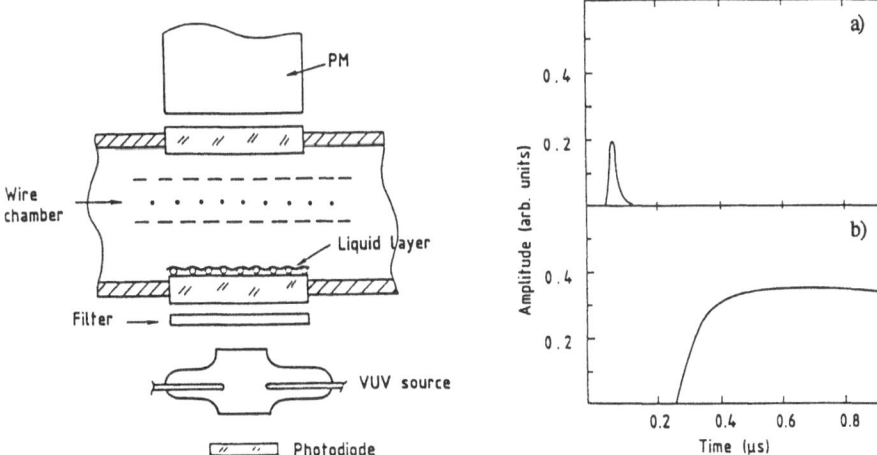

Fig. 3. Experimental set-up for the study of the semitransparent liquid photocathodes.

Fig. 4. Typical signals from a) the photomultiplier, b) the wire chamber.

the thickness of the layers was estimated by the volume of EF deposited on the surface of the window. At small thicknesses the layer was estimated by the absorption of VUV light in the layer deposited on the surface. Figure 4 shows the typical signals obtained simultaneously from the PM and from the wire chamber. It can be seen that there is a delay between them of about 0.2 μs, which is due to the electron drift in the chamber. The efficiency of the EF liquid photocathode as a function of its thickness is given in Fig. 5. We interpret these results in the following way. In the case of a very thin layer, d < 100 nm (Fig. 5a), the extraction of electrons is due to the excess kinetic energy of the photoelectrons, because in the photoionization process ϵ = $h\nu - E_{liq}$, where E_{liq} is the ionization threshold of the liquid. If $\epsilon > V_0$, then some fraction of the photoelectrons can escape from the layer into the gas phase. At small thicknesses (i.e. smaller than or equal to the photoelectron's mean free path in the

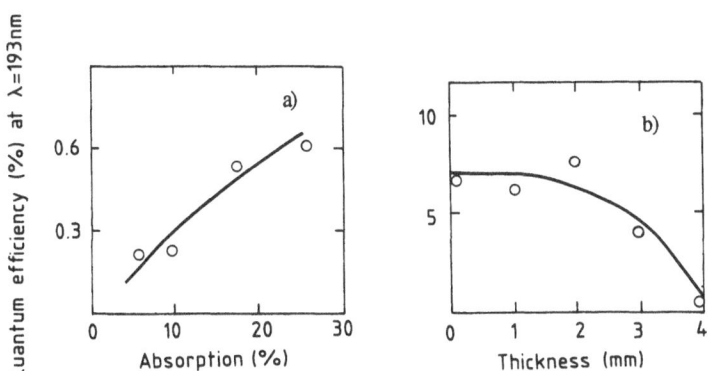

Fig. 5. Quantum efficiency of semitransparent liquid photocathodes as a function of the EF thickness for a) a thin layer, b) a thick layer.

layer), the efficiency increases with the thickness, simply because more VUV photons are absorbed. In the case of thick layers (d > 100 nm, Fig. 5b), the photoelectrons lose their kinetic energy quickly, but then drift, under the influence of the electric field, to the liquid–gas boundary. They gain the kinetic energy $kT_e = eD/\mu$, where k is the Boltzmann constant, μ is the electron mobility, e is the electron charge, and D is the lateral diffusion coefficient. This kinetic energy depends on the strength of the applied electric field, but the maximum possible value is $kT_e < Ed$. This is why the photoemission is high when the voltage in the layer drops by more than 10 V. At a thickness of more than 1 mm, the drop in efficiency is probably due to the electron trapping by impurities.

b) Reflective liquid photocathodes

The results of the investigation of reflective liquid photocathodes are presented in Fig. 6. The efficiency depends slightly on the material used for the support. This means that this device can be employed as the photosensitive element in an SSPC. The efficiency is smaller than the one obtained using TMAE vapour in the traditional SSPC approach. However, with adsorbed layers instead of vapours as the photosensitive element in SSPCs, it is possible to replace the low-pressure MWPC with a compact PPAC without having an intermediate conversion gap [17]. It was suggested that one of the mechanisms of photoelectron emission is the excess kinetic energy $\epsilon = h\nu - V_0$

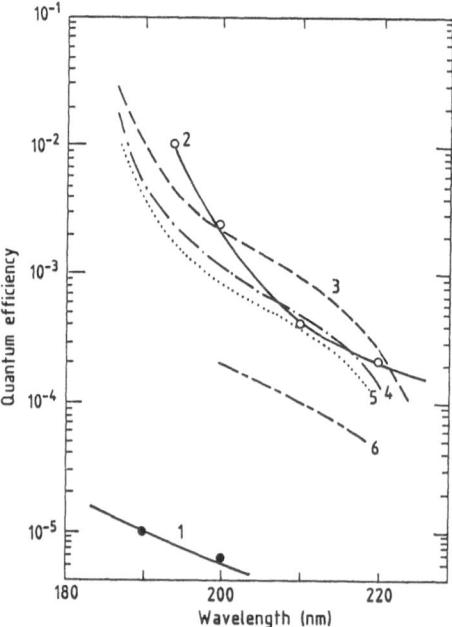

Fig. 6. Quantum efficiency of photocathodes as a function of the photon wavelength λ:
(1) Clean copper cathode.
(2) An EF adsorbed layer deposited on a copper cathode. The EF was produced in the USSR.
An EF adsorbed layer (synthetized at the University of Geneva):
(3) deposited on a copper cathode;
(4) deposited on an indium cathode;
(5) deposited on an aluminium cathode;
(6) when the metallic cathodes (Cu, In, and Al) were kept at 80 °C.

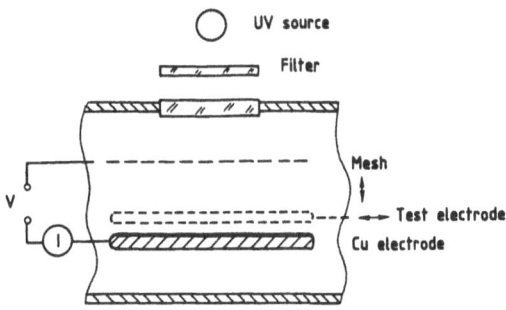

Fig. 7. Experimental set-up for the study of the contribution of the vapour gradient. The test electrode can be inserted as shown by the arrows.

mentioned above. But we have observed that a very thin intermediate layer of vapour between the liquid photocathode and the gas can also contribute to the efficiency. The apparatus used in this investigation is presented schematically in Fig. 7. It consists of a continuous VUV source and an experimental cell containing an ionization chamber. The ionization chamber is made of two electrodes and works at 1 atm. One of the electrodes is a mesh; the other one is a copper plate, which could be covered by a liquid. A typical voltage–current (U·I) characteristic of this chamber, measured in the spectral interval $\lambda = 180\text{–}200$ nm and when only clean metallic electrodes are used, is shown in Fig. 8 (dashed line). The left-hand curve (1) corresponds to the current from the mesh, and the right-hand one (2) to the current from the copper plate. Since the area of the mesh is only 20% of the surface of this plate, the corresponding current is also different. When the copper plate was covered with EF, the U·I characteristic changed dramatically: the current increased by three orders of magnitude, and the U·I curve became symmetric and more steep in the region of low voltage (solid line in Fig. 8). The latter fact indicates that the ions are collected basically from the volume. But when the distance between the two electrodes was increased by a factor of 4, no change in the current was observed. When the copper photocathode was covered by another metallic plate (test electrode), connected electrically to the electrode, the current dropped by a factor of 10, this residual current coming only from the gas volume. When this test plate was placed close to the mesh, the signal also dropped by a factor of 10, which shows that the current indicated by curve 1′ in Fig. 8 was not due to the photoemission process from the mesh. However, it was not due to the volume ionization of the EF vapour either. Therefore, the results of this experiment

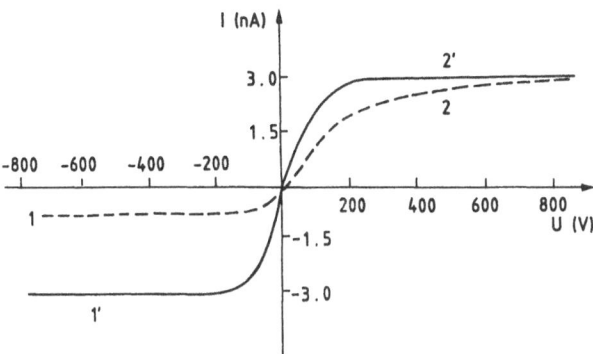

Fig. 8. Typical U·I characteristics of the ionization chamber. Dashed line: for a clean metallic electrode; solid line: value (divided by 10^3) of the current measurement when the electrode was covered with EF.

indicate that a thin layer (< 1 mm) of EF vapour close to the copper electrode gives the most important contribution to the current. This could be due to the gradient of the vapour density.

A thick layer of vapour could slightly spoil the timing properties of the detector. This effect is much less in the case of an adsorbed layer, but the quantum efficiency is also less. The best solution to this problem would be to use solid photocathodes, and our research in this direction is described in the next subsection.

2.2.3 CsI photocathodes

a) Investigation of CsI photocathodes without gas gain

Among the solid photocathodes already investigated, the best result was obtained with pure reflective CsI photocathodes, with a measured quantum efficiency of about 10% below 190 nm [17]: this result was verified by Dangendorf et al. [20][*].

This year some of us, in collaboration with the Séguinot–Ypsilantis group, performed a systematic study of CsI photocathodes [21]. This study was made in the ionization-chamber mode under very clean gas conditions: a stainless-steel vessel and tubes, Oxisorb absorbers, and so on. The quantum efficiency achieved at $\lambda =$ 193 nm was much higher, about 30% (see Fig. 9). This efficiency is so high that it was possible to record VUV scintillation light from the BaF_2 with a deposited energy > 1 GeV. For comparison, Fig. 10 shows the signal from the ionization chamber with a deposited energy of 5 GeV.

Thus the ionization chamber has some advantages over other types of readout. Large surfaces can be obtained at a low price because no window is needed between the BaF_2 crystal and the ionization chamber. These detectors can be flushed, and even if the photocathode is damaged, full recovery is possible after flushing with a

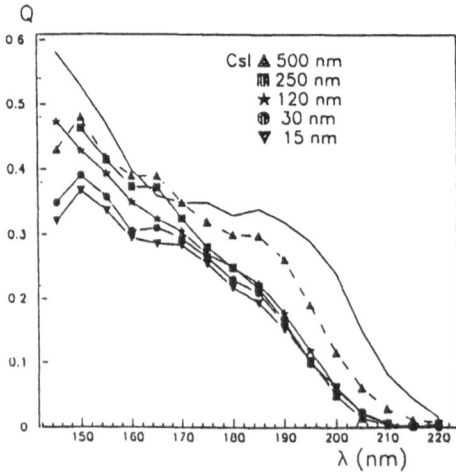

Fig. 9. The measured quantum yield Q versus the photon wavelength λ for five CsI photocathodes of different thicknesses. The solid line is the TMAE gas-phase quantum efficiency.

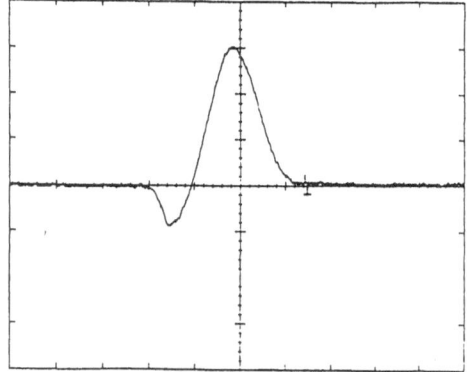

Fig. 10. Typical CsI photocathode signal observed, in an ionization chamber, at the output of the shaper, with a digital oscilloscope Tektronix 2440, for 5 GeV total energy deposited in the BaF_2 crystal. The horizontal and the vertical scales are 1 μs and 2 mV per division, respectively.

---

[*] Probably the first tests of CsI photocathodes in gaseous detectors were carried out by J. Séguinot and T. Ypsilantis in 1979.

gas mixture containing the same photosensitive vapour (TMAE or EF) as the one initially deposited on the CsI photocathode [21]. The transformation of the PPAC in a gaseous ionization chamber without amplification would allow a reduction of the gap and thus a decrease in the drift time of the photoelectrons. For a typical drift velocity of 5 cm/μs, the electron collection time will be only 10 ns in a 500 μm gap. The capacitance of such a detector will be about 2 pF/cm^2. If the typical size of a BaF$_2$ calorimeter module will be about 10 cm^2, the total capacitance of such a fast and large-size ionization chamber will be only 20 pF, which is compatible with fast electronics. Without any gain in the chamber, the mechanical tolerances needed for a PPAC are reduced, and a gain monitoring is no longer necessary. Note that as the photocathode cannot be damaged by feedback from the positive ions, the long-term stability of the chamber, which depends on the behaviour of the photocathode only, would be good. In addition, the ageing caused by polymerization due to the charge density in the avalanche does not exist since there is no gain. The influence of direct ionization is still negligible with respect to the required gap width (about 0.5–1 mm), particularly with a gas mixture based on helium. The capacitance of these gaseous ionization chambers is much lower than that of solid photodiodes.

b) Test of CsI photocathodes under the condition of gas gain

The measurements with the CsI photocathode described above were done in the ionization-chamber mode only. We have repeated them partly with high multiplication gaps, which are necessary for BaF$_2$ preshower counters. The results, presented here, are encouraging.

The set-up is similar to the one described in Ref. [14]. The same single-wire counter was used in the proportional mode, with high gain (up to 10^5), but the metallic cathode was not cooled. The CsI layer (500 nm thick) was evaporated onto the metallic cathode, which was then placed in the detector. A gas mixture of argon (91%) + methane (9%) was flushed into the detector through a bubbler of liquid EF or TMAE, so as to form the adsorbed layer on the CsI photocathode. The bubbler and the detector were kept at room temperature. In some measurements the gas mixture was flushed through the detector after the deposition of the adsorbed layer, without passing through the bubbler; this was done in order to remove the vapour.

Figure 11 shows the relative quantum efficiency versus the emission wavelengths for photocathodes of clean stainless-steel covered with a condensed layer of EF; of

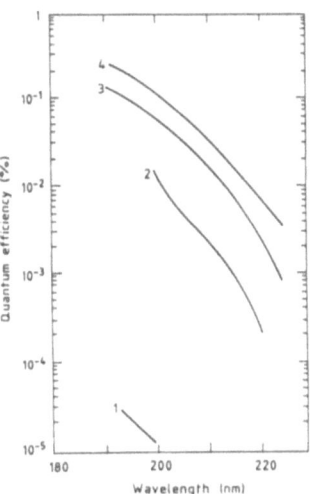

Fig. 11. Quantum yield as a function of wavelengths: 1) clean stainless-steel cathode; 2) EF layer condensed on the stainless-steel cathode; 3) pure CsI; 4) CsI + TMAE adsorbed layer.

pure CsI; and of CsI covered with a TMAE adsorbed layer. With EF + CsI, the quantum efficiency is about 15% at 193 nm; with TMAE + CsI, it is 20%. The main disadvantage of a pure CsI photocathode is that it cannot be in contact with air for more than 10 minutes without losing its sensitivity. We have observed that when the CsI is coated with an EF adsorbed layer, it can stay at least one hour in contact with air without any modification of the quantum efficiency. This effect was also seen with TMAE [21] and was explained by the strong attachment of these molecules to the surface, which changes the chemical properties of the molecule when it is adsorbed. Because of this property, it is probably not necessary to keep the CsI photocathode permanently in contact with the EF or TMAE vapours. Therefore CsI + EF or TMAE photocathodes can be used in the atmosphere of gases chosen for their excellent ageing properties, such as dimethyl ether [13]. These photocathodes are simple and easy to manipulate: they do not suffer from a brief contact with air; they are easy to deposit (vacuum deposition + adsorbed layers); they can achieve an efficiency that is comparable to or better than that of TMAE vapour, with a good. uniformity on large surfaces. For the readout of the BaF_2 fast scintillation, they can be combined with detectors such as PPACs, working at high gain (up to 10^5) without noticeable photon feedback and filled with gases having good ageing properties.

2.3 Investigation of VUV emission from the avalanches in parallel-plate avalanche chambers

In order to understand why we have little or no photon feedback even at a relatively high gain of $\sim 10^5$ (whilst the sensitivity of the photocathode is $> 10\%$), we have investigated the VUV emission from the avalanches. For these measurements, we used a PPAC coupled to a photosensitive wire chamber through a CaF_2 window (Fig. 12). The PPAC was flushed with pure argon or with a mixture of Ar + TEA vapour. The wire chamber was flushed with pure argon or with a mixture of Ar + 17 Torr of TEA. The avalanches in the PPAC were initiated by ^{55}Fe or ^{241}Am sources. The measurements were done in the PPAC at a gas gain of between 1 and 10^4. The VUV photons produced during the development of the avalanche in the PPAC were detected by the wire chamber with a probability of about 10% to 25%.

Some results of our measurements are presented in Fig. 13. One can see that the VUV emission from the PPAC is detected only for high charges collected on the anode: about 0.25 pC for the ^{55}Fe source, which corresponds to a gain of about 6 $\times 10^3$. Taking into account the efficiency of the wire chamber and the solid angle, the number of the VUV photons emitted in the avalanche is estimated to be about 10. With TEA, the number of inelastic collisions, which we estimate [22] from the intensity of the emission at 280 nm, is about 10^6. Therefore, during 10^6 inelastic collisions, only about 10 VUV photons are created. The reason is explained in Ref. [22]. According to our measurements, this probability decreases sharply when the

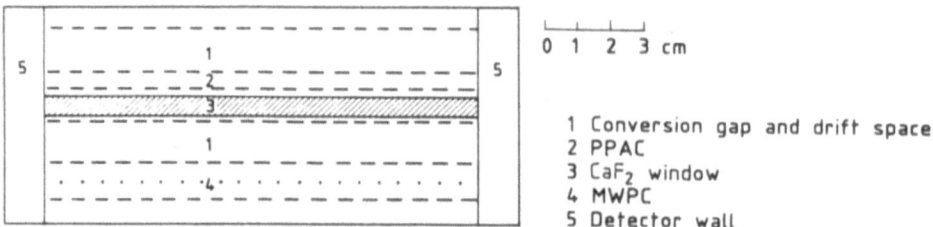

Fig. 12. Set-up for the investigation of the VUV emission from the avalanches in a PPAC.

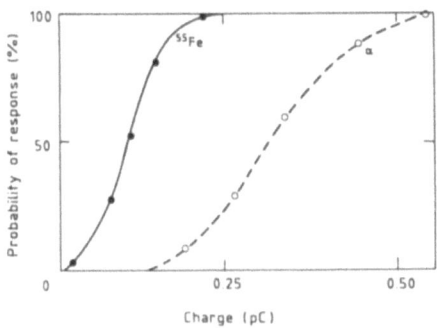

Fig. 13. Probability of detection of photoelectrons in the photosensitive wire chamber versus the collected charge in the PPAC, for α-particles and 5.9 keV photons from a ^{55}Fe source.

TEA concentration is increased. The same results are obtained with TMAE [22]. Our experiments show that solid photocathodes, which are highly efficient in the VUV region, can be combined with gaseous detectors that are being designed by us for the multiplication of the photoelectrons extracted from the photocathodes, even with large gains. These results are part of a systematic study, which is being pursued by us together with P. Fonte and F. Sauli, on the accurate characterization of photon emission in avalanches in the gas mixtures commonly used in various gaseous detectors.

3. PRELIMINARY BEAM TEST OF A BaF$_2$ CALORIMETER PROTOTYPE WITH A CsI PHOTOCATHODE

Looking at Figs. 9 and 11, one can see that a different quantum efficiency was obtained with and without gas gain: without gain it was a few times less. Earlier, the same results—lower efficiency at high gain—were systematically obtained by Dangendorf et al. [20]. This discrepancy should be carefully studied. Probably, the explanation could be the following: during ion bombardment, the implantation of impurities occurs. We should also take into account that all the measurements quoted in Ref. [21] were done under 'ideal' conditions, i.e. very clean gas, stainless-steel chamber and gas supply tubes, short exposition to air during installation of the CsI photocathode, and so on. Nevertheless, even the efficiency of about $\sim 10\%$ at $\lambda =$ 190 nm, achieved with a gas gain, is quite high enough for some applications. In order to have a clearer perception of other problems that can arise during CsI photocathode operation, we made a preliminary beam test of a small prototype BaF$_2$ calorimeter with such a photocathode*). The composition of the calorimeter prototype is shown in Fig. 14.

It consists of a BaF$_2$ crystal, ~ 20 cm long and ~ 5 cm in diameter, terminating in a PPAC with a CsI photocathode. The PPAC was formed by a mesh, touching the crystal, and by the CsI photocathode itself. The gap between the electrodes was about 4 mm.

The chamber was flushed with He + 10% CH$_4$ at a pressure of 1 atm. In order to imitate the real experimental conditions, no special precautions were taken to ensure that the system was clean. Also, because of a mishap, the chamber had sprung quite a big leak. Nevertheless the chamber worked in a very stable fashion up to a gain of

---

*) This work was done in collaboration with N. Zaganidis.

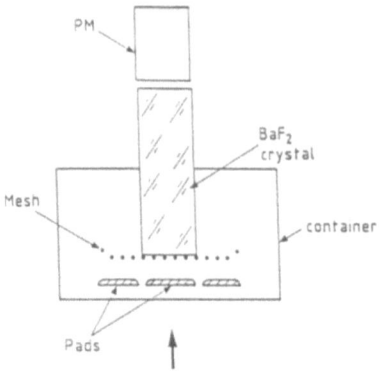

Fig. 14. Prototype of a BaF$_2$ calorimeter with a PPAC.

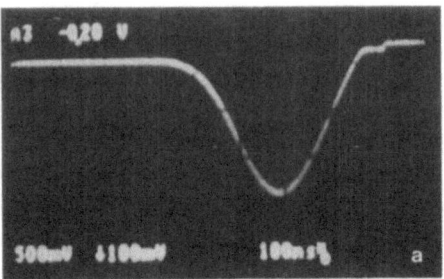 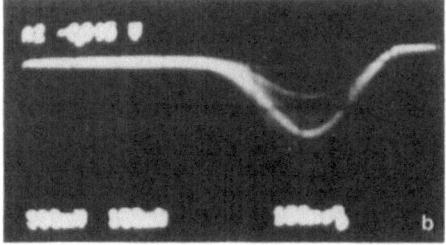

Fig. 15. Signals obtained at the output of the shaper in the PPAC for energy depositions, in the BaF$_2$ crystal, of a) 5 GeV from electrons, and b) 130 MeV from pions.

$\sim 10^3$–10^4. Typical signals obtained with 5 GeV deposited by electrons and 130 MeV by pions are presented in Fig. 15.

The energy resolution obtained was about 9% with 1 GeV energy deposited in the BaF$_2$. This is still a factor of 2 lower than that obtained with TMAE gas [5] or with a pure CsI photocathode without gain [21], and we are preparing another beam test using a CsI + EF photocathode, with a quantum efficiency that is at least twice as high. We are also planning to make a systematic study of a CsI photocathode working under the condition of gas gain in order to understand the factors that reduce the quantum efficiency.

4. BaF$_2$ CALORIMETER DESIGN

The final design of the experiments destined for the future large hadron colliders (LHC, SSC, and ELOISATRON) is not yet decided. But it is evident that the experiments need high-performance calorimeters. They have to be fast, be very resistant to radiation, and have a very good energy resolution and hermeticity and a good granularity [1]. Two families of detectors are being investigated. The traditional approach uses separate detectors to measure electromagnetic and hadronic showers. In the second concept, the same detector measures both the electromagnetic and hadronic showers. In the latter case, the electromagnetic energy resolution is degraded because it is necessary to use sampling techniques in order to obtain the compensation

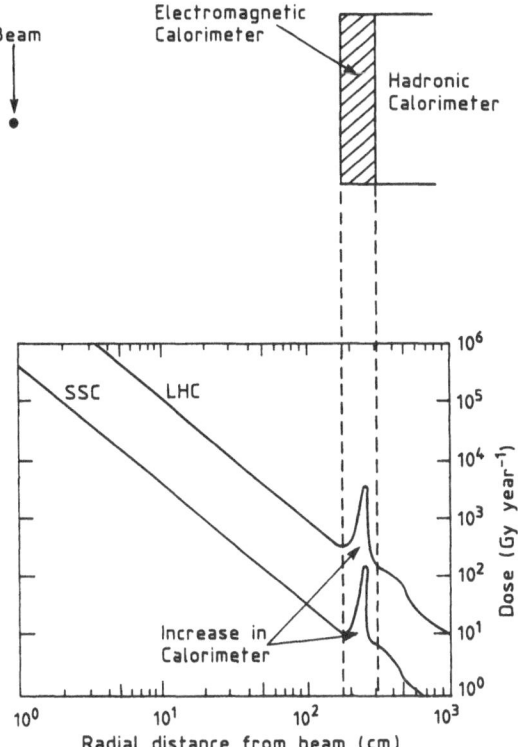

Fig. 16. Radiation dose in the calorimeter as a function of the distance from the beam.

(e/h = 1) for the hadronic measurements. The electromagnetic energy resolution measured with such detectors is about $15\%/\sqrt{E}$ (GeV), compared with 2–$4\%/\sqrt{E}$ (GeV) with homogeneous devices [23]. One of the main problems for detectors placed in the environment of future high-luminosity colliders is the high level of radiation. The radiation doses decrease when the distance between the detector and the beam increases, except in the electromagnetic part of the calorimeter where the π^0's produce a strong increase in the level of radiation (about a factor of 10) [24] (Fig. 16). This dose determines the radiation resistance needed by a detector that is used as the electromagnetic and the hadronic part of the calorimeter if the same technique is employed. If different detectors are chosen, the electromagnetic device will shield the hadronic one, in which case the radiation resistance will be 10 times greater than if the same detector were used for both the electromagnetic and the hadronic measurements.

We are developing an homogeneous electromagnetic BaF_2 calorimeter composed of a BaF_2 preshower counter with high granularity, followed by three or four layers of BaF_2 crystals, the total thickness being 25 radiation lengths. The geometry is presented in Fig. 17. The granularity for the BaF_2 calorimeter is determined only by the energy measurements and by a longitudinal sampling that is good enough for the e/π separation. The BaF_2 preshower counter placed in front of the calorimeter is about 2.5 radiation lengths thick, with a high-granularity readout (about 5 mm) to measure the position of the showers. Figure 18 shows a possible minimum-volume configuration using a BaF_2 calorimeter. The minimum distance from the beam is about 50 cm. The volume of BaF_2 needed for such a geometry is about 15 m^3.

A BaF_2 preshower counter, which appears to be a good candidate for the readout of the BaF_2 calorimeter, is described in the following subsections, together with the possibility of passive compensation with such an homogeneous device.

41

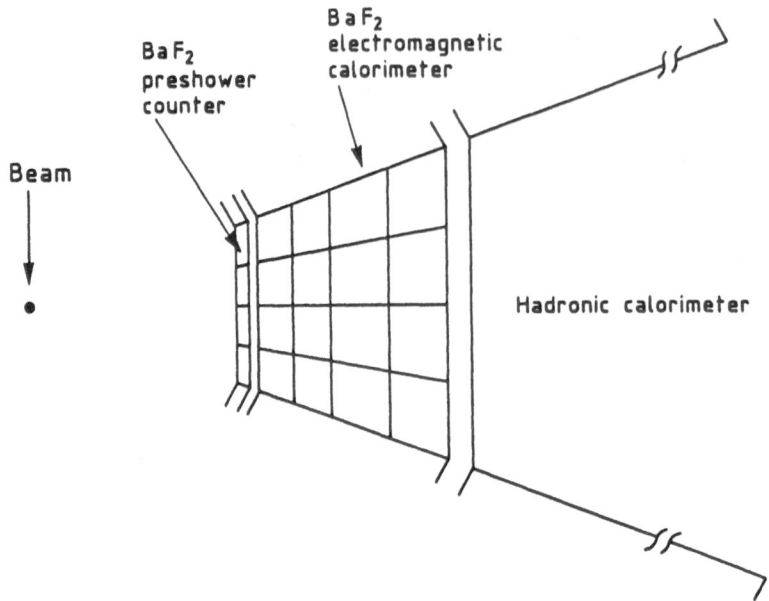

Fig. 17.   Design of the proposed BaF$_2$ calorimeter.

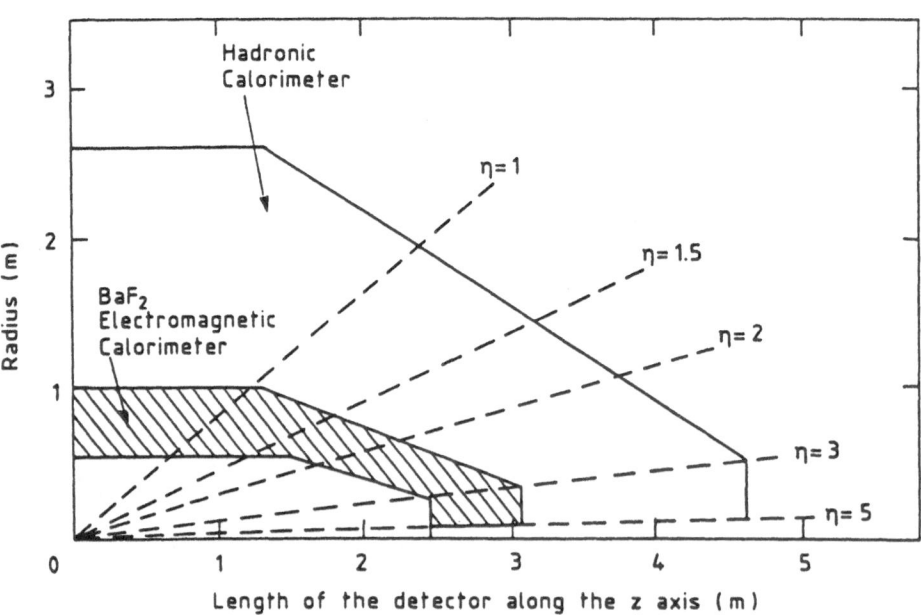

Fig. 18.   A possible minimum-volume configuration with a BaF$_2$ calorimeter for an experiment at the LHC, SSC, or ELOISATRON.

4.1   BaF$_2$ preshower counter

A preshower counter in front of a calorimeter is needed for high-precision position measurements of the showers, and to increase the e/π rejection capability. Such a device could be necessary not only for the BaF$_2$ calorimeter described in this paper, but also for many of the detector candidates for the calorimetry to be used in experiments

42

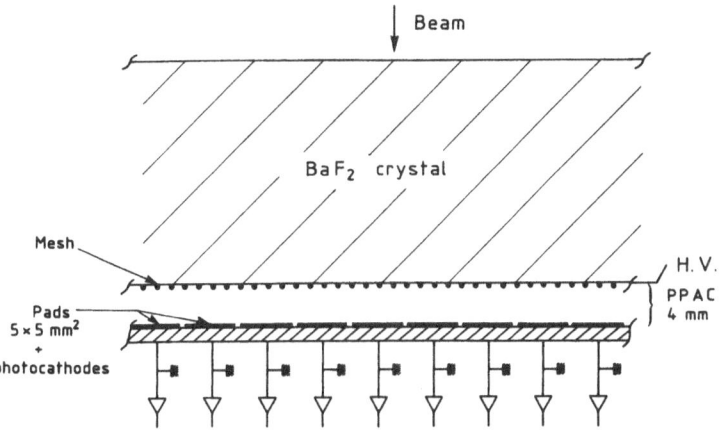

Fig. 19.   Construction of the BaF₂ preshower counter.

at the future high-luminosity colliders. We have made a preshower counter with BaF_2 crystals read out by PPACs combined with photocathodes. The design of the detector is presented in Fig. 19. Crystals of 2.5 radiation lengths (5 cm) are used, with the granularity of the calorimeter module following that of the preshower counter. Each crystal is coupled to a 3–4 mm thick PPAC having independent metallic pads, 5 × 5 mm², for reading out the signal produced in the chamber. Semitransparent or reflective photocathodes can be used, and the detector can work in either of two polarities simply by reversing the direction of the applied field. The best known candidate is the reflective photocathode made of CsI with an adsorbed layer of TMAE, deposited on the pads (see subsection 2.2.3 and Ref. [21]). Minimum-ionizing particles (MIPs) will deposit 32 MeV in the detector; an electromagnetic shower will deposit a few hundred MeV. For the photocathode efficiency estimated at about a few photoelectrons per MeV deposited in the BaF_2, the gain needed in the PPAC is about 10^3–10^4. The position of the shower, or of the MIPs, will be measured on the pads by the centre-of-gravity method, and the difference in the energy deposited will make it possible to measure a good e/π separation in the preshower counter. A prototype is being tested.

4.2   Readout of BaF₂ calorimeters

The relatively poor efficiency of SSPCs—about a few photoelectrons per MeV deposited in BaF₂—cannot really degrade the energy resolution of a calorimeter designed for experiments at the future large hadron colliders. The main limitation on the energy resolution of all the calorimeter candidates in such experiments comes from the constant term, which depends essentially on the calibration and the stability of these detectors. This term should be less than 1%, and readout systems with gain—such as SSPCs and PMs—involve complex technologies for controlling the stability of huge detectors at this level. The average thickness of the BaF₂ crystals in the calorimeter proposed in this paper is about 10–15 cm for a few layers. In this case, the energy deposited per crystal by an electromagnetic shower is high. At the energies that will be investigated in experiments where such a detector can be used, there will probably be no need for any gain in the SSPCs of the BaF₂ calorimeter. In order to simplify the readout of the BaF₂ calorimeter, we are attempting to transform PPACs combined with photocathodes into photosensitive gaseous ionization chambers working with a very small gain or without gain.

5. NEW DEVELOPMENTS

This year, in parallel with the construction and testing of the BaF_2 preshower counter and BaF_2 calorimeter, we have started to develop i) new photocathodes, ii) new crystals.

5.1 New photocathodes

The aim of our collaboration with the Laboratory of Organometallic Compounds of the Moscow State University is to make a systematic study of the chemical and physical properties of new compounds to be used as efficient photocathodes in the VUV spectral region for gaseous detectors, and particularly to read out BaF_2 scintillation. This laboratory has already supplied us with very good quality ethyl ferrocene, and has synthesized for us new organometallic compounds that are estimated to be good candidates for photocathodes: 1,1'-trimethyleneferrocenophene, biferrocenyl, (dimethylamino)methylferrocene, ferrocenyl phosphine, and ferrocenyl methyl alcohol and such derivatives. The photocathodes made from these products are now being tested. The laboratory also plans to carry out a systematic study of bent-sandwich complexes and square-planar complexes. It is expected that such molecules can combine small chemical activity with low ionization potential.

5.2 New scintillators

The efficiency of all the photocathodes which we have studied increases greatly in the wavelength region below 200 nm. This means that SSPCs using these photocathodes will be much more efficient with scintillators that emit in this region and thus achieve better energy resolution. However, BaF_2 was until recently the only known scintillator to possess adequate properties for SSPCs.

The discovery that in lanthanum fluoride (LaF_3) crystals doped with Nd^{3+} the scintillation peaks at 173 nm and has a decay time of 6.3 ns, allows the use of this scintillator in an SSPC filled with TMAE vapour [25]. The main advantages of LaF_3 over BaF_2 are its higher density (5.94 g/cm^3) and its shorter radiation length (X_0 = 1.7 cm), but the light yield is small (because it is difficult to obtain crystals with good transparency in the wavelength emission region) and the detection efficiency is no higher than that obtained with the BaF_2 device. On the other hand, the mechanism by which BaF_2 emits light with a fast component was not understood until the studies made by Valbis and his group [26] enabled them to explain it and to understand why BaF_2 has such a unique ensemble of properties as a scintillator. Based on this research, a new family of inorganic scintillators was discovered [26, 27]. Their properties are similar to those of BaF_2, but their maximum light emission is in a wavelength region between 140 nm and 200 nm, where the known photosensitive vapours and photocathodes achieve high efficiencies. The mechanism that explains the fast component of the BaF_2 scintillation, called cross-luminescence (CRL), was measured in CsCl, CsBr, RbF, and RbCl [26], as well as in KF [27]. It is possible to add a third component to these simple crystal bistructures in order to obtain crystals of higher densities, with the same light emission. The first crystals of this type were $KCaF_3$ and $KMgF_3$, and their emission spectra are shown in Fig. 20. An SSPC with a $KMgF_3$ crystal coupled to a wire chamber filled with TEA vapour achieved half the efficiency obtained with BaF_2 and TMAE vapour, i.e. four photoelectrons per MeV deposited in the crystal [28]. Then a prototype $KMgF_3$ crystal combined with a PPAC filled with TEA vapour was tested at Serpukhov and achieved the same efficiency [29]. But if the new scintillators, such as $KMgF_3$, have scintillation properties equivalent

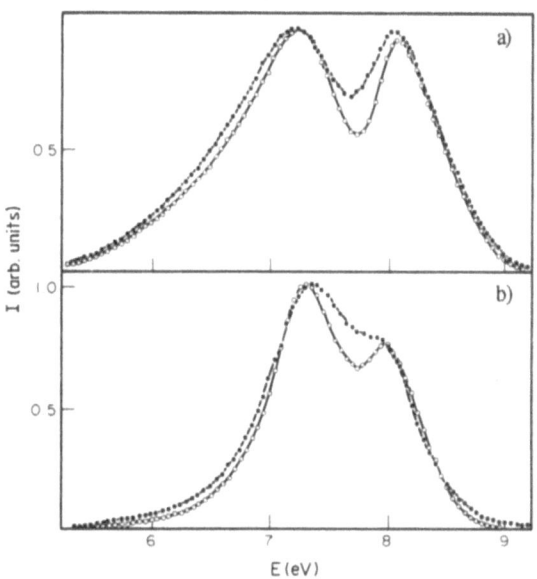

Fig. 20. Luminescence spectra of a) $KMgF_3$ and b) $KCaF_3$, both at 80 K (o) and 300 K ($\bullet$), as a function of photon energy.

to those of BaF_2 and a similar high resistance to radiation, they have the handicap of a lower density, 3.1 g/cm^3, and longer radiation lengths, $X_0 = 7\text{--}8$ cm. Valbis and his group [30], are now working on new crystals based on lutetium ($KLuF_4$) or yttrium (K_2YF_5), which have higher density and shorter radiation length, e.g. ~ 4.8 g/cm^3 and $X_0 \sim 2$ cm for $KLuF_4$. The emission spectrum of $KLuF_4$ is presented in Fig. 21. The maximum emission is peaked at 165 nm. But the main limitation comes from the fact that these new dense crystals probably cannot be obtained in large sizes, because their thermochemical properties do not allow the use of large autoclaves, similar to those used in producing large SiO_2 crystals for example, and they cannot be obtained from the melt by other methods.

However, different techniques are now being used to develop new crystals that are optically perfect, of a reasonable size, and have good properties as scintillators. We have also established a collaboration with the Institute for Physical Problems (Academy of Sciences of the USSR, Moscow), where some other new crystals will be synthesized for us.

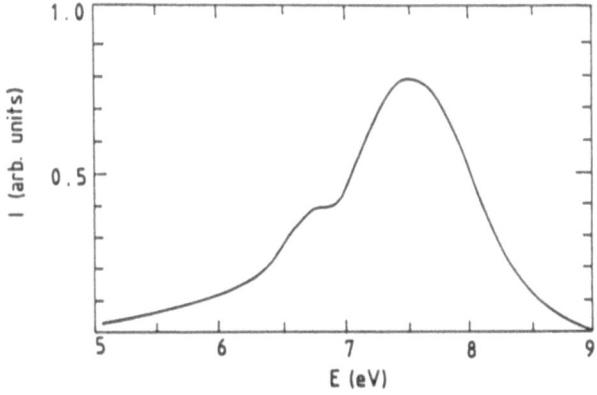

Fig. 21. Luminescence spectrum of $KLuF_4$ as a function of photon energy.

45

6. CONCLUSIONS

In studying the coupling of BaF_2 scintillator crystals to gaseous detectors, we have taken several steps towards showing the feasibility of BaF_2 high-granularity preshower detectors and electromagnetic calorimeters with properties matching the requirements of future colliders: fast timing, radiation hardness, and good granularity and energy resolution.

Our study reveals that simple amplifying structures made of solid photocathodes between parallel electrodes, in gases at atmospheric pressure, permit adequate amplification.

Steady progress is being made in the study of photocathodes that can be simply composed of adsorbed layers of photosensitive vapours or thin layers of photosensitive liquid mixtures, as well as solid photocathodes with adsorbed layers of photosensitive vapours with efficiencies of the same order of magnitude as that of photomultipliers in the VUV region.

Research is being carried out to replace BaF_2 with heavier scintillators emitting at a wavelength region that permits higher efficiencies to be achieved with gaseous detectors. But among the scintillators that will be available in the near future, it would seem that BaF_2 is the best candidate for the construction of calorimeters. Therefore, a thorough investigation should be undertaken to find a means of reducing the cost of the crystals for large-volume detectors to be used in future hadron colliders.

REFERENCES

[1] E. Fernandez and G. Jarlskog (eds.), Proc. ECFA Study Week on Instrumentation Technology for High-Luminosity Hadron Colliders, Barcelona, 1989 (CERN 89–10, ECFA 89–124, Geneva, 1989), 2 vols.

   T.J. Gourlay and J.G. Morfin (eds.), Proc. Symp. on Particle Identification at the High-Luminosity Hadron Colliders, Batavia, 1989 (Fermilab, Batavia, 1990).

   Proceedings of the Conference on Very High Resolution Electromagnetic Calorimetry, Sigtuna (Sweden), 1989.

[2] D.F. Anderson, Phys. Lett. **18** (1982) 230.

[3] D.F. Anderson, R. Bouclier, G. Charpak and S. Majewski, Nucl. Instrum. Methods **217** (1983) 217.

[4] D.F. Anderson, G. Charpak, C. von Gagern and S. Majewski, Nucl. Instrum. Methods **225** (1984) 8.

[5] R. Bouclier et al., Nucl. Instrum. Methods **A267** (1988) 69.

[6] C.L. Woody, IEEE Trans. Nucl. Sci. **NS–35** (1988) 493.

[7] E. Lorenz, G. Mageras and H. Vogel, Nucl. Instrum. Methods **A249** (1986) 235.

[8] M. Murashita et al., Nucl. Instrum. Methods **A243** (1986) 67.

[9] S. Majewski and D.F. Anderson, Nucl. Instrum. Methods **A241** (1985) 76.

[10] A.J. Caffrey et al., IEEE Trans. Nucl. Sci. **NS–33** (1986) 230.

[11] G. Charpak, V. Peskov, F. Sauli and D. Scigocki, CERN EP Internal Report 88–02 (1988).

[12] V. Peskov, G. Charpak, F. Sauli and D. Scigocki, Nucl. Instrum. Methods **A283** (1989) 786.

[13] M. Jibaly et al., Nucl. Instrum. Methods **A283** (1989) 692.

[14] V. Peskov et al., Nucl. Instrum. Methods **A269** (1988) 149.

[15] G. Charpak et al., Proc. Int. Workshop on Liquid-State Electronics, Berlin, 1988 (Hahn-Meitner-Institut, Berlin, 1988), p. 85.

[16] D. Astruc et al., same proc. as Ref. [15], p. 109.

[17] G. Charpak et al., Proc. Batavia Symposium in Ref. [1], p. 295.

[18] C.L. Woody, IEEE Trans. Nucl. Sci. **NS-35** (1988) 493.

[19] R.A. Holroyd, paper 6408, submitted to the 23rd Int. Conf. on High-Energy Physics, Berkeley, 1986.

[20] V. Dangendorf, A. Breskin and R. Chechik, A gas-filled UV photon detector with CsI photocathode for the detection of the xenon light, Weizmann Inst. report WIS-89/52 (1989).

[21] T. Ypsilantis, Proc. ECFA Study Week in Ref. [1], vol. 2, p. 661.
J. Séguinot, G. Charpak, V. Peskov, Y. Giomataris and T. Ypsilantis, article in preparation.

[22] V. Peskov, G. Charpak, W. Dominik and F. Sauli, Nucl. Instrum. Methods **A277** (1989) 547.

[23] C.W. Fabjan, *in* Experimental techniques in high-energy physics, ed. T. Ferbel (Addison-Wesley, Menlo Park, Calif., 1987), p. 257.

[24] E. Lorenz, Proc. ECFA Study Week in Ref. [1], vol. 2, p. 621.

[25] P. Schotanus, C.W. Van Eijk and R.W. Hollander, Nucl. Instrum. Methods **A272** (1988) 913.

[26] J.L. Jansons, V.J. Krumins, Z.A. Rachko and J.A. Valbis, Phys. Status Solidi (b): **144** (1987) 385.

[27] J.L. Jansons, V.J. Krumins, Z.A. Rachko and J.A. Valbis, Solid State Commun. (No. 2) **67** (1988) 183.

[28] A.F. Buzulutskov, L.K. Turchanovitch and V.G. Vasil'chenko, Nucl. Instrum. Methods **A281** (1989) 99.

[29] A.F. Buzulutskov, L.K. Turchanovitch and V.G. Vasil'chenko, private communication.

[30] A. Zichichi et al., Report on the LAA project (vol. 6), presented at the Third LAA Open Meeting, CERN, Geneva, 1989.

A NEW SCINTILLATOR, $LiBaF_3$, AND ITS BASIC PROPERTIES

V. Peskov and A. Zichichi

CERN
Geneva, Switzerland

World Laboratory
Lausanne, Switzerland

The purpose of this paper is to report on the existence of a new scintillator: $LiBaF_3$. Its properties are: density 4.9 gr/cm^3, radiation length $X_0 = 11.1$ g/cm^2. When a new scintillator is found a set of basic tests is needed in order to establish:

i) its emission spectrum;

ii) the coupling of the spectrum with a detector.

The first set of tests regarded the emission spectrum. This was done using a system of four different photomultipliers, in order to cover a large range of spectral sensitivities.

The photomultipliers (PMs) were: Hamamatsu R1460 (band of spectral sensitivity $\Delta\lambda = 185\text{-}300$ nm), Hamamatsu R324 ($\Delta\lambda = 200\text{-}400$ nm), Philips XP2020Q ($\Delta\lambda = 250\text{-}500$ nm) and 56DVP ($\Delta\lambda = 300\text{--}600$ nm). The range of λ to be observed was therefore: $\Delta\lambda = 185\text{-}600$ nm. For all measurements the crystals were placed in optical contact with the PM. Notice that all PMs have comparable rise-time ($\sim 2\text{--}5$ ns), but the Philips XP2020Q has a very fast decay time: ~ 3 ns. This allows a fast component of the scintillation light to be detected.

The samples studied were cylinder-shaped crystals: height 13 mm and diameter 12 mm. For reasons related to the production process the crystals studied were not very clean. The first step was the study of their purity: an X-ray energy dispersive analysis clearly showed some Pb contamination. Furthermore, transmission measurements were performed. They showed a strong absorption (about a factor of 10) of light for wavelengths $\lambda = 185\text{--}205$ nm, but the crystals were practically transparent for $\lambda > 210$ nm.

Nevertheless as the aim was to test the crystal as a scintillator, small Pb impurities could be neglected, because they did not have any effect for $\lambda > 210$ nm.

In order to measure the spectrum of the emitted light, a set of filters (WG 295, FNQ 023, FNQ 015) was used. The results showed that the scintillation spectrum of the $LiBaF_3$ is concentrated in the region $\Delta\lambda = 185\text{--}250$ nm.

The $LiBaF_3$ crystals were irradiated using three types of particles: β from ^{90}Sr, and α and X-rays from ^{241}Am. Time and spectral characteristics were measured with both radioactive sources and the results were all consistent. Let us quote some special case. A typical signal, obtained with the R1460 PM, when the crystal was irradiated using β from the ^{90}Sr, is shown

New Technologies for Supercolliders, Edited by L. Cifarelli and
T. Ypsilantis, Plenum Press, New York, 1991

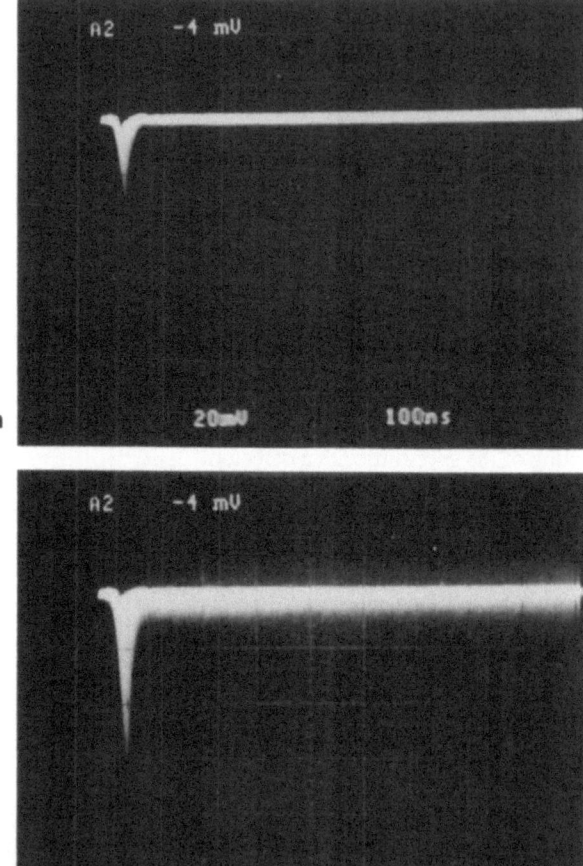

Fig.1. Scintillation light signal obtained via a β-source from ^{90}Sr:
   a) from LiBaF$_3$;
   b) from BaF$_2$.
Notice: in a) only the fast component is present, in b) the fast and the slow components
   are clearly visible.

in Fig.1a. For comparison, the signal from the scintillation light in a BaF$_2$ crystal of 2 x 2 x 4 cm^3, obtained using the same PM, at the same voltage, and with the same radioactive source, is presented in Fig.1b. Through a comparison between these two pictures, it is possible to conclude that only the fast scintillation light is present in the case of the LiBaF$_3$ crystal. Measurements made using the fast PM, XP2020Q, show that the decay time of this scintillation light is less than a few nanoseconds. After subtraction of the slow-component contribution, and taking into account the solid-angle differences (the solid angle at which the β radiation reached the BaF$_2$ crystal was wider than in the case of the LiBaF$_3$), it can be concluded that the fast component of the BaF$_2$ gives about 25% more scintillation light than the fast component of the LiBaF$_3$.

Let us denote the photon yield observed by the PM as [yield]$_{BaF_2}$ and [yield]$_{LiBaF_3}$ respectively. The result for the ratio is:

$$\text{[yield]}_{BaF_2} / \text{[yield]}_{LiBaF_3} \cong 1.25 \tag{1}$$

This ratio is suggestive of an interesting interpretation. In fact, it is well known that the fast component of the BaF$_2$ emission spectrum has two peaks whose photon yields are in the ratio:

$$\text{[yield]}_{BaF_2}^{\lambda=195} / \text{[yield]}_{BaF_2}^{\lambda=220} \cong 25\ \% \tag{2}$$

If the LiBaF$_3$ emission spectrum is like the BaF$_2$ one, then it is expected that

$$\text{[yield]}_{LiBaF_3}^{\lambda=195} / \text{[yield]}_{LiBaF_3}^{\lambda=220} \cong 25\ \% \tag{3}$$

The explanation of the measured ratio (1) is as follows: Pb impurities inside a LiBaF$_3$ crystal do absorb radiation for $\lambda < 200$ nm. Therefore the ratio (1) corresponds to the following

$$\frac{\text{[yield]}_{BaF_2}^{\lambda=195} + \text{[yield]}_{BaF_2}^{\lambda=220}}{\text{[yield]}_{LiBaF_3}^{\lambda=195} + \text{[yield]}_{LiBaF_3}^{\lambda=220}} \tag{4}$$

where [yield]$_{LiBaF_3}^{\lambda=195}$ is equal to zero because of the Pb impurities. This also means that

$$\text{[yield]}_{BaF_2}^{\lambda=220} / \text{[yield]}_{LiBaF_3}^{\lambda=220} \cong 1\ . \tag{5}$$

In other words the two emission spectra, of LiBaF$_3$ and of BaF$_2$ are nearly the same. The direct proof that this is indeed true is part of our project: in fact, the emission spectrum of LiBaF$_3$ will be measured as soon as Pb impurities free LiBaF$_3$ crystals will be available.

On the other hand, further evidence of the equivalence between the emission spectra of LiBaF$_3$ and BaF$_2$, comes from the second set of tests quoted above. Namely, the one where the coupling of the emission spectrum of LiBaF$_3$ with a detector has been studied. The aim of this second set of tests was to study the possibility of using a gas filled photosensitive detector instead of a PM. This new set of tests was done in two ways.

The first using a photosensitive gas, the second using a solid photosensitive layer. The advantage of the solid layer is that it offers the possibility of reducing the size of the detector, in the direction of the beam, down to a few mm [1,2]. Another advantage is the excellent time resolution: 0.5 ns [3].

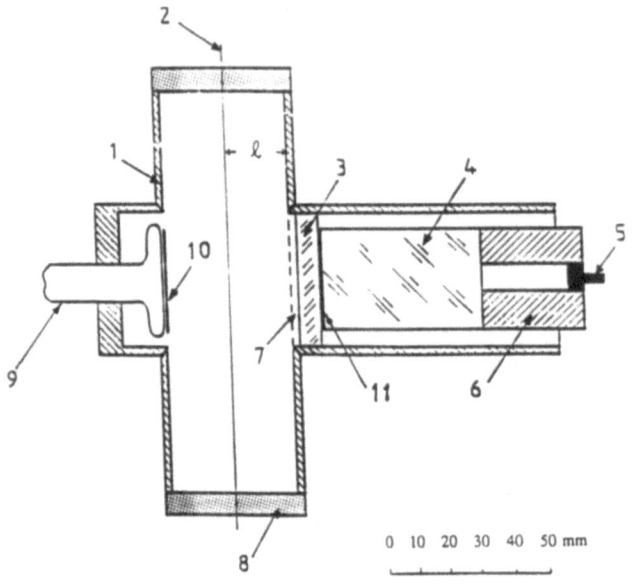

Fig.2. Set-up of the single-wire counter: 1) body of the counter (grounded), 2) anode wire to which the high voltage was applied, 3) CaF$_2$ window, 4) crystal to be tested, 5) radioactive source, 6) lead collimator, 7) cathode mesh, 8) ceramics, 9) aluminium cathode, 10) 0.5 µm thick layer of solid CsI with adsorbed TMAE, 11) optical contact.

The schematic drawing of the detector [4] is presented in Fig.2. It is a cylindrical single-wire counter of 40 mm diameter. The diameter of the (anode) wire is 50 µm.

Two transverse flanges were added in the middle of the cylinder. One of the flanges was equipped with a CaF$_2$ window (fully transparent for $\lambda > 130$ nm). The main reason for the CaF$_2$ window was to keep the gas system tight.

In the other flange a cathode, which could be covered by solid photosensitive layers, was mounted. The crystals were placed in contact with the CaF$_2$ window and were irradiated through a lead collimator by the radioactive sources mentioned above.

As specified above, two sets of measurements were made. In the first one [4], (photosensitive gas) a mixture of He + TMAE vapors at 25°C, at a total pressure $p = 1$ atm, was flushed through the detector. In this set of measurements the mixture acts as the photosensitive component of the detector. The photons from the crystal ionize the mixture, and the produced electrons generate the avalanche, detected in the wire. Note that for this set of measurements the metal cathode had no detection function. Its presence was necessary to shape the electric field.

In the second set of measurements (photosensitive solid layer) the photosensitive material—a CsI + TMAE layer [5]—was deposited over the cathode. The photons emitted by the crystal extracted electrons from the photosensitive solid layer, thus producing avalanches that were detected in the wire. A mixture of He + TMP was flushed at the same pressure as above (1 atm) in order to suppress spurious pulses and obtain a counting plateau.

In both cases the counter worked in the Geiger mode. The counting plateau was very clear and it was measured to be 100 V: from 1850 to 1950 V. Under these conditions the number of counts per second obtained from LiBaF$_3$ (N_{LiBaF_3}) and from BaF$_2$ (N_{BaF_2}) crystals was measured. In the case of the (He + TMAE) photosensitive gas, the ratio was

$$[N_{BaF_2}/N_{LiBaF_3}]_{TMAE\text{-}gas} \cong 8 \tag{6}$$

In the case of the CsI + (TMAE + TMP) photosensitive layer the same ratio was

$$[N_{BaF_2}/N_{LiBaF_3}]_{CsI\text{-}solid\ layer} \cong 6 \tag{7}$$

These results allow some additional conclusions to be drawn on the LiBaF$_3$ scintillation spectrum. The BaF$_2$ fast scintillation emission has two well separated peaks: at $\lambda_1 = 193$ nm and $\lambda_2 = 220$ nm. The intensity of the peak at $\lambda_2 = 220$ nm measured by PM is four times larger than the peak at $\lambda_1 = 193$ nm. But the photoelectron yield of TMAE gas irradiated by the BaF$_2$ fast scintillation light (this is in fact a convolution of the BaF$_2$ spectrum with the TMAE quantum efficiency) is calculated - see Fig.3 - to have a maximum at $\lambda \cong 193$ nm [6] and an enhancement at $\lambda \cong 210 \div 215$ nm. From Fig.3 one can see that the ratio of the first maximum to the second enhancement in the convolution is about

$$\left[N_{BaF_2}^{\lambda=190} / N_{BaF_2}^{\lambda=215} \right]_{TMAE\text{-}gas} \cong 6 \div 8. \tag{8}$$

The convolutions calculated for the CsI + (TMAE + TMP) photosensitive solid layer irradiated by the BaF$_2$ emission also show two maxima: one at $\lambda \cong 202$ nm, the other at $\lambda \cong 220$ nm. The ratio (in the convolution) of the emissions at $\lambda \cong 202$ nm with respect to $\lambda \cong 220$ nm is about

$$\left[N_{BaF_2}^{\lambda=202} / N_{BaF_2}^{\lambda=220} \right]_{CsI\text{-}solid\ layer} \cong 5 \div 6. \tag{9}$$

The fact that the $N_{BaF_2} \gg N_{LiBaF_3}$ whereas the total light yield is nearly equal in both cases (the difference being 25 %, as reported above) indicates that the main emission spectrum of the LiBaF$_3$ must be concentrated at $\lambda > 200$ nm. Indeed, the number of counts measured by a gas filled detector consists of contributions from those spectral intervals which the detector is sensitive to. For example, in the case of the BaF$_2$ scintillation emission

$$N_{BaF2} = N_{BaF_2}^{\lambda=190} + N_{BaF_2}^{\lambda=215} \tag{10}$$

The same relation can be written for the LiBaF$_3$ emission recorded by the gas filled detector

$$N_{LiBaF_3} = N_{LiBaF_3}^{\lambda=190} + N_{LiBaF_3}^{\lambda=215} \tag{11}$$

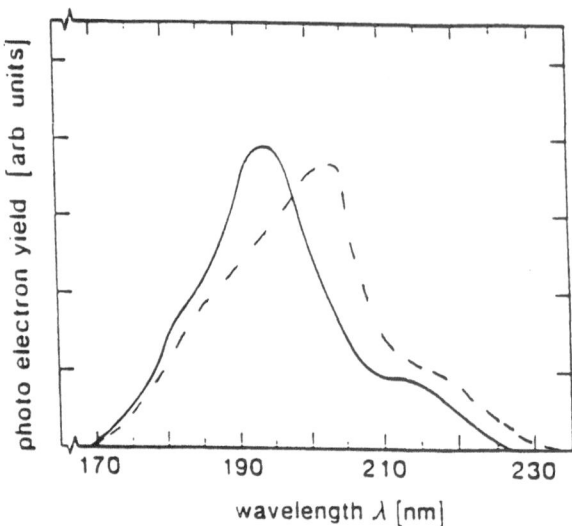

Fig.3. Wavelength-dependent photoelectron yield of a gaseous detector coupled to a BaF_2 crystal. The solid line corresponds to TMAE gas as the photosensitive element, the dashed line to the cathode covered with a CsI (+ TMAE) solid layer.

As mentioned above, the ratio between the first and the second maxima for the BaF_2 was estimated to be 6÷8 in the case of TMAE gas (formula 8 above) and 5÷6 in the case of the CsI photocathode (formula 9 above). Due to the Pb impurities mentioned before, the contribution $N_{LiBaF_3}^{\lambda=190}$ in formula 11 is missing.

Therefore the measured ratios quoted in formulas 6 and 7 have the straightforward interpretation in terms of the effective ratios:

$$\left[N_{BaF_2} \Big/ N_{LiBaF_3}^{\lambda=215}\right]_{\text{TMAE-gas}} \cong 8 \cong \left[N_{BaF_2} \Big/ N_{BaF_2}^{\lambda=215}\right]_{\text{TMAE-gas}} \tag{12}$$

and

$$\left[N_{BaF_2} \Big/ N_{LiBaF_3}^{\lambda=220}\right]_{\text{CsI - solid layer}} \cong 6 \cong \left[N_{BaF_2} \Big/ N_{BaF_2}^{\lambda=220}\right]_{\text{CsI - solid layer}} \tag{13}$$

being measured.

Conclusion: the $LiBaF_3$ spectrum, which produces counts in the gas filled detector, approximately coincides with the second emission peak at $\lambda \cong 220$ nm. This is perfectly consistent with the conclusions of the first set of tests: i.e. the emission spectrum of $LiBaF_3$ is very similar to that of BaF_2.

Tests of other solid photosensitive layers, more sensitive than CsI+TMAE, in the region 200–250 nm are under way. Such solid layers allow to record the $LiBaF_3$ scintillation light with few times more efficiency.

The plans for the near future are:

i) to measure the exact emission spectrum of $LiBaF_3$, using a powerful X-ray source and a monochromator;

ii) to investigate the possibility to record neutrons.

CONCLUSION

A new scintillator has been found: $LiBaF_3$. It has only a fast scintillation light emission (< few ns) and its photon yield is comparable with that of the BaF_2. The presence of Li may be interesting from the point of view of neutron detection, as well as for the study of compensation in a future calorimeter. This new scintillator has the advantage of being radiation hard and open to large scale production.

ACKNOWLEDGMENTS

We thank A. Borovik-Romanov and S. Petrov, from the USSR Academy of Science, and I. Valbis, from Riga University, who supplied us with the crystals

REFERENCES

1. G. Charpak, V. Peskov, D. Scigocki, J. Valbis, New scintillators for photosensitive gaseous detectors, preprint CERN–EP/89–66 (1989).

2. G. Charpak, D. Lamb, V. Peskov, D. Scigocki, J. Valbis, New developments in calorimetry based on VUV scintillators coupled to photosensitive gaseous detectors, in: Proc. ECFA Study Week on Instrumentation Technique for High-Luminosity Hadron Collider, Barcelona, 1989, Eds. E. Fernandez and G. Jarlskog (CERN 89–10, ECFA 89–124, Geneva, 1989), Vol. 2, p. 593.

3. G. Charpak, P. Fonte, V. Peskov, F. Sauli, D. Scigocki, D. Stuart, Investigation of operation of a parallel-plate avalanche chamber with a CsI photocathode under high gain condition, preprint CERN–PPE/91–47 (1991).

4. V. Peskov , G. Charpak, P. Mine, F. Sauli, D. Scigocki, J. Séguinot, W. F. Schmidt, T. Ypsilantis, Liquid and solid organic photocathodes, Nucl. Instrum. Methods **A269** 149 (1989).

5. J. Séguinot, G. Charpak, Y. Giomataris, V. Peskov, J. Tishhauser, T. Ypsilantis, Reflective UV photocathodes with gas-phase electron extraction: solid, liquid, and absorbed thin films, Nucl. Instrum. Methods **A297** 133 (1990).

6. P. Schotanus, C. W. van Eijk, R. W. Hollander, J. Pijpelink, Photoelectron production in BaFr-TMAI detectors, Nucl. Instrum. Methods **A259** 586 (1987).

HYBRID PHOTODIODE TUBE

Riccardo DeSalvo

CERN - LAA Project, Geneva, CH

Yunyong Wang

WORLDLAB, Lausanne, CH

ABSTRACT

We report the first test results on a new breed of light detector based on a photocathode followed by a silicon diode working in the bombarding mode.

1. INTRODUCTION

We need a new kind of light detector for the Spaghetti calorimeter capable of covering a span of at least 4 order of magnitude of linearity. The LAA project started a development program to develop such a light detector that led to the Hybrid Photodiode tube prototype production. The preliminary tests on these prototypes are reported here.

2. GENERAL DESCRIPTION

The Hybrid PhotoDiode tube (HPD tube) is made of an ordinary photocathode mounted in front of a planar semiconductor diode (Si diode or GaAs diode) anode. The photocathode and the diode at the anode are separated by a vacuum gap and held at a voltage difference of a few kilovolts. The diode is reverse-biassed to full depletion of the P-N junction. The ohmic contact facing the photocathode is only a few hundred angstrom thick. An incoming photon is converted on the photocathode into a photo-electron then accelerated and focussed toward the silicon diode by an electrostatic focussing system. The high energy photo-electrons stop in the depletion volume by generating electron-hole pairs. The gain obtained is equal to the number of electron-hole pairs created by an electron inside the silicon diode.

The collection of this charge on the electrodes produces the output current pulses of the hybrid photodiode tube.

In silicon, an average energy of 3.6 eV is required for generating an electron-hole pair. The theoretical gain is given by

$$G = (E - E_f) / 3.6$$

where E is the energy of a photoelectron after the accelerating voltage (10000-15000 V). E_f is the energy lost in the entrance window of the diode (in eV).

If $E = 10000$ eV and $E_f = 3000$ eV, then $G = 3 \times 10^3$. A typical curve of the gain as a function of the photocathode voltage is shown in Figure 1 and it is in reasonable agreement with the theoretical expectations.

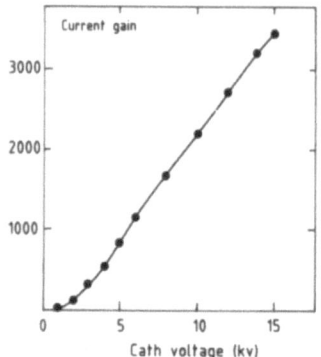

Figure1. Gain as function of photocathode voltage.

So far we received the prototypes from two companies: DEP and Philips.

The structure of one of the hybrid photodiode prototypes built is illustrated in Figure2.

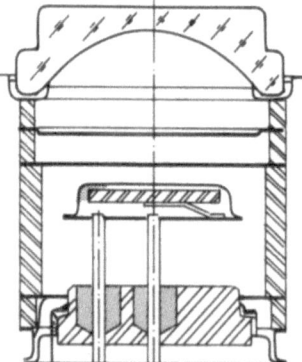

Figure2. Structure of one Hybrid photodiode tube prototype

The photocathode is a S20 (multi-alkali) optimized between 450 and 550 nm. The average sensitivity is 62 mA/W. The electrostatic optical design of this prototype is the so-called fountain tube was used because it creates a spherically symmetric electric field which allows focussing from a relatively large photocathode area onto a small anode. Glass has been chosen as the input window. The other HPD prototypes use a crossed focused electrostatic design.

3. CHARACTERISTICS

1) OUTPUT PULSES

The HPD is intrinsically fast. The flight time between the photocathode and the semiconductor anode for an electron depends on the operating voltage and the vacuum gap dimensions, but it is typically less than a nanosecond. The time-of-flight fluctuations are negligible. The electrons penetrate a short distance x into the silicon,

$$x = 3*10^{-10}*v^{1.4} \text{ cm}$$

where x is about 1.2 um at V = 10kV.

Since the typical electron drift speed in silicon is 50 um/ns, the charges is collected on the diode electrodes with a temporal spread of a small fraction of a nanosecond. The HPD speed is limited by the semiconductor diode rise-time.

The shape of the output pulse for a typical HPD is shown in Figure 3. In Figure 3 (left) the long tail comes from a LED and it is not present in Figure 3 (right) where a 200ps LASER source is used

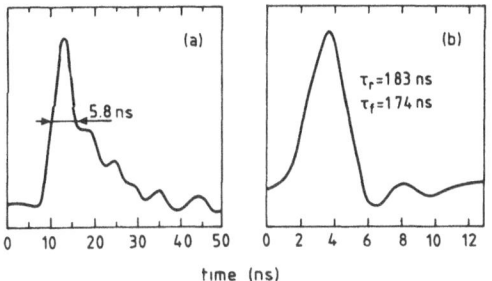

Figure3 Output pulse of HPD excited with fast LED by DEP (left) and with a 200ps LASER by Dr.Benetti,INFN Pavia (right)

2) GAIN STABILITY

Since each photoelectron generates a few thousands electron-hole pairs in a single step, the single electron response should be narrower than for a photomultiplier.

The gain of the HPD is directly proportional to the applied high voltage (not an exponential function like the photomultiplier) so the device has an intrinsically good gain stability as illustrated in Figure 4.

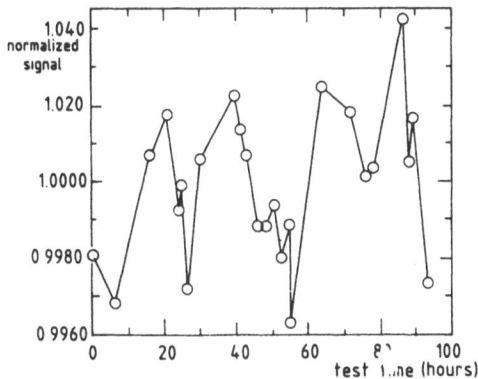

Figure4. Stability test results (temperature corrected)

3) UNIFORMITY

The uniformity of the prototype was measured by DEP scanning the photocathode surface with a 22° aperature light beam, having a cross-section of 1 mm-diameter at the input surface. The response is shown in Figure 5

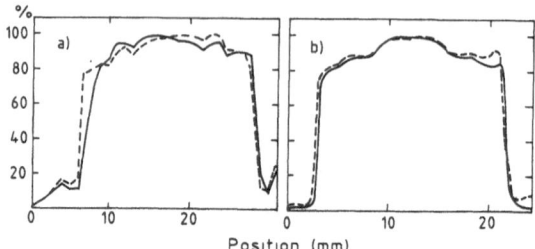

Figure5. Uniformity scan of a new photocathode (left) and after 3500 hours lifetime testing (right).

4) LINEARITY

The output signal of the HPD is the current given by the bombarded semiconductor diode. The semi-conductor diodes have a very large dynamic range and the HPD maintains good linearity over a similar range. The linearity of HPD has been measured with a set of LED as light sources. The result is shown in Figure 6 and the residuals between the expected and output signals are shown in figure 7.

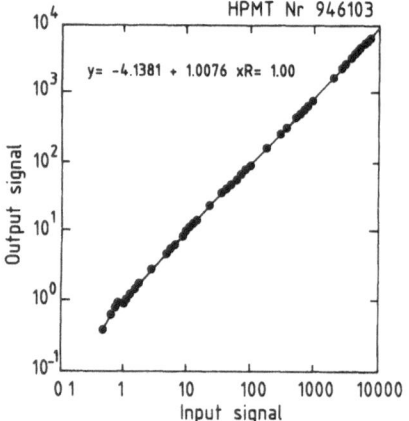

Figure6. Dynamic range of the HPD signal.

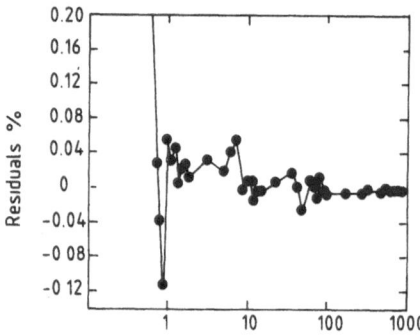

Figure7. Residuals of the dynamic r ange plot (%)

The schematic of the testing set-up is shown in Fig. 8. Two sets of 4 LEDs (4 mounted in front of the HPD and 4 outside the HPD black box) and a four channel light diode driver were used. The four LEDs outside the box were used only to load the light diode driver. when the LEDs in front of the HPD were not used.

The four channels of the LED driver were tuned to give an arbitrary amplitude roughly in the following ratio:

channel 1 = A
channel 2 = A/2
channel 3 = A/4
channel 4 = A/8

The HPD was read out with a normal ADC (Model 2249A) gated with the LED driver start signal. The four LEDs were fired one by one and their signal was recorded as reference. Then they were fired in all twelve possible combinations and the responce was plotted against the sum of same combination of the recorded signal. This allowed us to explore the linearity over about one order of magnitude; a small correction proportional to the number of used LEDs was also applied. Any deviation from a straight line is then attributable to non linearities. Once the linearity was tested in a short range one or more arbitrary neutral density filters (transmission 10%) were put between the LEDs and the HPD to probe the linearity into another order of magnitude.

5) POWER REQUIREMENTS

The HPD is a mechanically simple and compact device: it relies on electrostatic focussing and requires no voltage divider. The high voltage generator has practically no load (1 electron per converted photon). The silicon diode bias power supply has to provide the signal current at low voltage only. In the tests reported here, external power supplies were used. In the future each tube will be supplied with its own local miniature low power (10mW) high voltage supply developed for the image intensifiers market. This will eliminate the cabling and the voltage divider heat dissipation problems especially felt when large numbers of tubes are required.

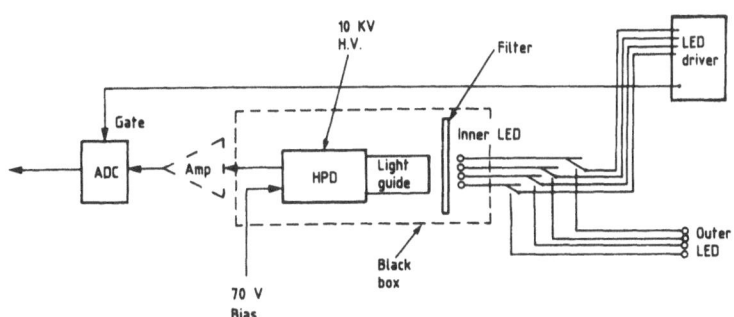

Figure8. Schematic diagram of the set-up for linearity testing.

6) LIFETIME

Lifetime testing has been made for 3500 hours at a 10 kV cathode voltage and 100 V diode bias. The HPD was tested continuously with a pulse of 6 mA peak output current roughly 10 ns wide. The frequency is 10 kHz, equivalent to 0.6 uA DC output. The pulse height as a function of testing time is shown in Figure 9 . The degradation of the pulse height is probably caused by the damage of the silicon diode which is shot at the same point by the incoming electrons during the testing period. The uniformity of the photocathode after 3500 hours is shown in Fig.5. (right) and shows no degradation while a hole is visible in the amplified signal (Figure 10). Tests on the HPD performances are still on going. A different kind of HPD under test seem not to show this problem.

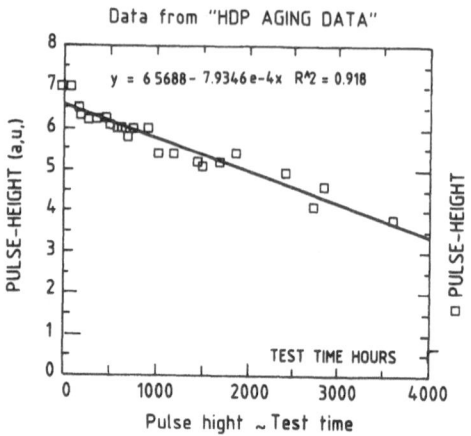

Data from "HDP AGING DATA"

Figure9.Pulse height as a function of testing time

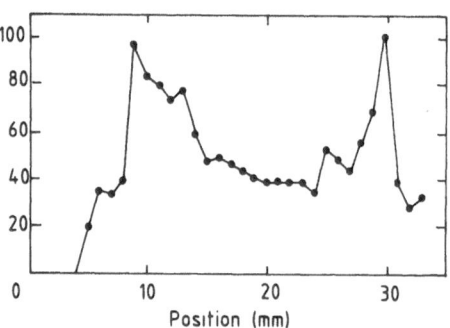

Figure10. Uniformity scan of the anode signal after 3500 hours.

4. FUTURE DEVELOPMENTS: MULTIPIXEL HYBRID PHOTODIODE

The HPD is built as an image intesifier and maintains the photocathode imagetopology on the silicon diode. If the anode is replac
with a segmented diode with independent segment readout and if the glass window is replaced by a fibre window, The HPD can be turned into a very efficient and fast imaging detector. We are in the process of designing such multipixel hybrid
photodiode tube to suit the SPACAL readout requirements. This development is intended to improve the already impressive space resolution and e/ separation of SPACAL. Developed for the Spaghetti Calorimeter by LAA, the HPD might be very useful in many other fields especially in medical imaging systems and nuclear detectors.

ACKNOWLEGEMENTS

Part of the measurements reported here were performed by the DEP company. The HPD speed measurements were performed with the pulsed LASERs of F.Hartjies at NIKHEF Amsterdam and of Dr. Benetti at INFN Pavia.

REFERENCES

1. L. G. Wolfyang, J. M. Abraham and C. N. Inskeep
IEEE Trans. on Nucl. Sci. NS-13 no.3   46(June, 1966)
2. P. Chevalier   Nucl. Instr. and Meth. 50(1967) 346-348
3. J. Fertin , B. Lach, J. Meuleman, J. Dupuis, L'Hermite and R. Petit
IEEE Trans. on Nucl. Sci. Ns-15 no.3   (June, 1968)
4. R. Desalvo   CLNS 87-92
Cornnel University Ithaca NY 14853

CALORIMETERS SAMPLED WITH WARM LIQUIDS

Jean Pierre Mendiburu

Collège de France
Paris, France

ABSTRACT

We report on the R&D undertaken since 1987 by the WALIC* collaboration (composed of CEA-Saclay, Collège de France-PARIS, LBL-Berkeley, LAPP-Annecy) on sampling calorimeters using room temperature liquid hydrocarbons like Tetra-Methyl-Silane (TMS) or Tetra-Methyl-Pentane (TMP).

We have found that several metals, glasses, plastics and glues, very promising for their mechanical, electrical and radiation hardness properties, can be put in contact without observing, during the drift time, any significant decrease in the life time of the electrons created by ionisation.

We are constructing a prototype, the frame of which is made of plastic (VECTRA) using lead as absorber and TMS as ionising medium.

1. INTRODUCTION

The room temperature dielectric liquids as active media for ionization detectors are studied since a decade (Reference. 1 to 10). Nowadays, the building of calorimeters sampled by warm liquids (Reference 11 to 13) is explored by several very active R&D groups, using Tetramethyl-Silane (TMS) or Tetramethyl-Pentane (TMP).

One of the main difficulties of this technique arises from the level of purity at which the liquids have to be obtained and kept .Typically, to maintain the trapping on electronegative components under a few per cent, the life time of the free electrons has to be in the order of one microsecond or more, and at least one order of magnitude more than the drift time (figure 1).

Drift chambers filled with Tetramethyl Germanium (TMG) are also foreseen for neutrino or double beta decay experiments. For such purposes, the lifetime has to reach several hundreds of microseconds, corresponding to a level of impurity as low as 10^{-9}.

All these detectors have to be built with non contaminating materials and to be extremely clean and leak tight.

The present paper reports on the results obtained, in search for new compatible materials with TMS at College de France (CDF) and TMP at Lawrence Berkeley Laboratory (LBL), and on the project of plastic calorimeter.

Section 2 describes the tools and their preparation, section 3 discusses the methods used to measure the compatibility, the section 4 gives our resultson compatibility and section 5 describes the prototype in construction.

2. THE INSTRUMENTS

2.1 The purification station

The CDF station,which has a capacity of 1 liter, has been developed at P.S.I. (Reference 11). For purification, the TMS is evaporated by heating a start container, passes through a first molecular sieve filter filled with 4A grade, through a second one filled with a mixture of 3A, 4A and 5A grades and then the vapor is condensed by cooling a storage container. This process is reproduced as many times as necessary and at each step, the progress in life time is measured by a purity monitor device mounted in parallel on the storage container (Reference 14). This one

New Technologies for Supercolliders, Edited by L. Cifarelli and
T. Ypsilantis, Plenum Press, New York, 1991

consists of a ionization chamber, lightened by rectangular UV light pulses produced by a low pressure mercury vapor lamp. The shape of the current generated by the drifting electrons is analyzed on a digital scope and this gives directly the life time with an error of a few per cent. The agreement of this method with the test cell measurements has been verified.

Further details on this station and on purification procedure are described in reference 15.

2.2 Test cells

For CDF tests, we have built a set of 10 cells (figure 2) made of ultra-high vacuum stainless steel and ceramic elements, already known for their compatibility with TMS.

Each cell has been put through a " standard cleaning cycle " :

- washing for 10 minutes in a heated with ultrasonic bath successively with *i)* acetone at 40°C, *ii)* alcohol at 60°C and *iii)* ultra-pure water (18 MΩ / cm) at 80°C.

- assembling, storage in a dust free room, under a lamina flow of air.

- As a final washing, flushing of ultra pure water in the cell which is immersed in a ultrasonic bath at 80°C until the resistivity of the water coming out has reached 18 MΩ / cm.

Each cell is dried, for 6 hours at room temperature in a gas flow of ultra pure helium .

2.3 Samples preparation

We first standardize the size of all the samples to $30*10*2$ mm^3, giving a contact surface of 760 mm^2. Each sample follows as close as possible the " standard cleaning cycle " described in 2.2 and is dried for at least 3 hours under the lamina flow of air at room temperature. When possible (for conductive materials), we systematically test its cleanliness by Auger spectroscopic analysis .

2.4 Final assembling and filling

The clean sample is hung up to a ultra-high vacuum flange and this one is mounted to close the top of the cell. This cell is connected to the filling station and baked out at 150°C (or at the maximum temperature allowed for the material in test) down to a pressure of 10^{-8} millibars.

After a last vacuum tightness test of the cell, we transfer by evaporation 10 cm^3 of liquid in the cell, which is then ready for life time measurements.

3. COMPATIBILITY MEASUREMENTS

To test the compatibility, we make a first lifetime measurement before any contact between the liquid and the sample. Then we turn the cell upside down and leave the cell in this position for the time of contact. We turn it back to its normal position to stop the contact and measure the new life time. We repeat this process with increasing contact times. For most of the samples.

To evaluate the life time, the liquid of the cell is ionized with beta rays from a Ru ^{106} radioactivesource.

3.1 Electronic setup and data taking

The electronic set up is represented on figure 3.
To reduce the noiseof the preamplifier, the traditional resistance in the feed-back loop of its first stage is supressed. This feed-back is purely capacitive (capacitance of .2 pf) and to avoid the output signal to grow up to saturation, a very slow (several ms) base-line restoration is produced by an LED, driven by the output signal and opto-coupled to the input FET.

The output signal from this preamplifier is analyzed by an analog to digital transient recorder Lecroy TR- 8837 F. For each measurement, we have averaged the shape of the pulse for a few hundred events to improve the ratio signal to noise by a few tens.

3.2 Signal parametrisation

Due to the pure capacitive feed-back of its first stage, our preamplifier response is proportional to the integral of the current created by ionization between the electrodes of the chamber; we thus ideally expect a parabolic shape output in absence of electronegative contaminants.

TABLE 1 . COMPATIBILITY OF METALS.
compatibilities quoted (1) have been measured with TMS at College de France
and those quoted (2) have been measured with TMP at LBL

Material	Contact time	Life time (errors) (µs)	Drift Voltage (Volts)
copper (1)	5 mn	42.0, (-6.8,+7.8)	20 and 40
	2.5 hours	49.4, (-6.4,+7.3)	20 and 40
	8 days	40.6, (-6.0,+6.0)	40
	21 days	88.0, (-5.0,+5.0)	20 and 70
lead 1 (2)	0.	2.1 (.4)	
	6 hours	3.8 (8)	
	28 hours	5.9 (1.2)	
	1.5 days	9.3 (1.9)	
	6 days	27 (5.4)	
	29 days	33 (6.6)	
lead 2 (1)	0	41.7 (-10.8,+19.5)	70
	1 hour	45.7 (-7.6,+10.7)	70
	13 days	49.5 (-7.8,+10.8)	70
	27 days	45.5 (-6.0,+7.8)	70
lead 3 (1) (see text)	0	48.1, (-12.,+24.9)	103
	15 mn	38.7, (-8.6,+14.5)	103
	3 days	61.7, (-14. 6,+25.6)	70
	10 days	63.7, (-17.4,+34.6)	70
	17 days	54.5, (-10.6,+16.4)	70
	44 days	76.7, (-17.3,+29.6)	70
	61 days	111.4, (-11.0,+13.4)	40
tungsten (1)	0	54.5, (-15.8,+27.9)	30 and 50
	2 hours	54.6, (-11.2,+16.7)	40 and 70
	3 days	96.1, (-17.1,+23.6)	20
	10 days	243, (-53,+83)	10
	21 days	383, (-75,+114)	10
	33 days	381, (-90,+155)	5 and 10
aluminum (2)	0.	9.6 (1.9)	
	5 hours	9.6 (1.9)	
	61 days	7.2 (1.5)	
	120 days	9.5 (1.9)	
	270 days	8.5 (1.7)	
	360 days	7.7 (1.5)	

TABLE 2 . COMPATIBILITY OF INSULATING MATERIALS.
all these compatibilities have been measured with TMS at College de France.

Material	Contact time	Life time (errors) (µs)	Drift Voltage (Volts)
TEFLON	0	100	12
	2 mn	0	1500
PEEK	0	14.7,(-2.2,+3.)	800 and 1000
	5 mn	6.1,(-1.8,+2.)	1000
	49 hours	19.9,(-4.5,+8.)	1000
	14 days	15.6,(-3.,+4.7)	500 and 600
	31 days	7.2,(-2.8,+8.0)	300 and 250
VECTRA (metalized)	0	18.9,(-4,+6.3)	100.7
	5 mn	18.5,(-5.3,+11.6)	131.3
	2 hours	27.1,(-12.0,+15.4)	116.5
	7 days	25.2,(-6.1,+9.7)	70
	16 days	23.9,(-3.5,+4.5)	143
	37 days	28.1,(-4.9,+6.6)	50
	55 days	27.7,(-6.7,+10.9)	70
ZERODUR	0	40.1,(-6.0,+8.3)	100
	1 hour	29.4,(-7.5,+13.4)	100
	11 days	32.8,(-6.8,+10.7)	100
	28 days	45.2,(-9.7,+16.0)	70
	46 days	55.5,(-15.5,+23.7)	70
SUPTAX	0	117.6,(-26.0,+43.4)	35 and 70
	1 hour	78.4,(-22.0,+45.8)	70
	15 days	64.1,(-8.9,+11.8)	70
	35 days	74.4,(-11.5,+16.1)	70
Kerimid 500	0	14.9,(-2.2,+2.9)	70, 105 and 150.4
	15 mn	16.2,(-3.0,+4.5)	151.7
	8 days	17.3,(-2.9,+4.1)	151.5
	17 days	6.8,(-1.5,+2.)	153.4
	28 days	15.4,(-3.,+4.4)	153.5
	41 days	18.4,(-4.3,+7.4)	153

TABLE 3 . MAIN PROPERTIES OF ELECTRIC RESISTIVE MATERIALS .

Commercial name	ZERODUR	SUPRAX	KERIMID	PEEK	VECTRA
Technical name	Vitroceramic	Borosolicat	Polyamide-Imide Glue	Poly ether sulfone (450GL30)	Liquid cryst. polymer (C150)
Produced by (Company)	Schott	Schott	Rhone Poulenk	Ici	Celanese
Traction resistance (MPa)	≈ 100	≈ 100		157	200
Lengthening before break (%)				2.2	1.2
Flexing module (GPa)				10.3	19.
Working temperature (°C)	> 600	> 600	250	250	252
Dilatation coef. x 10^{-6} (°C^{-1})	5.7	3.3		22	7
Density (g / cm^3)	2.6	2.2		1.49	1.80
Dielectric constant at 1 MHz	7.4	5.6		3.2	2.9
Dielectric rigidity (kV / mm)				19	35
Resistivity at 20 °C (Ω.cm)	3.10^{13}	10^{14}		$6\ 10^{15}$	$>5\ 10^{15}$

TABLE 4. COMPARISON OF MECHANICAL PROPERTIES
OF METALS AND PLASTICS [17]

Materials	Breaking load: (R) at 25°C (MPa)	Merit factor at 25°C =(R / density)	Breaking load: (R at 300°C) (MPa)	Merit factor at 300°C =(R / density)
Soft steel	440	5.6	420	5.4
stainless steel: 18/8	700	9	550	7
Super resistant steel	2000	22.5	1750	22.5
alloy with titanium	1250	28.5	1000	23
Light alloy	460	6.5	45	1.6
PEEK 450 GL30	157	105	1.8 at 315 °C	1.2 at 315 °C
VECTRA C150	220	122	1.8 at 250°C	1.0 at 250°C

Two distortions are introduced by the electronics: one by the base line restoration which is not proportional to the amplitude of the signal, and a second one by the decay time of the second stage. We have gathered these two contributions in a single exponential decay time RC.

The following formula has been used to fit the signal with three parameters : the total charge produced by the ionizing primary particle, $Q_{tot} = 2\ n_e$, the electron lifetime τ and the global decay time RC:

$$V(t) = -\ G\ \frac{n_e}{T_d^2}\ \frac{RC\ \tau}{\tau - RC} \left\{ \left[\frac{RC\ \tau}{\tau - RC} + T_d - t \right] e^{-t/\tau} - \left[\frac{RC\ \tau}{\tau - RC} + T_d \right] e^{-t/RC} \right\}$$

G is the gain of the preamplifier and T_d is the electron drift time.

3.3 Signal fit

The analysis time is chosen long enough to constrain RC by the shape of the signal still present after the drift time. In our fit, this part of the signal is almost independent of the 2 other parameters n_e and τ.

We have to note that our results rely on the knowledge of electron mobility, taken from published values. It is not possible to exclude the unlikely case where the presence of poluants should have preserved the initial number of electrons but modified their mobility.

4. THE RESULTS

This R&D being oriented to the construction of a large calorimeter, we have first selected the materials for their mechanical strength, resistance to high temperature and to ultra-cleaning, very high or very low electric resistivity, radiation hardness, and also the facility to be managed on a large scale by industry (materials that could be molded, extruded, metallized, welded, glued etc.).

Tables 1, 2 and figures 5a,5b and 6 summarize our results.

4.1 METALS

a) Materials for containers, structures and connections:

The untreated copper from in our store (purity 99%), prepared as described in (2.2) has been found to be compatible with TMS. It can be used for connections, containers, or pipes with the advantage on stainless steel that these pipes can be crimped.

Aluminum is also compatible; TMP has still a good life time after one year of contact. Aluminum can be used for structures and containers.

b) Materials for absorbers :

The lead that we have tested was hardened with .06% of Calcium and 1.3 % of tin.

The cleaning process has been adapted to preserve the surface of lead from the cavitation effects of ultra sonic cleaning. The last washing by ultra pure water was done without using ultra sonic treatment; instead the washing time was doubled. The Auger spectrometry analysis has shown that the cleanliness was not significantly different.

The sample 3 is an element of the warm liquid calorimeter prototype being constructed at LBL (reference 12). It consists of a piece of lead, glued with Kerimid 500 (see section 4.2) to a sheet of Kapton, metallized on the other face.

c) Getter effect

As shown in Table 1, the life time has significantly increased during the contact between liquid and stainless steel, tungsten or lead. We interpret this as a trapping at the surface of the impurities, moved by Brownian movement. This effect could be similar to the one operating in ultra vacuum getter pumps; It depends strongly of the processing of the sample.

This effects could help to maintain the liquid purity in large devices or even be used in specific getter purifiers.

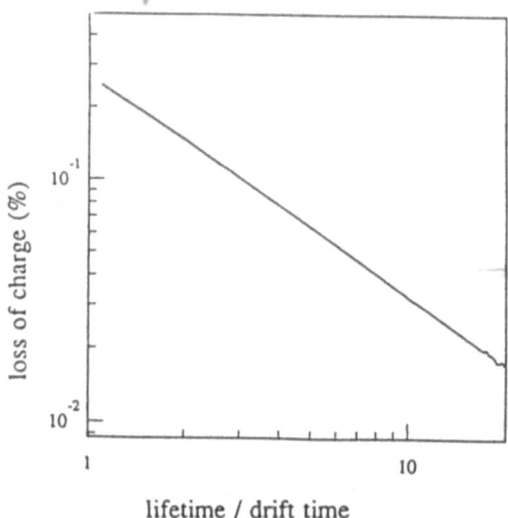

Figure 1   Proportion of trapped electrons versus ratio [lifetime / drift time]
(The measurement time is supposed equal to the drift time
and the electronic decay time infinite).

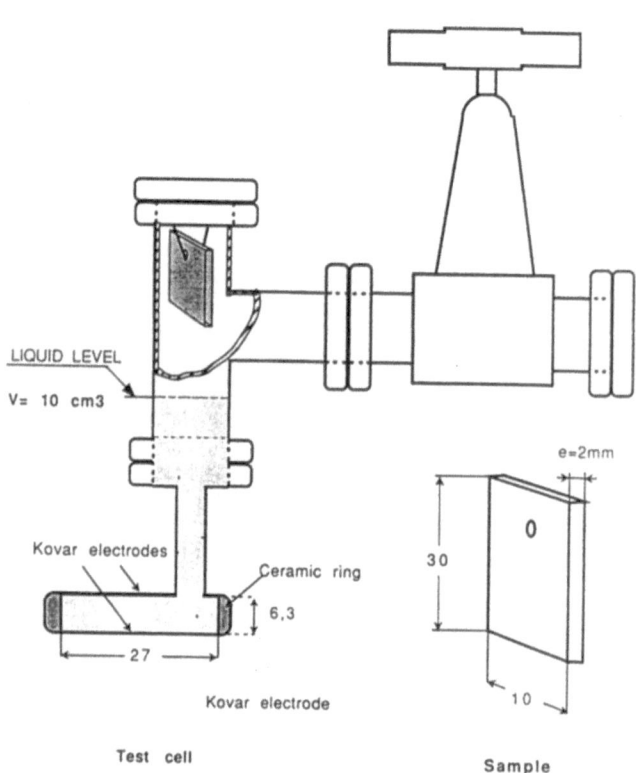

Figure 2   Test cell and sample

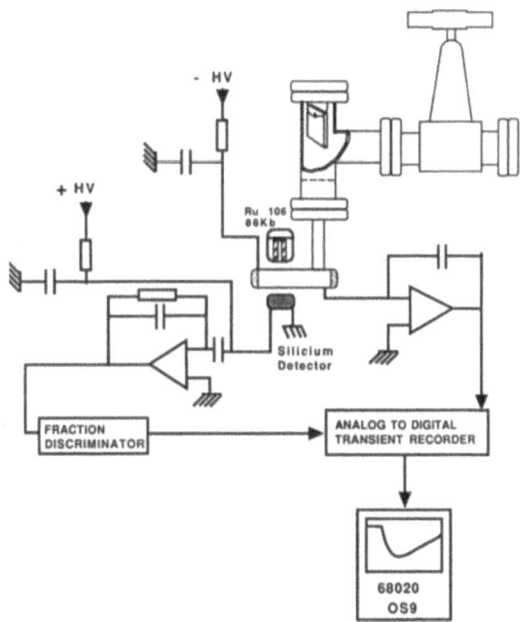

Figure 3   Electronic set up for life time measurements

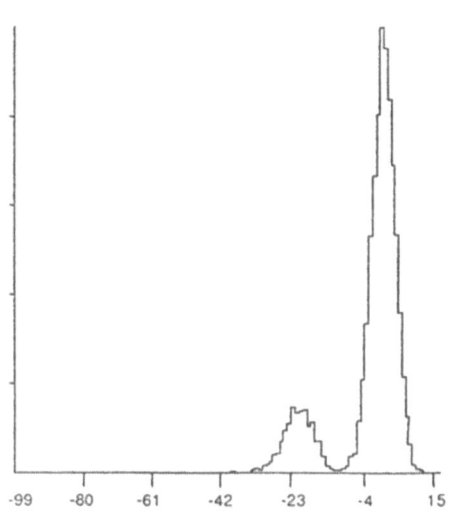

Figure 4 a   Total charge collected on a cell for a lifetime of 20 microseconds.

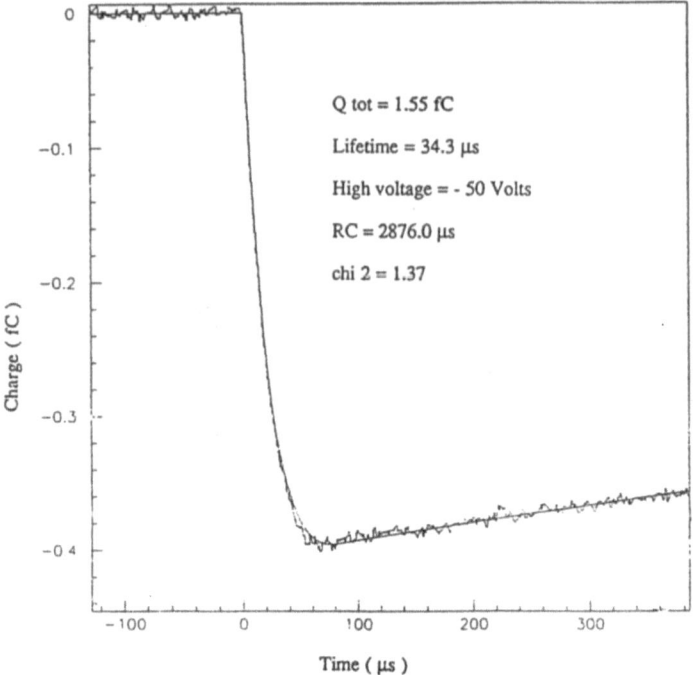

Figure 4 b   Example of shape of signal (data and fit)

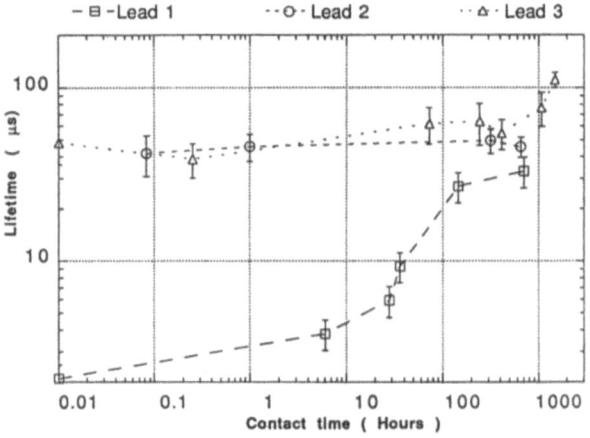

Figure 5a   Compatibility of lead

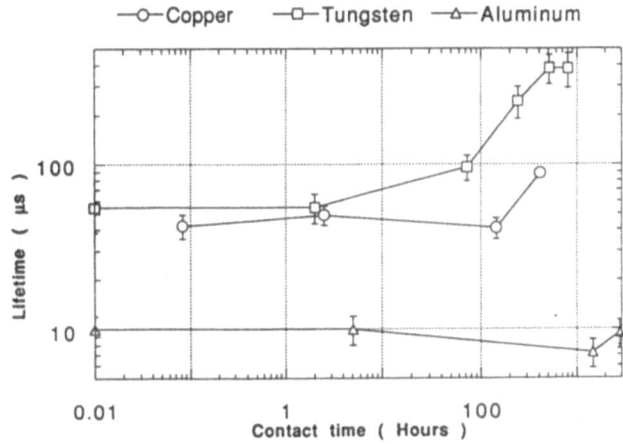

Figure 5b   Compatibility of other metals

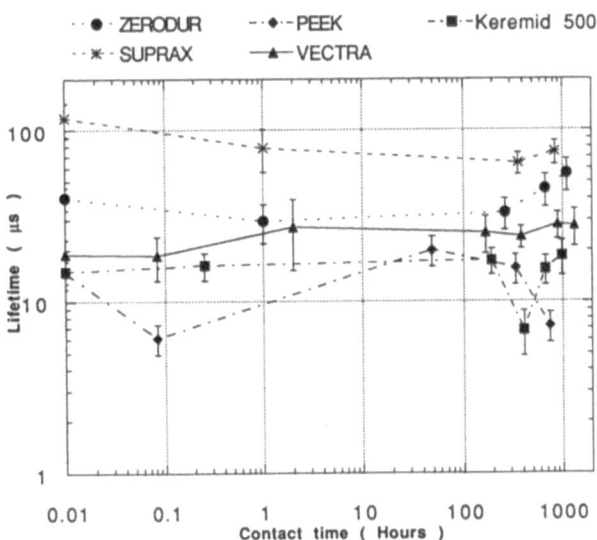

Figure 6   Compatibility of electric isulating materials

73

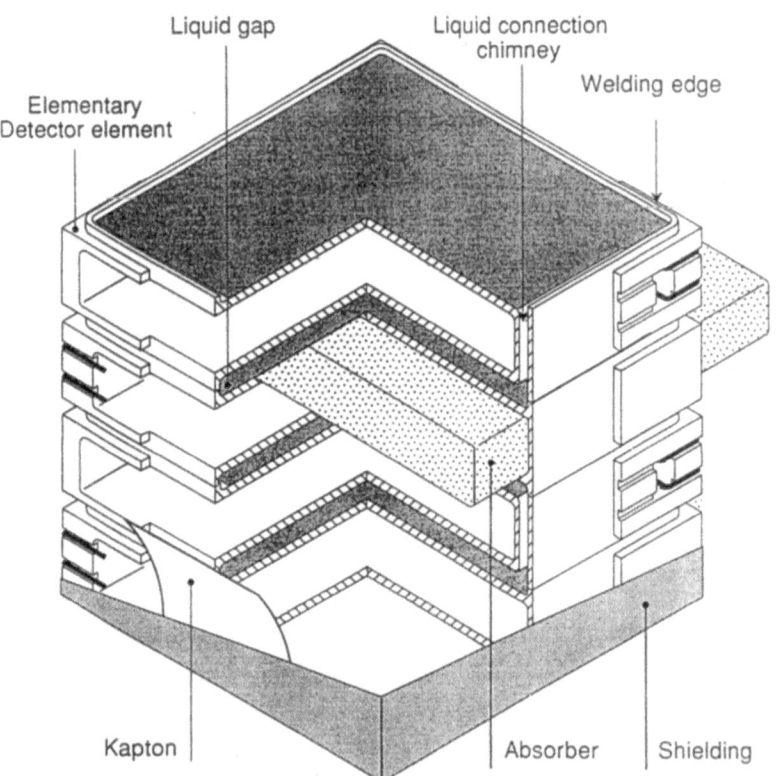

Figure 7   Prototype of plastic calorimeter

4.2   INSULATING MATERIALS

Table 3 and 4 show the mechanical and electrical properties of isulating materials that we have tested.

Zerodur and Suprax are glasses that can be machined like metal and welded by thermo-compression, the polyamide-imide " Kerimid 500 " is a multi-purpose adhesive, Teflon, which is the commercial name of Fluored-Ethylen-Propylen is not compatible,.perhaps due to its fluorine content. The PEEK is the comercial name of the polyether-ether-ketone; it has the required mechanical and electric properties to build frames (except that it is difficult to glue or metallize). The Vectra is the most interesting isulating material that we have found. This plastic is a mixture of several elements that can be adapted to desired characteristics. Its ratio strength to density is between aluminum and stainless steel, but with lower interaction and radiation lengths (λ_T = 21.7 cm, λ_a = 62.7 cm, d=1.57 g/cm^3); it has a low porosity, ; it can be baked out, molded, extruded, machined, metallized, glued and welded ; It is an excellent insulator, radiation hard and non flammable.

A calorimeter, based on a Vectra frame is under investigation for LHC (Reference 13).

5. THE PLASTIC CALOREMETER

A prototype is being constructed at CDF, in which the liquid is only in contact with copper and VECTRA. This technique, based on welded elements is forseen at the start in the aim of an easy industrialization. The liquid circulates in a sort of maze, and ionizing chambers face sheets of lead absorber (figure 7). The read out of these ionizing chambers are grouped into a small number of samples. The actual size of this prototype is of 9 prisms of 20 elements each, the depth being of 30 radiation length.It will contain about one litter of TMS and his total weigth will be 35 Kg.

The technique of electrostatic transformer has been adapted in a charge coupled device that allows a low capacity equivalent at the input of the preamplifier and a very simple design, in particular, a low drift power supply.(reference 17). A realistic design for large hadron collider is being studied.
It is planed to put this set up in a test beam of the SPS to study its response to electrons and hadrons during 1991.

6. CONCLUSIONS

All the tested materials that do not contain electronegative elements such as Chlorine, Fluorine, Sulfur etc. are compatible. The main difficulty in using warm liquids is more in the cleaning and the tightness of the detectors than in the choice of their building materials.

We have observed a very significant " gettering " effect of several metals (lead, tungsten, stainless steel) which is promissing in maintaining (or even improve) the life time of the warm liquids.

We are constructing a plastic calorimeter prototype and forsee to put it in a beam in 1991.

We thank F. Moreau, Delataille from LPNHE, Ecole Polytechnique (Paris) for their very fruitful collaboration, A.Grillot and D.Latorre from CERN for Auger analysis.

REFERENCES

1	W.F.Schmidt	Can.J.Chem. 55 (1977), 2197
2	T.G.Ryan, G.R.Freeman	J.Chem.Phys 68 (1978), 5144
3	H.Jungblut,W.F.Schmidt	NIM A241 (1985), 1128
4	R. Holroyd et al.	NIM A236 (1985) 294
5	J.Engler et al.	NIM A252 (1986), 29
6	I.Lopez,W.F.Schmidt	Z.Naturforsch 45 A (1990),832
7	S.Geer et al.	NIM A287 (1990), 447
8	Lopes et al	NIM A271 (1988), 464

9	M.G. Albrow et al.	NIM A265 (1988), 303
10	S.Ochsenbein et al.	NIM A273 (1988), 654
11	B. Aubert at al.	Fast hermetic calorimetry using warm liquids: Proposal for generic detector R&D for SSC/LHC Walic collaboration, LBL report, Sept. 1987.
12	B.Aubert et al.	Warm Liquid Calorimeter Large Subsystem Proposal DOE/SSC
	J. Busenitz et al	Warm liquid Calorimeter Large Subsystem R&D Progress Report
13	L.Dobrzynski et al.	The Prism Plastic Calorimeter SubmittedtoCERN/DRDC?90-72;DRDC-P18
14	F.Moreau et all.	NIM A278 (1989) 417 and "Purity measurement in TMS" to be published in NIM
15	L.Dobrzynski et al.	Purification of Warm liquid : station and processes to be published
16	Delataille	Thesis Paris
17	Mendiburu Jean Pierre	Charge transfer device for sampling caliorimetry, LPC 90.48

ELECTRON IDENTIFICATION WITH TRD AT FUTURE COLLIDERS

P. Nevski

Moscow Physical Engineering Institute

Abstract
A Tracking Transition Radiation Detector for the future collider experiments is discussed. It consists of foam radiator in which a large number of small diameter cylindrical chambers are imbedded.

1 Introduction

The physics to be studied with the new generation of particle accelerators (LHC, SSC, Eloisotron) is mainly accessible via the leptonic decays of new particles. These particles which should occur at the TeV scale of electro-weak theory breaking and unitarity violation are expected to have significant branching ratio into leptonic channels. Highly efficient lepton identification over a large solid angle will play a crucial role in the experiments planned for these machines.

Since the typical lepton energies will be several dozen GeV and higher, transition radiation (TR) is a powerful if not unique technique available for electron identification at those energies. The radiation process is not destructive in the sense that it does not significantly affect the particle energy and direction. These measurements can therefore be carried out in addition to calorimetry.

The techniques of electron identification with transition radiation is used in different experiments (see, for example review [1]). The cluster counting method seems to be one of the most promising tools due to a better hadron rejection and the simplicity of electronics [2]. The main problems are known to arise in a multi-particle environment, where ionization losses from overlapping particles imitate TR quanta, thus deteriorating the hadron rejection.

One way to avoid rejection deterioration is a simultaneous dE/dx measurement in order to reject overlapping particles. This method requires multiple measurements of the transition radiation together with ionization losses along particle tracks. This solution permits also to use a TRD not only as an electron identifier, but at the same time as a tracking device sensitive mainly to high γ particles.

The new accelerators are planned to have a high collision rate and aim to register very rare process. This requires a small registration time, the ability to provide a trigger and efficient electron detection over the full solid angle. Study of multilepton decays such as $H^0 \rightarrow 2e^+ + 2e^-$ implies additional requirements of the absence of dead zones in a detector.

New Technologies for Supercolliders, Edited by L. Cifarelli and
T. Ypsilantis, Plenum Press, New York, 1991

Much efforts have recently been made to meet this challenge of electron identification at future colliders. In this paper we will present mainly results of the R&D programs [3],[4] that are developing an integrated transition radiation detector and tracking device for the LHC and the SSC.

2 New radiators

Transition radiation is produced when a charged particle crosses the interface between two media with different dielectric constants [5],[6]. The probability of a single quantum emission being about $\alpha = 1/137$ per crossing, a large number of surfaces is needed to produce a substantial signal.

One of the problems to resolve while constructing a TRD with a full coverage of solid angle is a choice of radiator. Traditional radiators, usually made of thick (10-20 μm) foils stretched with 200-300 μm spacing between them, need some supporting structure. This leads to appearance of dead zones in a detector. Moreover the radiator performance is changing for different particle trajectories.

A very interesting solution is the use of foams as radiators. The foam itself has a good mechanical properties to allow a self-supporting construction. And for many types of plastic foam the bubbles dimensions are remarkably constant (figure 1). The performance of different foams as radiators was studied by several groups [7],[8]. The most comprehensive study, which includes a comparison with regular foils as well as a minimization of the radiator thickness, is in the reference [9].

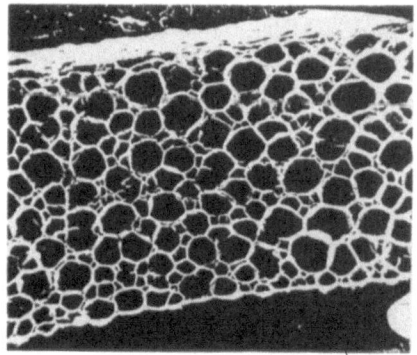

Fig. 1. Microphotograph of a foam

The radiator material has to be transparent to X-rays, therefore different carbon and hydrocarbon foams were selected as potential radiators. For a given total radiator thickness the TR quanta yield was studied. Figure 2 presents the radiator performance as the probability for 3 Gev electrons to deposit an energy above some threshold in a registration chamber behind the radiator versus the same probability for 3 Gev pions. The higher the curve is, the better is the radiator.

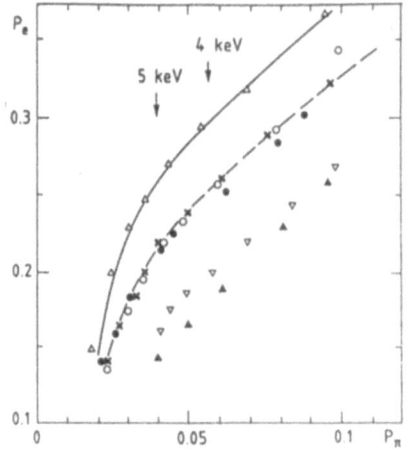

× - polyethylene $\rho = 0.059$
o - carbon $\rho = 0.080$
• - polystyrene $\rho = 0.110$
▽ - polystyrene $\rho = 0.040$
▲ - polyurethene $\rho = 0.055$
△ - optimized regular foils
ρ - foam density $[g/cm^3]$

Fig. 2. Performance of different foam radiators.

The polyethylene foam with a density of 59 Kg/m^3 was found to be one of the most promising radiators. The pore diameters of the foam ($\approx 200\mu$m) and the wall thickness ($\approx 5\mu$m) between pores are fairly uniform. The mean pass inside the polyethylene for any trajectory in this foam is $13\mu m$. Tests of this foam indicated a TR X-ray yield reaching 85% of that of a regular stack of polyethylene foils, making it very attractive to use this foam for a TR detector.

3 TR quanta detection

Taking into account well known disadvantages of traditional flat drift chambers, a new technique is also recently developed to detect TR quanta. This technique makes use of thin walled straw tubes as a detector for TR quanta and ionization. Research is being carried out in different laboratories to develop such straw chambers as an industrially produced multi-purpose device (see for example [10],[11]).

The advantages of straw chambers are:

a) Higher reliability due to the independence of channels and to the absence of cathode wires - a break of one anode does not affect the operation of other anodes.

b) Due to reduced drift space the total accumulated charge is small. Continuous cathodes also reduce the danger of cathode polymerization.

c) The small cell size simplifies the tracking algorithms due to low multi-hit probability.

d) The mechanical properties of a straw imbedded in a foam allow reliable operation with a gas mixture at 3-5 atm pressure, making the system more compact without loss of momentum resolution in case of measurements in a magnetic field.

The comprehensive study of gas mixtures for straw chambers was done in a paper [12]. The straws operating with a gas mixture $Xe : CO_2$ (50:50) were found to have both an efficient TR absorption and a short drift time. This time was measured to be 38 ns for a 4 mm tube diameter and a 50 μm anode wire diameter. Another attractive mixture is $Xe : CF_4$, which has even a shorter drift time with similar energy resolution. This mixture also enhances the robustness of the chamber with regard to breakdowns.

The uniformity of the gas gain of the straws has been measured and the rms deviation of the gains was found to be very small ($\approx 1\%$). The first results of the radiation hardness test indicates that this mixture also provides very long lifetime for the straws [13]. The use of straw chambers together with foam radiators opens a fascinating possibility of creating an effective integrated TRD-tracker with a large solid angle coverage.

4 TRD-tracker prototype

The RD6 collaboration at CERN has recently performed a detailed test of such a TRD-tracker prototype in order to study its rejection and tracking power [14].

The tested prototype consisted of 192 straws arranged in 64 rows each containing 3 straw tubes, 4 mm in diameter, 200 mm in length. The position of three straw-sets in each row was randomly displaced with respect to the preceding one in order to achieve uniform sensitivity of the detector regardless of the beam angle and impact point. The straw tubes were made of 64μm thick Kapton$^{©}$ and were placed in holes drilled into a polyethylene foam with 8 mm between their centers. The prototype was installed in the X5 test beam of the CERN Super Proton Synchrotron (SPS) (figure 3).

The measurements were done with a mixed beam of pions and electrons. The energy was varied from 10 to 110 Gev. The beam was incident at angles from 90° to 48° to the straws. A lead class calorimeter as well as two Cerenkov counters C_1 and C_2 (not

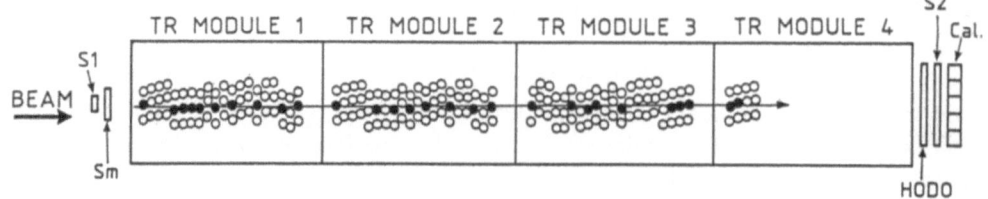

Fig. 3. TRD-tracker prototype test layout

shown) were used to identify electrons and pions. Scintillator counters S_1, S_2 and S_m were used to select particles in the TRD.

The absorption of a TR photon by a Xe atom produces several primary electrons due to the Auger effect. Their range is much smaller than that of a single δ-electron of the same energy as the TR photon. Thus, the ionization cluster produced by the photon is more localized. The difference should even be bigger in case of ionization losses from overlapping particles. One could expect to improve the discrimination between TR photons and ionization clusters by a very fast signal shaping. To check this possibility the time shape of the TRD signal from one of the straws was registered during the test, providing about 40 amplitude digitizings with a time step of 2.5 ns.

5   Results of test measurements

Figure 3 shows an event with a single particle crossing the detector. Black circles indicate "hit" straws, i.e. straws in which more than 0.2 KeV were deposited. A "track" was defined as a band of $\pm$ 2 mm width in which more than 10 straws are hit. Deposited energy spectra for a straw crossed by tracks of 10 Gev electrons and pions at a 63° angle are shown in figure 4. There is a large excess of deposited energy above 4-5 keV for electrons, as expected from TR X-rays. The energy depositions above 5 KeV were defined as TR "clusters". This TR yield, presented in figure 5 as a function of particle γ-factor in the interval $70 < \gamma < 6 \times 10^4$, contains both the contribution of the relativistic rise in the ionization losses for $\gamma < 350$ and of the TR photons for $\gamma \geq 350$.

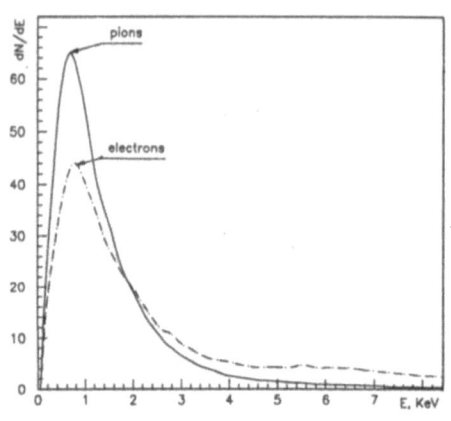

Fig. 4. Deposited energy spectra for pions and electrons.

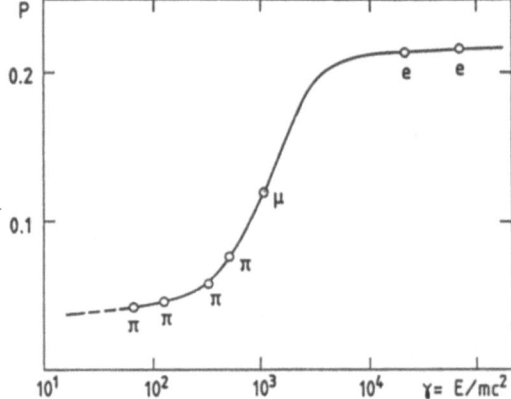

Fig. 5. γ-dependance of the probability of energy deposition in a straw above 5 keV.

Figure 6 shows the single pion rejection factor corresponding to 90% electron detection efficiency as a function of the number of straws crossed.

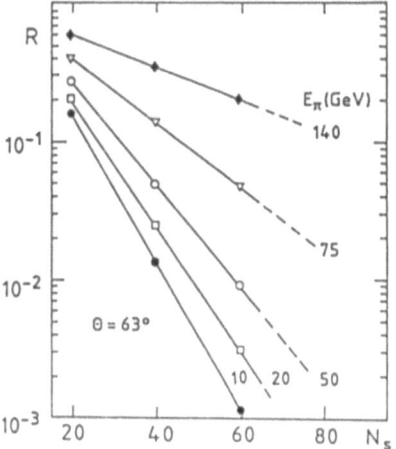

Different curves cover the pion energy range 10-140 GeV for the 63° angle between straw and beam. Instead of 140 GeV pions muons of 110 GeV were really measured.

The tracking performance of the detector has been studied with the following procedure: one track was reconstructed using the odd numbered straws of the original track, whereas the second used the even numbered straws.

Straight line fits were made to each of these "tracks" and fitted parameters were assumed to be independent measurements of the same original track. The difference in the slopes and intercepts of these tracks was taken as an estimate of the reconstruction accuracy.

Fig. 6. Pion rejection as a function of number of crossed straw.

The distribution of these differences, divided by 2, is shown in figure 7. The errors on track reconstruction are 1.5 mrad in angular and 0.4 mm in position resolution for 30 crossed straws.

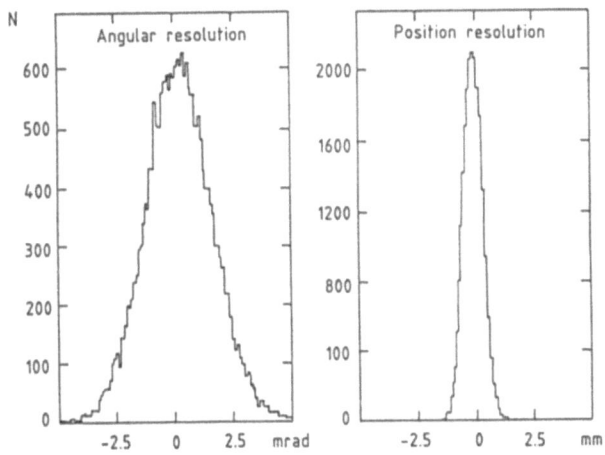

Fig. 7. Reconstruction accuracy of the straw TRD prototype.

As mentioned above a new facility of the proposed TRD is its possibility to discriminate electrons not only against single pions, but also against overlapping particles. A two dimensional analysis was used for this purpose. A traditional "cluster" counting (N_3 - number of energy depositions above some energy threshold, typically about 5 Kev) for transition radiation was completed by counting of "single particle hits" N_{12} - number of energy deposition close to the most probable energy loss for a minimum ionizing particle from about 0.25 MIPs to 1.5 MIPs. The energy thresholds for this algorithm can be optimized. Figure 8 demonstrates the scatter plots of the fraction of single particle hits N_{12} versus the fraction of high energy clusters N_3. The results are given for a 63° angle between beam and straw axis and particle energy of 10 GeV. The efficiency of the two-dimensional algorithm in the overlapping particle rejection is evident.

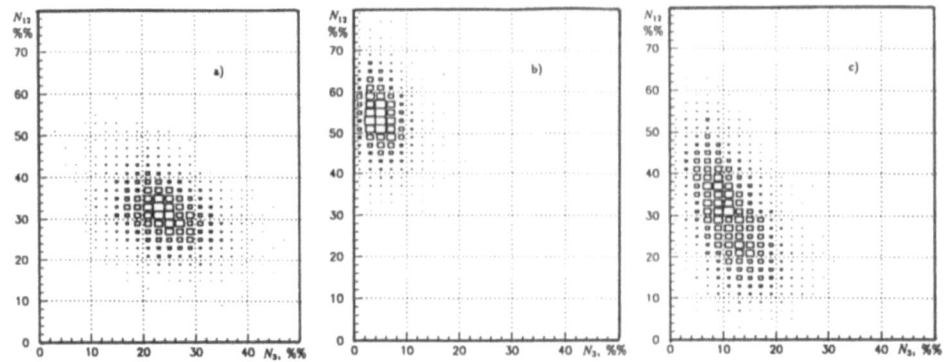

Fig. 8. Scatter plot of the fraction of single particle hits
against the fraction of TR clusters on a track
of electrons (a), pions (b), and two pions (c).

It is interesting to compare these 2-dimensional scatter plots with the same plots obtained with a very short integration time (figure 9) to understand the nature of the rejection improvement. One can see that the number of clusters and single particle hits become less by the same amount both for pions and electrons when the integration time become short. But the number of clusters from overlapping particles is reduced much more.

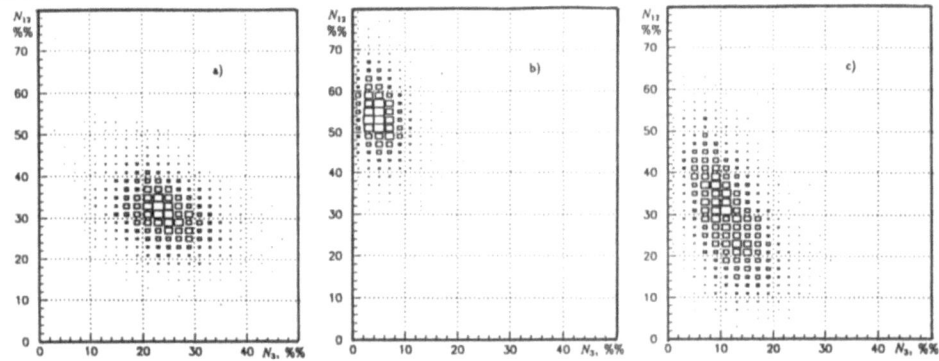

Fig. 9. Same as figure 8, but with 7.5 ns integration time.

Figure 10 shows the dependence of the two pion rejection factor on the signal integration time.

This rejection improvement can be explained by the difference between local ($\approx 300\mu$m), very fast pulses from TR clusters and relatively slow dE/dx pulses from two or more pions.

Saturation below 7 ns reflects the finite extent of the point like TR clusters due to diffusion and residual electronic integration time.

The single pion refection power found to be the same up to this physical time limit.

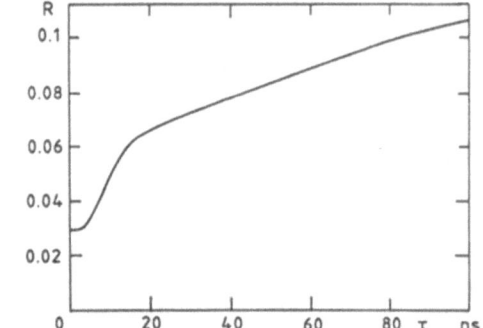

Fig. 10. Two pion rejection factor as a function
of the signal integration time.

6 Detector construction for colliders

Using the data from the prototype test we can simulate TRD-tracker response for any detector geometry and any length. Two possible examples of this geometry are shown on figure 11. Both geometries have the advantages that the electronics can be located outside the detector and that the detectors have no dead zone. The average angle between straw axes and particles for these geometries varies from 48° to 63°.

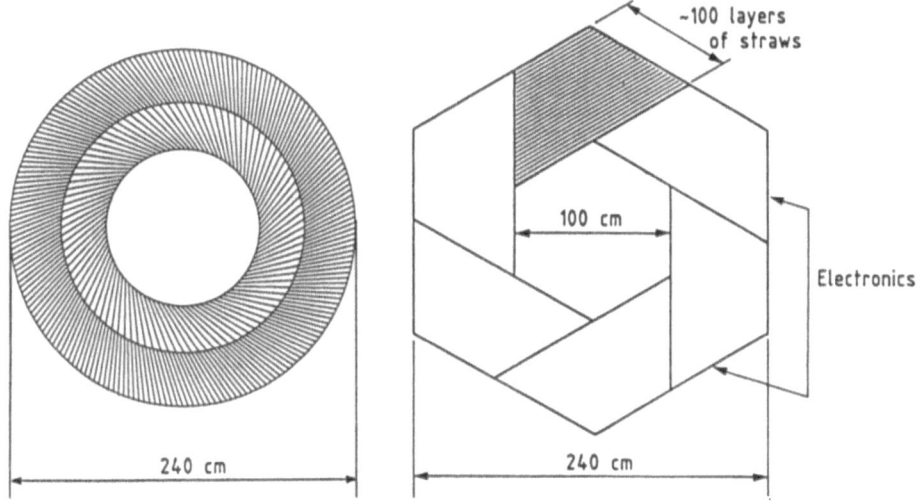

Fig. 11. Cross-section of possible straw TRD-trackers.
a) *halo* design; b) *pinwheel* design

As a reasonable example of future experiment we consider a detector with about 100 layers of straws. The average number of crossed straw will be 50 and the detector length along a particle track at 90° to colliding beams will be 80 cm. A Monte Carlo simulation shows that such a detector can provide a rejection of false electrons at a level of 10^{-1} - 10^{-3} (figure 12), depending on the collider luminosity and the multiplicity of overlapping collisions as well as on the electron rapidity and the number of crossed straws.

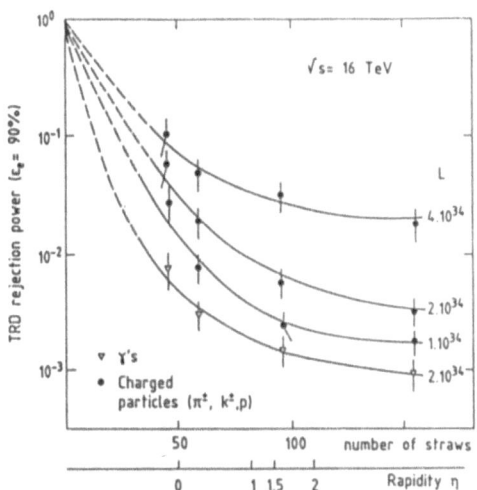

Fig. 12. Monte Carlo evaluated rejection of the TRD-tracker as a function of rapidity for different collider luminosity.

7 Conclusion

A possibility to develop a combined, self supporting TRD-tracker with uniform structure based on polyethylene foam and thin-walled straw chambers looks very promising. Simultaneous dE/dx measurement allows not only to discriminate single pions against electrons, but also to effectively reject multiple particles (hadrons in jet events and conversions). This detector has a good tracking capability. The short registration time of straws and their radiation resistance make this TRD-tracker suitable for operation in a high-luminosity environment.

References

[1] B. Dolgoshein, ECFA study week on Instrumentation Technology for High-Luminosity Hadron Colliders, CERN 89-10, Barcelona **v.2** (1989) 650.

[2] C.W. Fabjan *et al.*, Nucl. Instrum. Methods **185** (1981) 119.

[3] V.A. Polychronakos *et al.*, Integrated high-rate Transition Radiation Detector and tracking chamber for the LHC, CERN/DRDC/90-38 (1990).

[4] S. Ahlen *et al.*, Proposal to develop an Integrated High-Rate Transition Radiation Detector and tracking chamber for the SSC (1989).

[5] V.L. Ginzburg and I.M. Frank, Zh. Eksper. Teor. Fiz. **16** (1946) 15.

[6] G.M. Garibyan, Sov.Phys.-JETP **10** (1960) 372.

[7] J. Shank *et al.*, Proc. Symposium on Particle Identification at High Luminosity Colliders, Batavia, USA (1989) 399.

[8] R. Ansari and B. Merkel, Proc. Symposium on Particle Identification at High Luminosity Colliders, Batavia, USA (1989) 359.

[9] V. Cherniatin *et al.*, Foam radiators for transition radiation detectors, Submitted to NIM.

[10] B. Zhou *et al.*, Performance of small radius thin-wall drift tubes in an SSC radiation environment at MIT research reactor. Preprint BUHEP-90-2 (1989).

[11] H. Schonbacher, ECFA study week on Instrumentation Technology for High-Luminosity Hadron Colliders, CERN 89-10, Barcelona **v.1** (1989) 359.

[12] B. Dolgoshein *et al.*, Nucl. Instrum. Methods **A294** (1990) 473.

[13] V. Bondarenko *et al.*, to be published.

[14] J.T. Shank *et al.*, Test beam performance of a tracking TRD prototype, Submitted to NIM.

TRIGGER AND DATA ACQUISITION PROSPECTS
FOR THE SPAGHETTI CALORIMETER [*]

Riccardo DeSalvo

EF division/LAA Project, CERN
CH-1211 Geneve 23

ABSTRACT

The Spaghetti Calorimeter is becoming a better and better candidate for calorimetry in the future supercolliders. Using R&D results obtained by the SpaCal group at CERN, I am attempting a partial review of some Spaghetti Calorimeter prospects and capabilities. This report is focussed principally on trigger and data acquisition topics. The Spaghetti Calorimeter will be capable of producing very high quality data in the supercolliders.

INTRODUCTION

One of the toughest requirements for a Supercollider Detector is to be able to produce a fast, efficient and tight trigger. If the trigger is not good enough the data taken can easily be biassed beyond recovery or a very high price may be paid in terms of trigger rate, software effort and/or effective luminosity. In fact in the past non-uniform, non-hermetic and unyielding detector design caused awkward readout systems, lousy triggers, loss of effective luminosity and diluted or biassed data samples. In a supercollider, given the increased event rate and complexity and the rarity of interesting events, we cannot afford any of these problems.

In supercollider experiments the trigger will be mostly provided by the calorimeter and the muon detectors. Triggering on tracking would presumably be useful only in specialized magnetic detectors (for example studying B physics). Thanks to its highly segmented, fully projective structure and to the fact that the Spaghetti Calorimeter (SpaCal) signal is naturally delivered on the calorimeter's outer surface, an experiment based on a spaghetti calorimeter lends itself to a particularly easy, fast and effective trigger and data acquisition structure. The data collected will be of uniform quality in azimuthal angle and in a very large range of rapidity. The aim of this report is to describe a possible readout structure for a SpaCal based experiment.

(*) Also presented in a preliminary version at the 9th Workshop of the INFN ELOISATRON Project on "Perspectives for new detectors in Future Supercolliders", Erice, Italy, 17-24 October 1989.

This report, together with the conceptual spaghetti calorimeter design presented last year at Tuscaloosa[1], is intended as a basis for montecarlo simulation and as a milestone for further design evolution. The conceptual calorimeter design mentioned above is used as the hardware base of the trigger and data acquisition system described here.

There can exist an infinite choice of detector configurations in a collider experiment and I will discuss just an arbitrary one. In the rest of the report I assume that tracking is limited to locating and counting vertices and that a PreShower Detector (PSD), capable of localizing particles and distinguishing between charged and neutrals, is an integral part of the calorimeter.

The calorimeter performance as particle identification device is very rapidly spoiled by the electrons produced by photon conversion. For this reason the material in front of the calorimeter must be kept to a bare minimum, possibly just the vertex locating device.

THE SPACAL ADVANTAGES

The signal of the SpaCal is fast, gate lengths of only 70-80 ns are required to collect 98 to 99% of the charge. The SpaCal signal can be shaped to 20-30 ns duration maintaining all the information necessary to the trigger. Very importantly all the signals are available at the outer surface of the calorimeter where cabling, electronics installation, power evacuation and access are easiest.

Thanks to the projective geometry it will be possible to make thin channels for the preshower detector signals without compromises in hermeticity or uniformity[1] (similar channels would be used for cobalt wire calibration sources). The signal of the preshower detector tiles will then arrive at the outer surface of the calorimeter at the same time and in the same place of the rest of the calorimeter signal for every shower. As a result, few tens of nanoseconds after a particle impact, all the information related to a given shower will be available in a geographically limited region at the calorimeter's outer surface.

Because of its fine transverse granularity, the region carrying the signal will match the natural dimensions of the shower process, the unitary region analyzed by the e.m. trigger will be only a few cm wide and even the hadronic trigger will have to look over a radius of only ~15 centimeters in order to obtain trigger energy resolutions of $60\text{-}70\%/\sqrt{E}$ [2].

Since the fibres run radially and deliver their signal to the outer surface an arbitrarily fine segmentation will be possible at the only cost of increasing the number of readout channels.

In projective modules, 75% of the fibres do not run across the whole calorimeter thickness; reading these short fibres separately will give information on the longitudinal shower profile without the need of a mechanical longitudinal segmentation.

Longitudinal channels along the modules corners will allow full volume calibration with radioactive sources.

The SpaCal has many other advantages over other calorimetric techniques like its high speed, high resolution, compensation, small constant term, etc.. Since these SpaCal performances are discussed elsewhere[3] I will not repeat the discussion here.

INFORMATION PIPELINING

During the trigger process the information delivered by any detector must be fully buffered. Trigger processing and information pipelining can be dealt with independently provided that the signal is split at the detector level into two distinct but synchronized streams[4].

With the technique described in this chapter the calorimeter and PSD information can be pipelined by means of Parallel In Serial Out CCD[5] (PISO CCD) analog and shift register memories which run at the Supercollider repetition

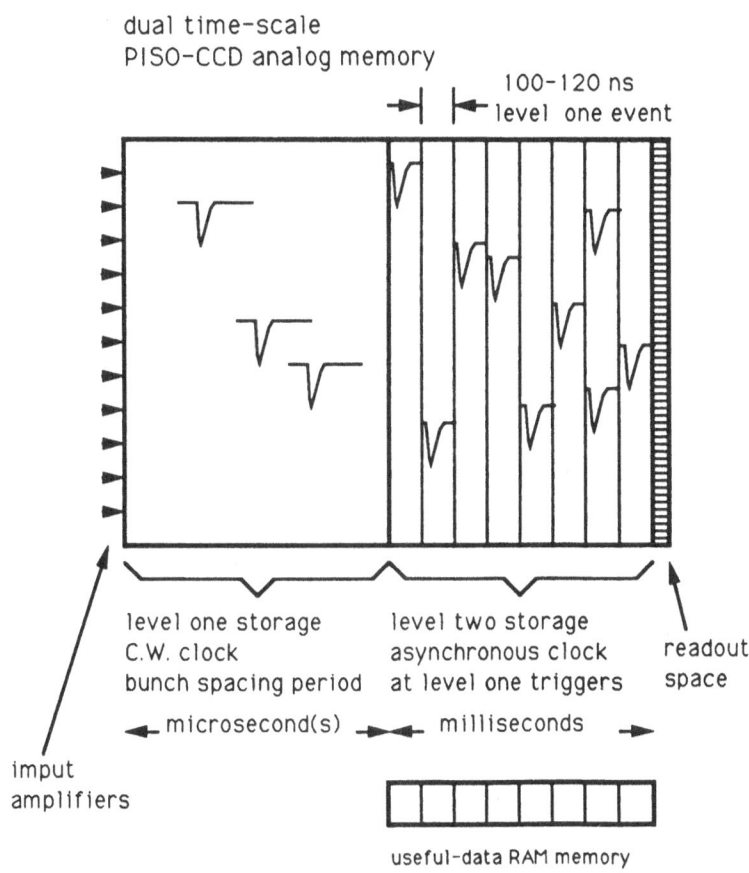

FIGURE 1 Analogic memory structure capable of recording the SpaCal information both during the first and the second level trigger. The full information is stored in the first half of the memory during the level 1 decision while only the preselected events are kept in the second half. If the CCD cannot store the full dynamic range of the signal the same signal can be injected in more than one channel with different volume.

rate for the time necessary to reach the second level trigger decision. Storing the calorimeter information into CCDs will spoil the fine time structure information peculiar of the SpaCal but this information can be extracted from the fast analog signal, converted in charge, and stored in the CCD as well.

The full information is stored during the processing of the first level trigger ($\sim$1 ms) and only the useful part is kept during the milliseconds needed for the second level decision. Storage in both time scales can be achieved on a single CCD chip of limited depth (figure 1). The PISO CCD is a few hundreds channels deep, its depth is divided in two main sections, the first one is shifted by a continuously running clock so that the information is continuously moving along. In absence of a level 1 trigger, the unwanted information is disposed of at the border between the first and the second section of the CCD. The second section of the CCD is divided into 100-120 ns subsection. In case of a level 1 trigger the clocks of this section are run for the time necessary to shift in the required event for storage during the level 2 decision period. Every CCD subsection will then contain a complete event including the history before and after the event.[*]

A pyramidal structure offers a natural modularity for the Spaghetti Calorimeter. The information from the PreShower Detector tiles, the calorimetric signals (electromagnetic plus hadronic from the long fibres and purely hadronic from the shorter fibres) and the time structure signals of each SpaCal pyramid will be stored on a PISO-CCD chip. One chip may store the information corresponding to one or several cells.

The trigger information and any other digital information locally generated could be stored in similarly structured shift register memories.

TRIGGER PROCESSING

When designing a trigger system and listing the different physics signatures one immediately recognizes two main classes of triggers; the first class has local structure while the second class has topological structure. Correspondingly, the decision of the first class of triggers can and will be taken locally while the overall picture of the signal in the full calorimeter will be needed for a decision of the second class.

The first class contains triggers on the production of high energy electrons, muons or isolated gammas and of jets at high transverse momentum. In first approximation the detection of any of the above particles in any point of the calorimeter is enough to retain the event with no need for further checks. These local triggers should be relatively rare and therefore the fake rejection and efficiency should be emphasized when designing them. Since SpaCal preserves the natural geometrical structure of the showers and presents the needed information in a small geographical region, these trigger decisions will not require complicated cabling and can hence be made fast.

The second class of triggers, the topological ones, includes the large total transverse momentum and the missing transverse momentum triggers, often in conjunction with one of the first class triggers. They are much more difficult to deal with because the complete calorimeter signal is required. The following problems arise,
1   the information from the 60-80 m^2 of the calorimeter outer surface must reach a single point to be processed

---

[*]   It may be important to keep track of the baseline in order to be able to reject triggers generated by overlap of different events or to correct for the effects of such overlaps. The baseline information can be dealt with at the time of data digitalization by the local data acquisition processor.

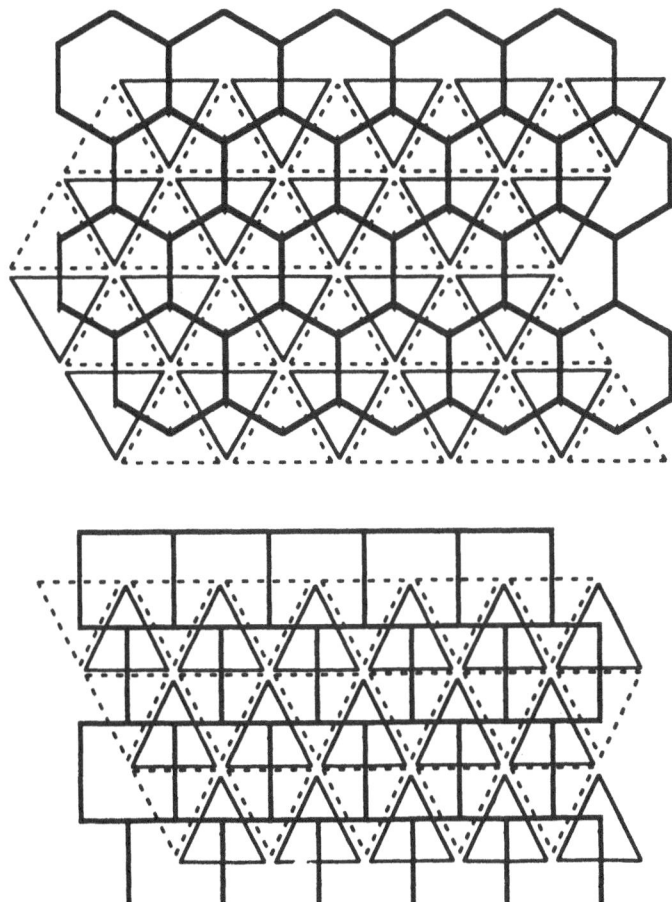

FIGURE 2   The topological equivalence of the hexagonal geometry (top) used in our prototypes and the brick-wall structure (bottom) that will be used in the 4 p projective calorimeters is shown here.   The triangles represent the 3 by 3 ganging of the modules in the e.m. trigger.   The solid and dashed triangles represent the two fully overlapping sets of trigger channels that cover the full surface without border effects.

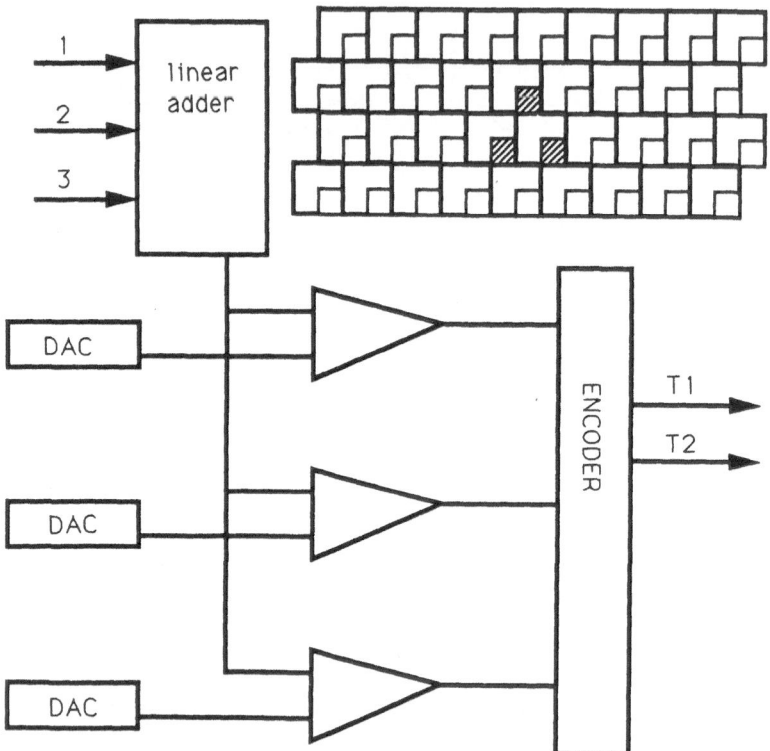

FIGURE 3   Signal mixing in an electron trigger (3A)   and of its hadronic activity veto (3B and C).

FIGURE 3 A   The signal from the long fibres of three pyramids (shaded squares) are added and discriminated.   Three programmable thresholds are applied on the sum signal and encoded on a two bit line for further processing

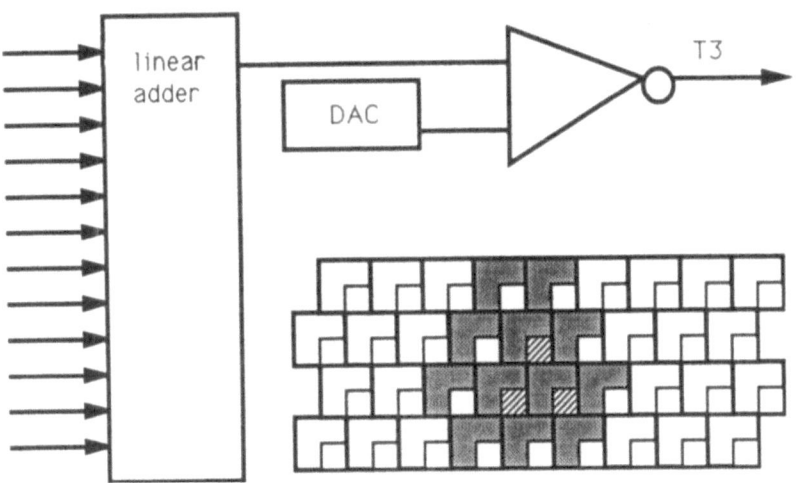

FIGURE 3 B   With an isolated electron signal   all the hadronic shower
sectors (grey notched squares) surrounding an electron triplet should be
silent.   The signal from these channels is used as veto in electron triggers

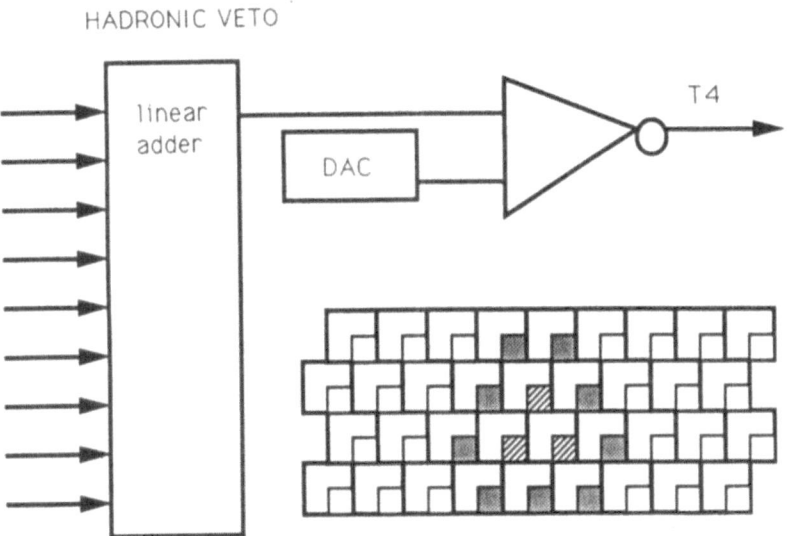

FIGURE 3 C   Since at least a triplet will fully contain an e.m. shower
vetoing on neighboring e.m. channels (grey squares) will help reject electron
like pions

2   the large mass of information (from more than 100.000 channels) must be reduced to a few variables of interest in a fixed and narrow time slot

3   the electronics noise from more than 100,000 channels, if not properly treated, may sum up to large trigger noise

4   there may be a large background from uninteresting events occurring close in time.

For the missing transverse energy trigger, there is the additional problem of dealing with detector inefficiencies and broken channels. The trigger will have keep track of changing electronics noise and breakdown patterns to identify and reject the pathological behavior in order to avoid trigger biases by these spurious effects.

The central problem of any trigger is to build the correct tree so that the information is correctly organized. In the following paragraphs we will indicate how this could be done for the triggers listed above.

ELECTRON TRIGGER

The electron showers are very small and localized; in first approximation making an electron trigger at the level of a single tower would be sufficient. In practice electrons hitting the border region between modules would then be detected with lower efficiency. Ganging many towers together to overcome this problem would not be an appropriate solution because, if the electron trigger unitary surface becomes too wide, many hadrons will be misidentified as electrons. This problem can be solved taking advantage of the bricklayer structure described in ref 1 and shown in figure 2B. This structure is fully equivalent to a hexagonal structure[*] and as a consequence it has no four neighbors corners. For the trigger the pyramids can be ganged in two fully overlapping sets of channels each reading three pyramids sharing a common corner. Given the typical pyramid front section of 35 by 35 mm and the short Moliere radius (17 mm), any given electron will be fully contained inside a e.m. trigger channel. In this scheme the number of electron trigger channels is twice the number of pyramids, but all electrons will be treated with the same efficiency, independently of their impact point. Also every tower will be checked in six channels thus providing a healthy level of trigger redundancy. The loss of a limited number of channels would result in lower trigger performance at the borders of the bad cells. To completely loose trigger at any given position a breakdown should occur in three contiguous channels in each independent sets of trigger channels.

SpaCal offers several independent ways of identifying electrons; some of these are correlated and maybe not all of them need to be implemented. The optimal selection can be done only after extensive prototype tests and some level of redundancy will need to be be kept. We list here some of the known electron identifier algorithms and sketch how they could be implemented in practice.

The first requirement is obviously a minimal energy deposit. Possibly a single threshold would not be sufficient in the trigger processing and, in order to leave enough flexibility, the thresholds should be programmable. A three level threshold accommodated on a two bit line is shown in figure 3A. Only the signal from the long fibres (flat top section of the pyramids) would be used in the electron trigger while the signal from the shorter fibres (side wedges of the pyramids) containing purely hadronic signals would be used in veto as shown in figure 3B. Since electrons are always contained inside a triplet of pyramids pions could be rejected by requiring negligible energy deposit in the nine neighboring electronic and hadronic channels (figure 3C).

_____

[*]   it can be obtained from a hexagonal lattice as in figure 2A by opening to 180^0 two opposed 120^0 angles without changing the topology

peak/charge comparator block diagram

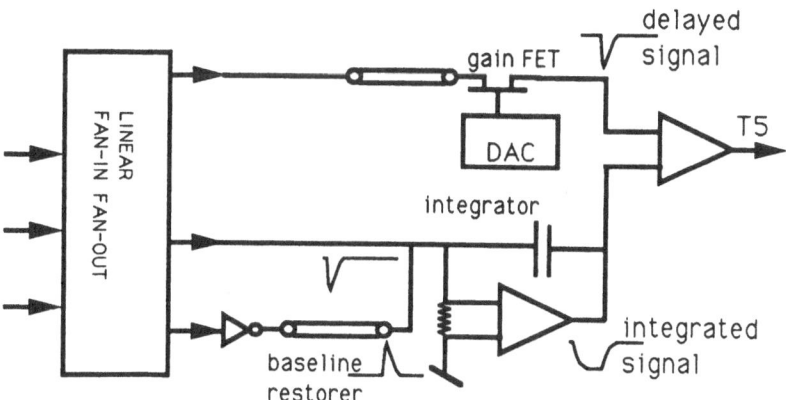

FIGURE 4   The relevant pulse-shape characteristics are transformed into charge to be stored on the PISO-CCDs.

FIGURE 4 A   A peak/charge comparator is obtained comparing the signal with its own integration.   the ratio is adjusted changing the attenuation of the delayed signal line.

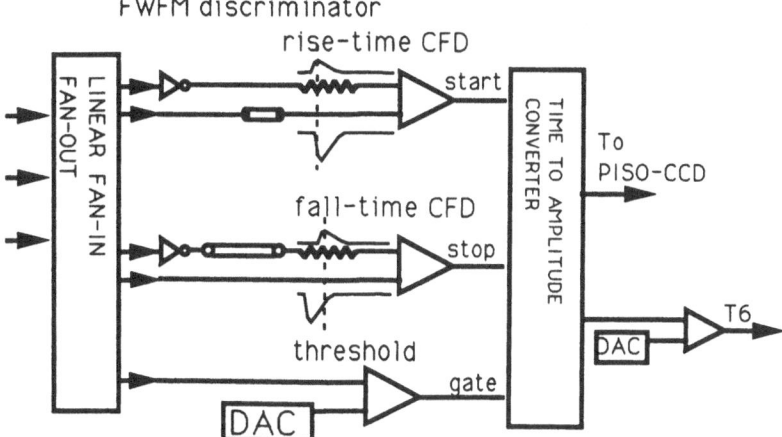

FIGURE 4 B   A constant fraction over threshold signal is built with two CFDs and a minimal threshold gate.   The.electrons produce short light pulses that will give small pulses in the time to amplitude converter thus allowing discrimination against the longer pion pulses.

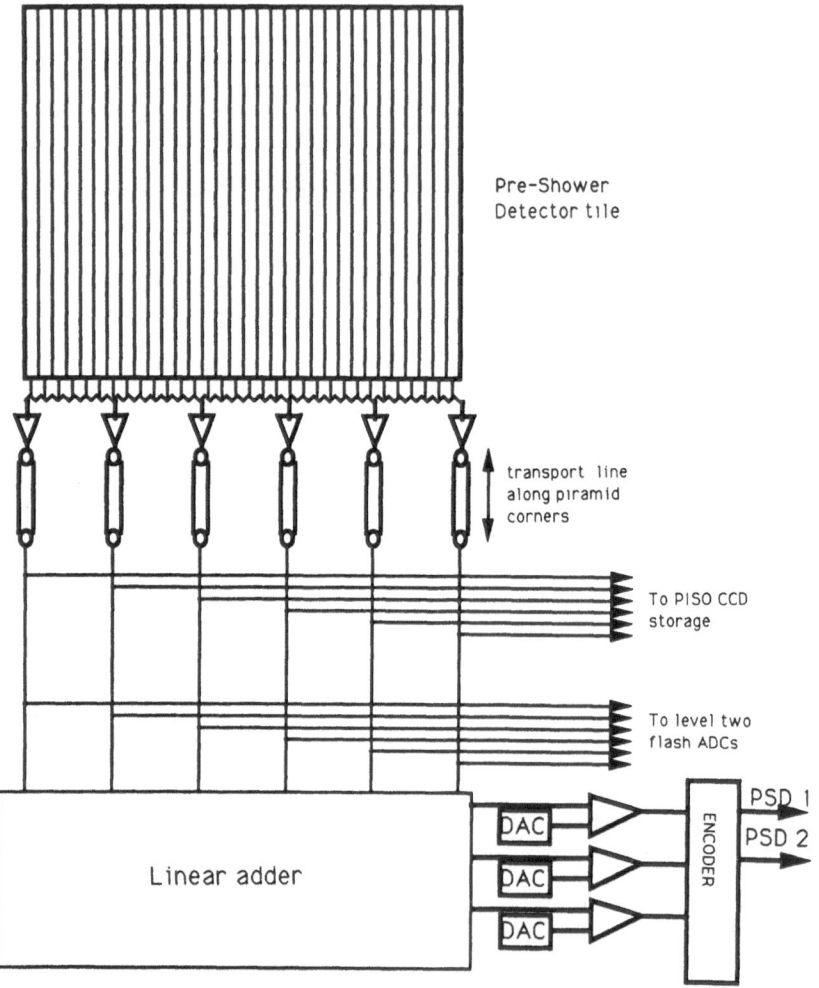

FIGURE 5   Preshower Detector trigger and data acquisition architecture based on a microstrip silicon detector

FIGURE 5 A   Possible architecture of a PSD trigger. A minimum of two readout layers before the converter and two after it will be necessary.   The signal is split between the level 1, level 2 and buffering streams.

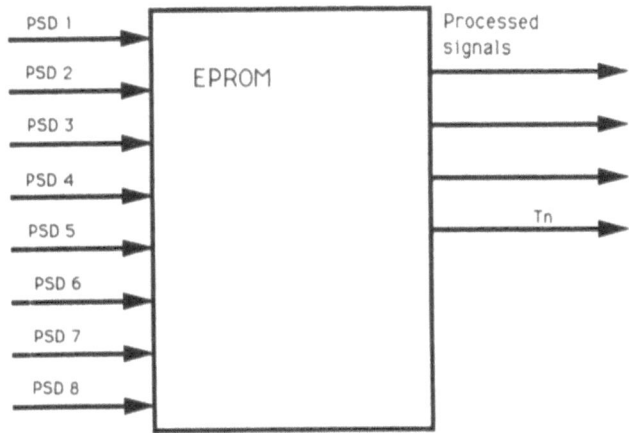

FIGURE 5 B   Processing of the PSD signal to match the signal of different readout layers. The EPROM memory is preloaded with all the possible outcomes for all the possible PSD threshold combinations.

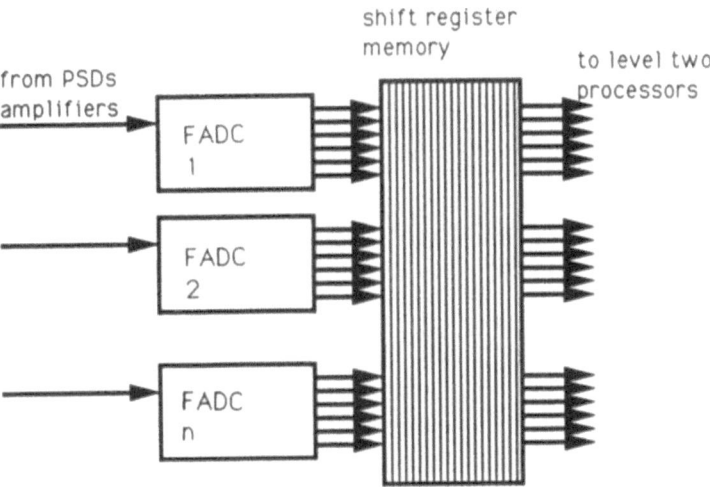

FIGURE 5 C   The full information from the PSD is digitized and stored while waiting for the level 2 trigger

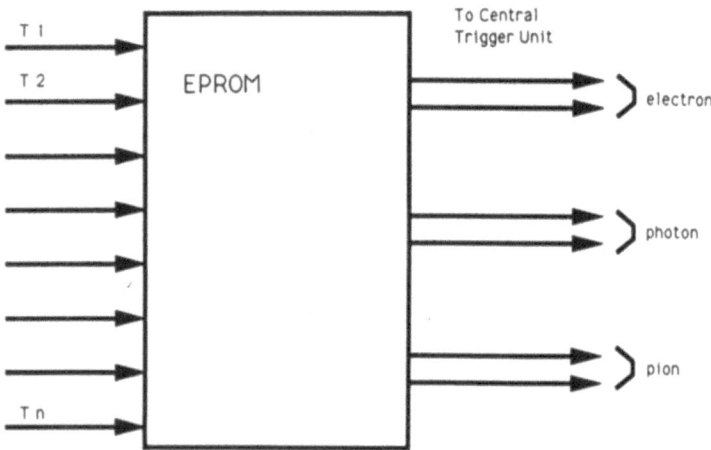

FIGURE 6   All the information coming from the different electron identifiers are analyzed in a way similar to the analysis of the PSD signals.   The trigger signal are sent to the central trigger unit and/or start a local level 2 process.

Next, the pulse shape may serve for identifying electrons[6]. The pulse shape analysis could be performed either on the triplet signal or on the signals of each single pyramid.

We have demonstrated that electrons and pions can be distinguished through amplitude-to-charge ratio and the Full-Width-Fifth-Maximum (FWFM) width. These quantities can be measured with circuits conceptually shown in figure 4A and 4B. Similar circuits can also be used to transform any other interesting feature of the pulse into a charge signal stored in the PISO CCD.

In the SpaCal collaboration we are also studying the use of a "perfect" filter identifying the pulse shape of an electron and rejecting any other shape. Such a filter can never work completely perfectly since some pions will fake an electron, e.g. through a charge exchange reaction in the PSD. A "quasi-perfect" filter with variable thresholds could improve the pion rejection factor.

Next and possibly the most important component for the electron trigger is the PSD signal. The choice of the PSD complementing the SpaCal is critical. It must be fast enough to match the fibres' speed. It should be segmented in tiles matching the SpaCal modules' front section and should have millimetric resolution in two dimensions. A few candidates are available, of which the microstrip silicon detector is currently considered the most promising.

In the PSD electron first level trigger only the total deposited charge before and after the converter would be looked at ignoring the signal distribution inside the tile. A minimum ionizing signal before the converter and multiple tracks after will be the indicator of the presence of an electron. The required information is obtained tuning the thresholds of figure 5A.

The PSD information needed for the level 2 trigger is more detailed and will require a readout and storage of segmented information on several points across the tile. Digitalization could be done by flash ADCs followed by a storage shift register memory (figure 5 C). The tile could be readout in the charge division mode thus requiring only a small number of fast analog lines running from the PSD along each pyramid. Since the charge division technique usually yields percent type resolution a few reading points on a 40 mm tile will allow sufficient multiple track resolution to reject neutral pions trough cluster size and fragmentation of the signal after the radiator.

Rejection of a p^0 accompanied by a soft $p^{+/-}$ is obtained checking the alignment of the preshower signal and of the calorimetric signal in the second level trigger. This operation could also be done locally.

Any other electron identifying information could be added to the above mentioned techniques. The information bits from all the different level 1 electron trigger qualifiers would be preprocessed entering them as an address into an EPROM as shown in figure 6.

It is often asked how close to a jet center it is possible to identify an electron. With the SpaCal interaction length of 20 cm a hadronic shower at 99% containment will extend up to half a meter of radius (encompassing maybe 200 pyramids). SpaCal has a segmentation much finer than the typical hadronic shower dimension (typically 35 by 35 mm front face pyramids) and the hadronic shower shapes are quite irregular. Each hadronic shower will hit only a fraction of the pyramids inside the half meter maximum shower radius. In SpaCal an electron can be identified even within the typical hadronic (or jet) shower radius if it happens to fall in a region with negligible hadronic activity. Many pyramids inside a maximum hadronic shower radius will be virtually empty; in a 150 GeV pion beam test of a two meter long spaghetti calorimeter prototype (twenty non-projective hexagonal modules 85 mm apex to apex.) modules placed 30 cm from the pion impact point were found over threshold less than 30% of the time and, when hit, were showing a signal in average of the order of the percent of the total particle energy. With the smaller cross section foreseen for the SpaCal pyramidal modules an even larger fraction of the channels will be empty. The quantitative estimate of the detection efficiency will require beam tests of projective prototypes or extensive montecarlos. To boost the electron identification inside a jet the readout of the e.m. section of the

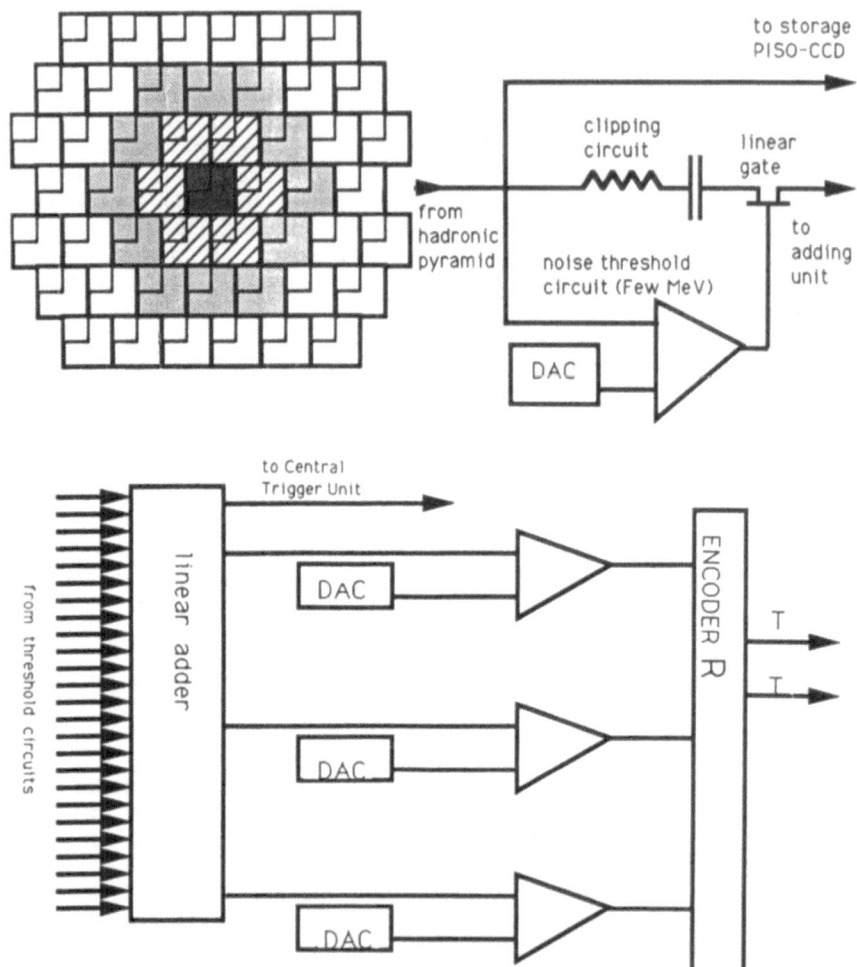

FIGURE 7   To form an hadronic domain the signals from 19 modules (shaded squares in top left corner) are filtered to reduce the noise and shorten the pulse (top right corner)   and added (bottom).   The signal of a hadronic domain is then discriminated for trigger processing.   A simple analogic gate system applied on the raw signals can mask electronics noise in the trigger sum circuit and will be an extremely useful tool to deal with pathological channels.

pyramids could be further segmented to dimensions comparable with the radiation length. Such a fine segmentation would allow a sampling and fitting of the e.m. shower shape inside a jet in a second level trigger. Due to the favorable geometry of the spaghetti this finer segmentation could be easily achieved with segmented cathode photomultipliers or with an imaging version of the new Hybrid Photo Diodes that we are testing[7]. If gammas from neutral pions can be converted by material mounted in front of the PreShower Detector, the sample of the electrons selected with the trigger system will be correspondingly polluted.

ISOLATED GAMMA TRIGGER

Similar to the electron trigger but much more difficult to achieve is the isolated gamma trigger. It can be obtained by requiring an electron trigger with no charge before the pre-radiator. Because of the background from neutral pions, extremely tight isolation requirements will have to be applied in order to keep the level 1 trigger rates at a reasonable level. This trigger could be obtained from different output lines of the EPROMs of figure 6 while requiring the surrounding hadronic domains to show negligible activity.

MUON TRIGGER

The measurement and triggering of muons will be largely independent from the calorimeter because the passage of a muon leaves on average a signal of just a few GeVs which is too low for triggering. At the level 2 trigger a matching minimum ionizing track traversing the calorimeter and pointing from the interaction region to the muon hit could help reject cosmic rays. This will improve the knowledge of the muon energy; at 225 GeV 3% of the muons showed energy deposits larger than 50 GeV in the SpaCal prototypes.

VERY HIGH ENERGY SINGLE JET TRIGGER

Because of the larger dimensions of hadronic showers the scenario is quite different for the hadronic or jet trigger even in the simple case of a single isolated jet. Given the large dimensions of hadronic showers, several modules have to be ganged together to get the full jet energy. Nineteen SpaCal pyramids can be analogically added together (figure 7) to make a hadronic trigger domain of 15-20 cm radius[(*)].

The domains could be further ganged in triplets like the electrons towers in order to get rid of border effects.

The SpaCal signals last for 70-80 ns, however, the trigger could work on clipped signals in order to reduce the signal width and hence the occupancy.

When selecting events with a single very high energy jet it would be enough to discriminate the added signal like in the case of the electrons.

TOTAL ENERGY AND MISSING TRANSVERSE ENERGY TRIGGER

In general the trigger on hadronic signals will be done after analyzing the amplitude and distribution of the energy deposition in the entire calorimeter. The added signal of the hadronic domain would be routed to the central trigger unit to be processed into the topological triggers. Since the transverse momentum has a strong dependence on the geographical distribution (cos(q)) of the deposited energy, the pulse height information from

---

(*) With domains of this dimensions, hadronic resolutions of $60-70\%/\sqrt{E}$ have been measured with pions in Spa Cal prototypes. For better trigger resolutions larger domains should be used.

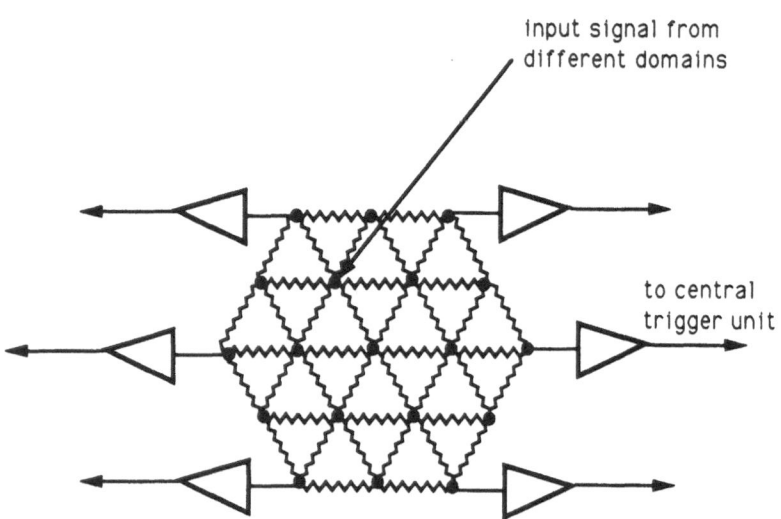

input signal from
different domains

to central
trigger unit

FIGURE 8   The signal from several hadronic domains can be analyzed
with the help of a resistive network in order to extract the geographical
coordinates of a jet.   Injecting the signal from each domain with different
weight the Lorentz boost correction can be taken into account directly.

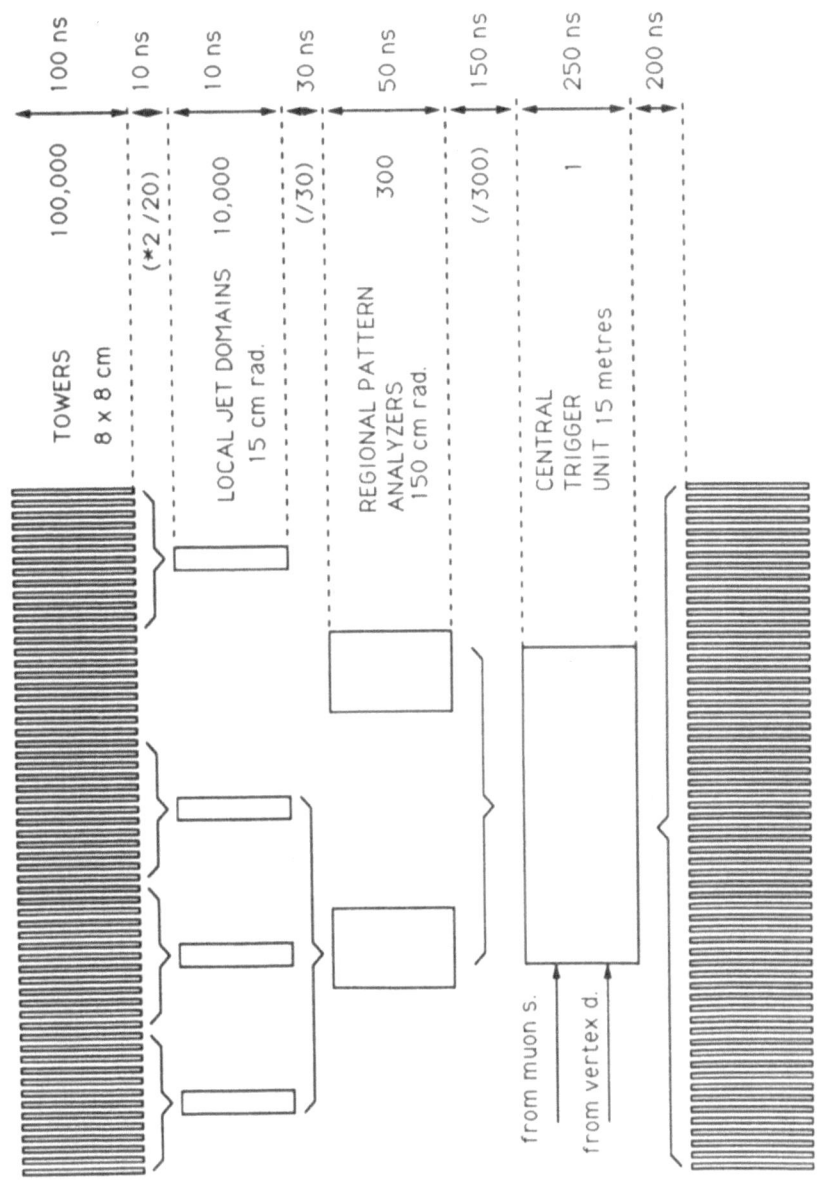

FIGURE 9 Flow chart of the information from the pyramids to the central trigger unit and back.

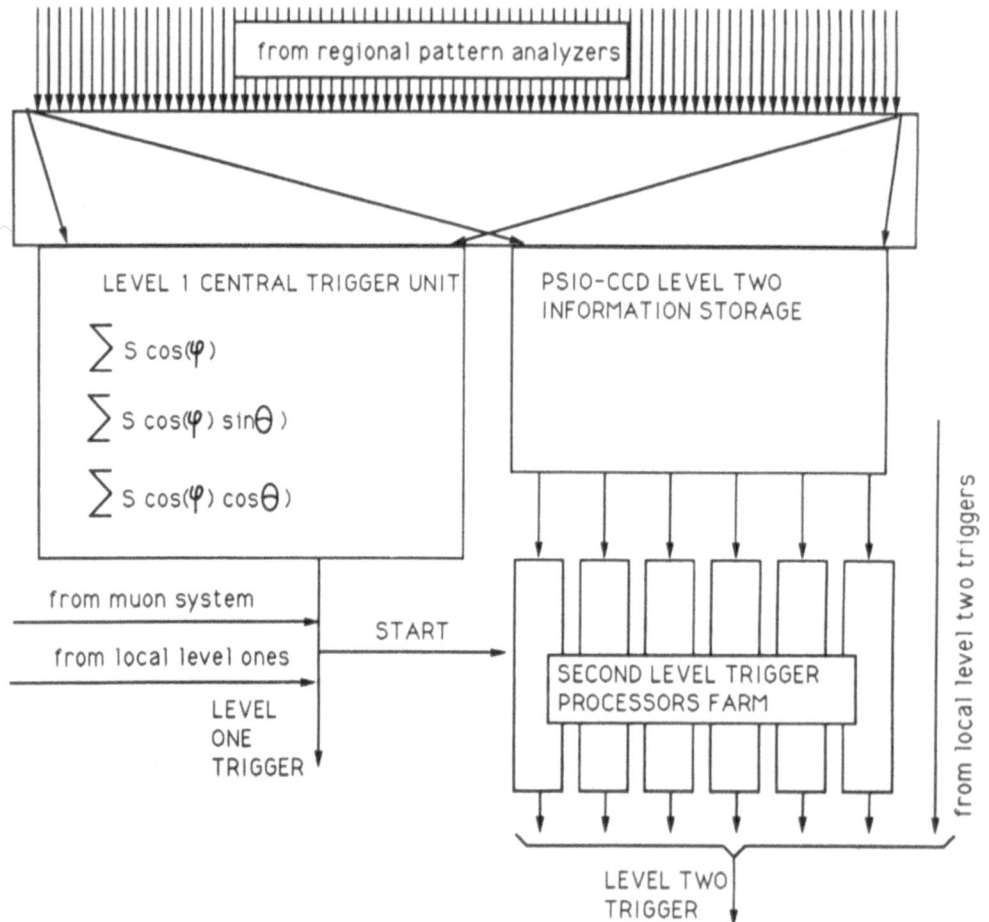

FIGURE 10   Possible block diagram of the central trigger unit.   The information is stored   into PISO-CCDs during the level 1 trigger coarse processing so that it is also available for the level 2 digitization and processing.

each hadronic domain should be brought to the central trigger unit on separate lines. In order to reduce the number of cables without loosing the geographical information regional pattern analyzer could pre-weight the hadronic signal of several domains and compact it into a higher hierarchy signals. The pattern analyzers could either be analogic as in figure 8 or digital. The signals from the regional pattern analyzer, containing the amplitude and position information, is then routed to the central trigger unit for further correlation.

The central trigger unit has several functions;

it receives and sorts the local triggers and the signals from th: local domains

it correlates the signals from the few hundreds regional pattern analyzers to form the total transverse momentum and the missing momentum level 1 trigger.

it broadcasts the level 1 trigger and data storage signals starting the level 2 signal processing.

it processes the level 2 hadronic trigger and sorts the local level 2 triggers.

The information flow from the pyramids to the central trigger unit and back to the towers is shown in figure 9. Since the level 1 trigger in this scheme is a synchronous trigger all the analysis must be done in a way that maintains the temporal coherence of the signal. This means that the central trigger unit is either a large analogic signal analyzer or that the information of the regional pattern analyzer must be digitized synchronously with the machine repetition rate and processed in a conveyor-belt-like digital processor where the data advances at (or at a multiple of) the collider repetition rate. In both cases it will not be possible to implement complex algorithms for the level 1 trigger. After an event passes the level 1 trigger the same information is re-analyzed by the level 2 trigger processors. These processors will have milliseconds to take their decision. For a schematic diagram of the data flow in the level 1 and 2 central trigger unit see figure 10. Note that the level 2 trigger analysis for the local triggers will be done locally on distributed processors triggered by the level 1 signal.

CONCLUSIONS

The Spaghetti Calorimeter is made with lead and plastic fibres which have a fast response to particle showers and fast compensation. The fast light signal can be converted into electronics signals at corresponding speed. The whole trigger and data acquisition scenario here discussed can work at a supercollider 100 MHz collision rate. As a result the Spaghetti Calorimeter can deal with supercollider's event rates and the event overlap problem is minimized.

The Spaghetti Calorimeter has fine segmentation, high resolution, high uniformity and can be made with perfect hermeticity. It has the capability of delivering very high quality data in supercollider experiments and should be chosen for one of the interaction regions.

Development in view of Supercollider experiments has just started and there is a long way still to go. Focussed R&D is vigorously pushed forward by the SpaCal group at CERN and elsewhere and the program is well advanced on the fibre and lead side. Much more emphasis should be put on the trigger and data acquisition problems.

REFERENCES

1   R. DeSalvo, "Dream on a supercollider spaghetti calorimeter", CERN/LAA/HC/89-004 and proceedings of the Tuscaloosa 1989 SCC calorimetry workshop

2   December 1989 SpaCal beam test results, to be published

3   F. Hartjes, R. Wigmans, "Scintillating plastic fibers for hadron calorimetry", Nuclear Instruments and Methods, A277 (1989)379-385
      D. Acosta et al., "Result of prototype studies for a spaghetti calorimeter", submitted to Nuclear Instruments and Methods, 1990

4   this technique is already used in a few experimental data acquisition systems, one of the latest being in the Zeus experiment

5   D. Daub, H. Effing: Philips solid state special products. "Micro Strip detector read out using charge coupled devices", CERN/EF 85-10, July 1985

6   R. DeSalvo et al., "A novel way of electron identification in calorimeters", Nuclear Instruments and Methods, A279 (1989), 467-472

7   R. DeSalvo, "The Hybrid Photodiode Tube", CLNS 87-92, Newman laboratory of nuclear studies, Cornell University, Ithaca, NY 14853, USA

DEVELOPMENT OF FRONT END ELECTRONICS FOR FUTURE SUPERCOLLIDER EXPERIMENTS

Francis Anghinolfi, Paul Aspell, Michael Campbell, Erik Heijne, Pierre Jarron, Gerrit Meddeler

CERN, CH-1211 Geneva 23, Switzerland

ABSTRACT

Key issues of front end electronics for future supercollider experiments at high luminosity are discussed. Advances in microelectronics engineering allow to fabricate high density, high speed, and low power VLSI CMOS circuit customised for particle detector readout. Demonstrator chips fabricated in CMOS technology using a pipeline architecture are described and experimental results are presented.

1. INTRODUCTION

The need for large detection volume and high precision event reconstruction in future supercollider experiments leads to readout systems with a very large number of electronic channels. Consequently, the vast quantity of data to be read out at the high rate expected at these high luminosity experiments will imply the development of complex and high density readout electronics, only feasible when using the potential of modern microelectronics technology.

Micron scale CMOS technology and modern VLSI CAD tools make now possible the fabrication and design of a very complex customised readout system integrated on a single monolithic chip. Customised VLSI chips can contain the low noise amplifier, the discriminator, the analog storage pipeline element, the level 1 trigger filtering and the amplitude digitizer.

The high level of integration of VLSI circuits offers the opportunity to conceive new approaches to detector readout architecture in which advanced signal processing technique will play an important role. The CERN-LAA R&D project supported by Italian funds was specifically established to initiate effort on such new techniques using modern microelectronic technology.

A front end electronic system built with custom VLSI chips can be custom designed to meet detector characteristics. In particular, trackers and calorimeters have different needs in channel density, speed, noise and dynamic range, leading to different design trade off and circuit optimization.

Thanks to miniaturization, a novel pipeline architecture can be conceived in which analog storage capability of switched capacitors circuits can be exploited to create an analog delay matching the trigger decision latency. In sect. 2 this novel architecture we developed in the framework of the LAA R&D [1] program; the Hierarchical Analog Pipeline Processor (HARP) for analog readout application, is described.

New Technologies for Supercolliders, Edited by L. Cifarelli and
T. Ypsilantis, Plenum Press, New York, 1991

In section 3, we report the experimental results of test chips designed to prove the feasibility of this approach. In section 4 is presented another readout processor principle; FASTPLEX in which the signal processing is digital .These two readout processors have a pipeline architecture with no dead time, which is an important key feature for high luminosity collider running at 15 ns bunch interval, whereas the level 1 trigger latency is in the range of 1 µs.

2. HARP READOUT SYSTEM

The readout of particle detector employs basically four analog signal processing functions:

- Low noise amplification.
- Pulse shaping.
- Delay.
- Digitization, a discriminator or an analog to digital converter.

The miniaturization of modern VLSI circuits allows to accommodate on a single monolithic CMOS chip several identical channels (16-128), each containing those basic analog processing functions with in addition the necessary "glue" logic to control the readout system, common for all channels.

Obviously, such an approach considerably improves system performance in packing density but also in speed and power consumption because of the reduction of parasitic capacitances. A simple 1 cm connection in printed circuit board has a parasitic capacitance in the range of one pF whilst a 10 µm connection in a VLSI circuit has less than ten fF. Moreover, the fact that the analog delay can be integrated on a silicon microchip allows trigger filtering before digitization, thus saving considerable power .

2.1 HARP architecture principle

Based on this approach, we have developed a novel readout concept [2], named 'Hierarchical Analog Readout Pipeline Processor' (HARP) which is shown in fig. 1. Under this scheme, analog information is stored locally until a trigger decision tells the readout system to digitize only the analog signals which belong to the triggered sampling time. Consequently, this on chip data selection allows to digitize and transmit a very small fraction of the raw data which considerably simplifies and speeds up the readout operation.

The key of this readout system is the analog delay element. It is designed as a multicell analog memory of which the write cycle is controlled in pipeline mode.

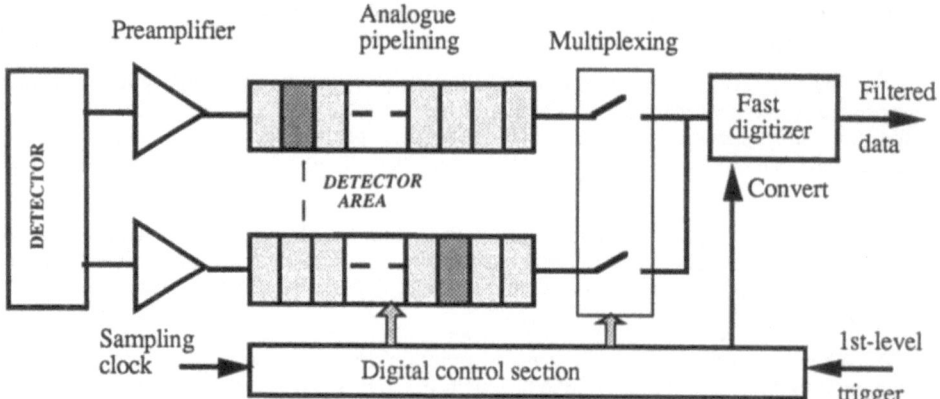

Fig. 1 New approach uses analog pipeline on the detector as signal delay element. Only data selected by first level trigger are digitized, also on the detector. The data filtered by this trigger are transmitted in digital form to the remote electronics area.

2.2 Analog storage and pipelining readout operation

Charge Coupled Devices (CCDs) are well-known analog signal processing elements [4]. Their precision is usually limited to a precision in the order of 8 bit, which could be not sufficient for the physics experiments. Data stored on a CCD can only be read out serially, when the transferred charge reaches the end of the pipeline. Moreover it is difficult to make radiation hard CCD technology.

The switched capacitor technique offers the possibility of storing analog signals on good quality capacitors without moving charges. The advantage is that analog signals stay in place on the storage capacitors, whilst the memory addressing is done by read and write pointers driven by digital shift registers. Using this technique, analog signals can be stored with 11 or 12 bit precision [5] and data can be randomly accessed through addresses or pointers.

Signals from detector elements, after amplification if necessary, are passed into a switched-capacitor analog pipeline memory which allows storage for the time needed to build the first level trigger decision T1. The memory is organised in such a way that reading and writing are performed simultaneously, thus eliminating system deadtime.

The principle is illustrated in fig. 2 for a matrix of memory cells. The write and read operations use separate addressing registers, bus lines, switches and amplifiers. The delay time T1 between write and read can be adjusted to the latency of the level 1 trigger.

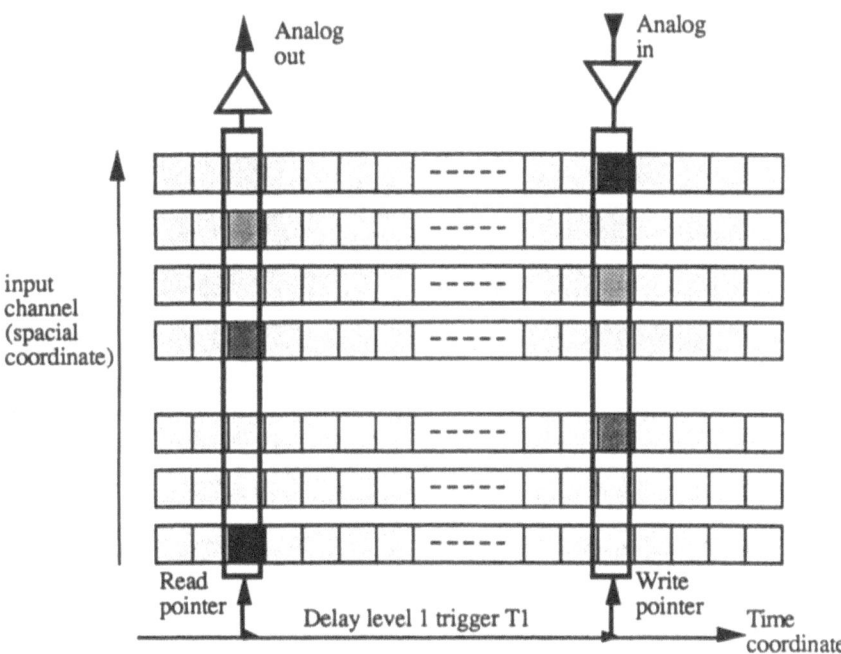

Fig.2 Matrix of analog signal storage cells. Each row contains the consecutive data from a single detector element. The signal of the present event is being written in the column of cells indicated by the write pointer. At the same time data from another event, T1 seconds ago indicated by the read pointer, can be read out if selected by the trigger.

A simplified circuit scheme of the addressing of the memory cells is shown in fig. 3 for a 3 cell array. Each cell consists of a capacitor C_S, a write switch and a read switch. The write register WSR addresses the write switch defining the cell to be charged while the read register defines the cell to be read out. Both operations can be carried out simultaneously within the same clock cycle.

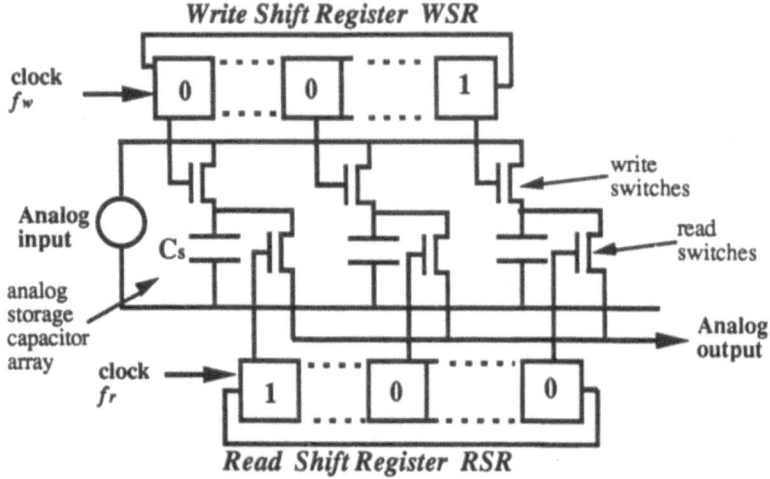

Fig. 3 Schematic of a circuit which allows to write and read simultaneously in different storage capacitors.

2.3 Signal sampling technique

The input to the analog memory can be configured as a voltage sampler or a current integrating sampler. These configurations are illustrated in figs 4 and 5 for a 3 capacitor array. The voltage sampling technique of fig. 4 is the most widely used. The write amplifier W_{amp} in the voltage follower configuration simply stores the instantaneous input voltage on capacitor C_{s1} when the write switch opens after closure. In this example C_{sn} is simultaneously being read by the amplifier R_{amp}, and C_{si} is disconnected.

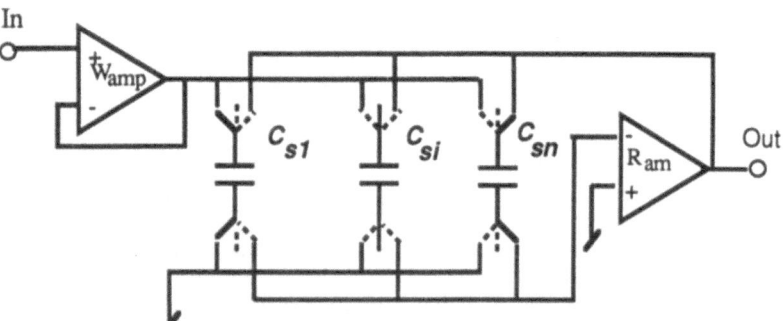

Fig. 4 Voltage sampling input circuit

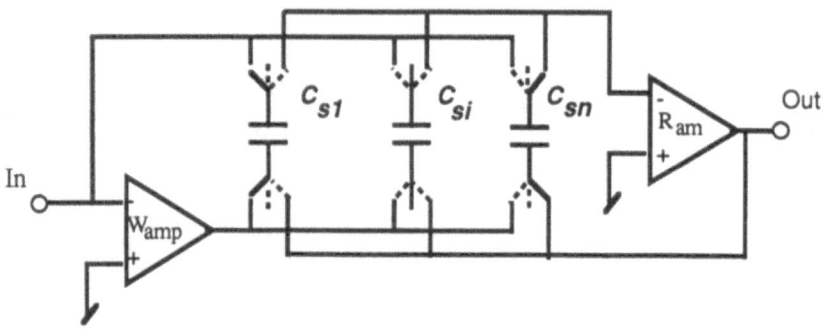

Fig. 5 Current sampling input circuit.

The current integrating configuration of fig. 5 uses memory capacitor C_{si} as a feedback capacitor of the write amplifier W_{amp} so that this circuit operates like a traditional charge amplifier. The resulting voltage on the capacitor is proportional to the integral of the input signal current from the detector during the sampling interval $Ts = 1/fw$. We call this circuit configuration the Charge Sampling Integrator (CSI).

In particle physics detector applications, the CSI configuration has several advantages over voltage sampling technique, mainly because the charge is the relevant information of most particle detectors based on ionization. In supercolliders the detector signal may spread over several clock periods of 15 ns. Assuming there is no pile-up, the simple sum over several cycles would yield the detector charge with a good tolerance regarding jitter. To get it with voltage sampling, a relatively complicated algorithm involving knowledge of the pulse shape would be needed . Moreover, when using short pulse shaping tunes e.g. 10 ns, the sampled pulse shape may well be sensitive to local variations, as the detector geometry may lead to different input capacitances, the charge collection speed may vary, e.g. after radiation damage etc.

The current sampling integrator has the sensitivity of the charge amplifier, thus it may turn out superfluous to have a pre-amplifier between detector and analog pipeline circuit. Also, the CSI circuit acts as a gated integrator equivalent to a time variant filter on each sampling period, followed by a time invariant filter consisting of the transfer function of the write amplifier. In principle, no further analog signal processing should be needed to enhance the signal to noise ratio.

In order to reduce clock feedthrough effects, identical write and read switches are placed symmetrically on the two nodes of the storage capacitor as illustrated in fig. 6. The Analog Memory Element (AME) is completed by a reset switch which discharges the capacitor C_S after reading, when all other switches around the capacitor are turned off. The injected parasitic charge is then equipartitioned on both capacitor electrodes.

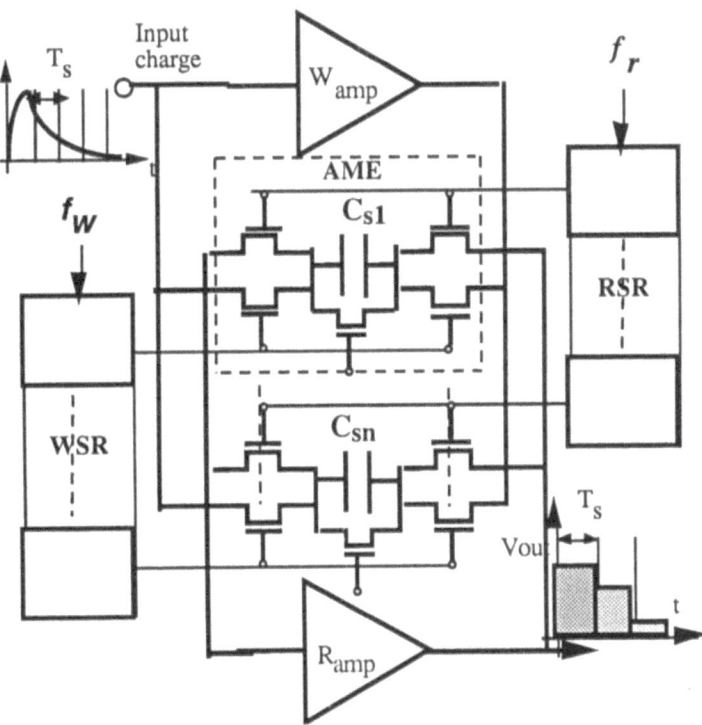

Fig. 6 Schematic diagram of the analog memory elements (AME) in the current sampling integrator circuit. The symmetric write switch and read switch transistors are actuated by the respective shift registers WSR and RSR.

The total equivalent Noise Charge (ENC) of the CSI circuit is expected to have three dominant noise contributions. First the ENC of the gated integrator W_{amp} which can be calculated following the approach of V. Radeka for signal filtering for Ge detectors [6]. Second, the KTC noise caused by the reset switch, which is 400 e⁻ for C_S= 1pF. The third contribution is the noise voltage of the read amplifier. Further analysis is expected to show that the gated integration is the dominant noise contribution, at least in the ideal case.

One limitation of the CSI circuit is the mismatch between capacitors in a single pipeline channel which may cause a gain spread. If double polysilicon capacitors are used, the spread may be limited to 0.5-1%. If capacitor signal is read as a charge, capacitor mismatch has no effect. Voltage sampling configuration does not suffer from this limitation.

3. EXPERIMENTAL ANALOG PIPELINE TEST CHIPS

Based on the design circuit technique described in section 2 employing the analog memory principle, we designed and fabricated two experimental VLSI chips called 'Simultaneous Analog Pipeline Element ' (SAPE) . The first one, SAPE 3, has a double pipeline architecture and was fabricated in a 3 μm n-well CMOS technology with double poly[(*)]. The second chip, SAPE 1.5 is a single pipeline circuit. In CMOS 1.5 μm in a double poly, double metal process. It is similar to, but much faster than, the SAPE 3 circuit.

3.1 Design of SAPE 3

The block diagram in fig. 7 shows the double pipeline structure with 64 memory cells in the first pipeline and 8 cells in the second. This is also clearly visible on the chip photograph in fig. 8. The shift registers WSR and RSR act as pointers to the 0.6 pF storage capacitors to be written or read. The size of each AME cell is 110 μm × 65 μm. The total length of each channel is 4140 μm. Two of the channels have voltage sampling inputs and the other two have a CSI configuration. The write amplifier is a folded cascode Operational Transconductance Amplifier (OTA) with n-channel input transistor and ~ 0.5 pF input capacitance, open loop gain of 75 dB, unity gain at 117 MHz, an integral dynamic range of ± 2 V. The power dissipation of the OTA is 5 mW and the non-linearity in the range ± 1.5 V is < 0.2%.

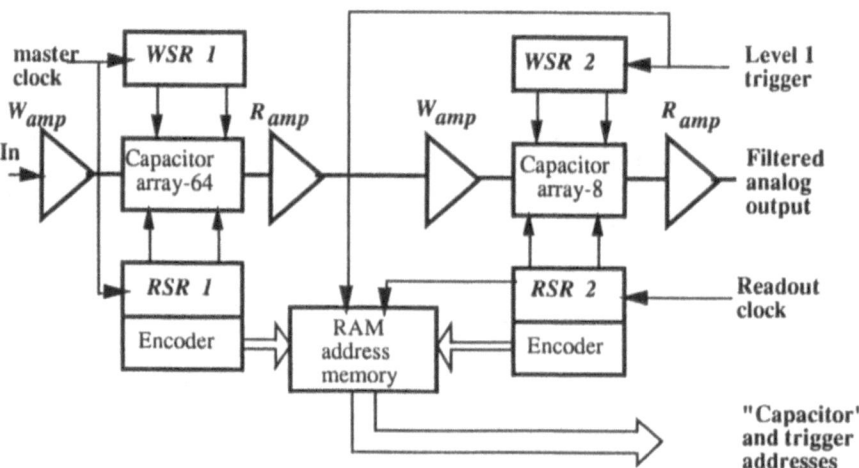

Fig. 7 Block diagram of the double SAPE circuit made in a 3 μm CMOS technology.

(*)   Mietec N.V., Oudenaarde, Belgium

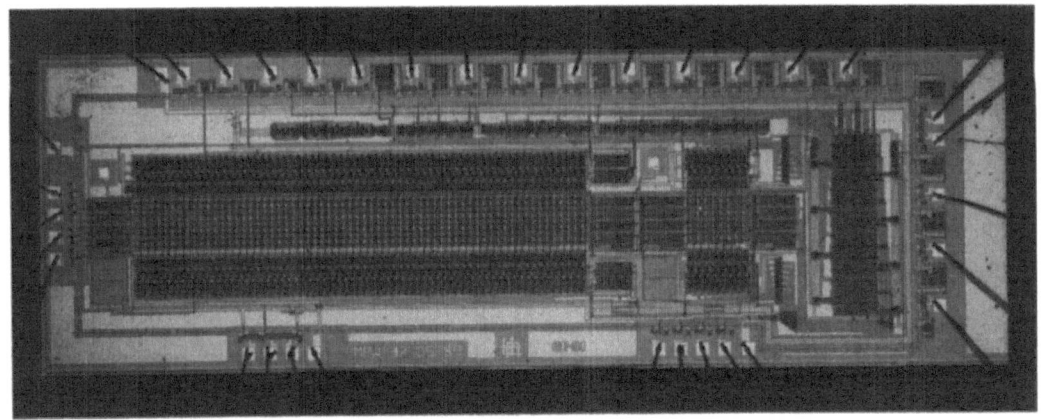

Fig. 8 SAPE 3 view.

The transfer of a signal from a cell in the first array to the second array is controlled by an external trigger signal and the addresses of the original cell and the destination cell are externally available in order to allow pedestal correction and gain equalization, if needed. A 4-bit counter keeps track of the trigger number. The operation of this circuit has been visualised in fig. 9, where two turning wheels represent the pipelines containing the memory locations, which are written or read as they pass the respective pointers. The contents of the second pipeline are designed to represent at most 1/8th of the data in the first. The transfer of a signal from the left wheel to the right wheel only becomes effective if an external trigger signal moves the write pointer in the second wheel to the following position. The cells in the right wheel contain 'good' data which can be read out at any desired moment. Therefore, this pipeline acts as a derandomizer element.

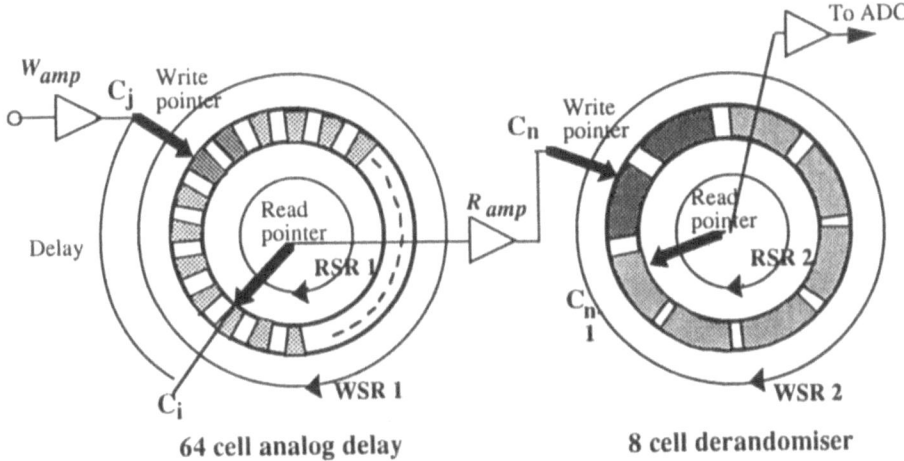

64 cell analog delay   8 cell derandomiser

Fig. 9 Visualization of the operation of the double pipeline circuit. The write pointer in the derandomizer array at the right will only move to a new cell after a 'good' event. The 'good' events can be read out of the second array at any desired moment.

3.2 Experimental results of SAPE 3

The SAPE circuit performs several functions in an integrated manner: low-noise amplification, analog signal storage and data filtering. One would like to characterize each of these functions as in the case of separate circuits: noise, sensitivity, dynamic range of the input amplifier, sampling speed, pedestal and gain uniformity of the analog memory function. The evaluation of such a system is complicated by various summing and interference effects and it is not easy to identify the origin of a performance limitation like noise, non-linearity or non-uniformity. Cross-talk between the digital and the analog circuitry is obviously a major concern.

Our test set–up uses the commercially available programmable data acquisition system ASYST with LabMaster modules, running on an Olivetti M300 Personal Computer. Although this system provides ADC and DAC functions, another low-noise DAC is added. The programmed clocking sequences are stored in a RAM, which is mounted in a test box, together with an address decoder and an output signal track/hold circuit.

The operation of the SAPE 3 circuit as a delay line is illustrated in fig. 10 with a write clock frequency of 20 MHz in the first pipeline. The read clock (arbitrary trigger) frequency is fixed at 2 MHz so that 1/10th of the data are read out.Figure 11 shows an oscilloscope picture of a single pulse in voltage sampling, at the 20 MHz write clock frequency. Several tests have been made to verify that one can store signals in successive capacitors and retrieve them correctly, as shown in fig. 12 for 5 pulses. The current sampling input is shown in fig. 13 where a current is presented at the input via a 100 kΩ resistor. This picture is one of a sequence in which the input current pulse was gradually moved from one clock cycle to the next. The charge is either integrated in exactly one memory cell or distributed over the two storage capacitors proportionally to the time division.

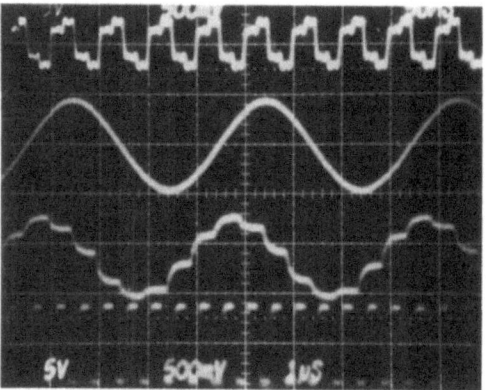

Fig. 10 Operation of SAPE 3 as delay line. The top trace is the 20 MHz write clock and the bottom trace the 2 MHz read clock. The second trace is a 250 kHz sinusoidal input signal, which is reproduced at the output, 3rd trace.

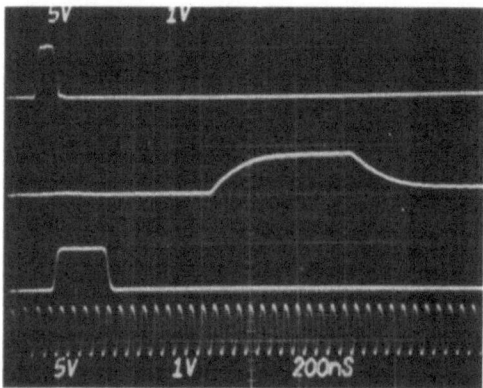

Fig. 11 Top trace: input voltage signal of 1 V. The second trace shows the corresponding output signal. The 3rd trace gives the trigger signal for readout and the 20 MHz write clock is shown at the bottom. The read clock is a +10 derivation of the write clock frequency.

112

The sensitivity and the linearity of the current input are shown in fig. 14 for positive input charge. A similar curve is obtained for negative signals. The sensitivity is 1.62 V/pC. The linear range is ± 1.2 pC. The system has ± 2 V dynamic and 2 mV of pattern noise which correspond to a precision of 11 bits. Similar values will be shown for the SAPE 1.5. The writing speed of the SAPE 3 is limited by the digital circuitry, and the settling time of the output signal is limiting the readout frequency.

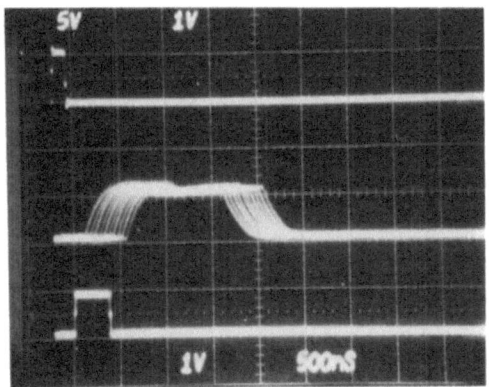

Fig. 12 An input signal with a duration of 5 clock cycles at 10 MHz (top) in voltage sampling is stored in consecutive storage cells and read out (middle trace). The bottom trace is the trigger signal.

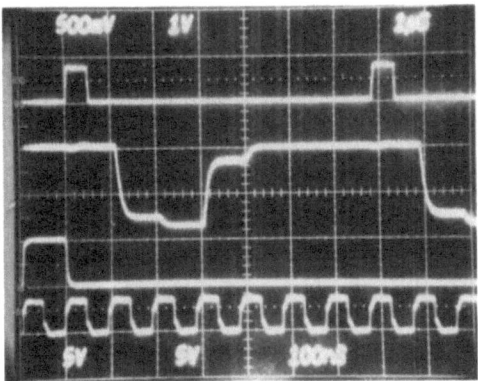

Fig. 13 Charge sampling of a signal, top, in a 100 kΩ resistor. A delay is applied in order to shift the 100 ns long signal across two read cycles, so that the integrated charge appears divided over two storage elements, second trace. Trigger signal and 10 MHz clock are as before.

Cross talk problems appeared in the measurement set-up, in particular digital clock feedthrough was serious at first. It was found that the main path to the sensitive inputs passes via the ceramic Dual-In-Line package, because the pedestal uniformity improved strongly after cutting the input wire bonds. Disconnecting the pins from the board contacts did not help. One will have to use low-level, differential input logic signals like ECL as well as sophisticated packaging for this type of device.

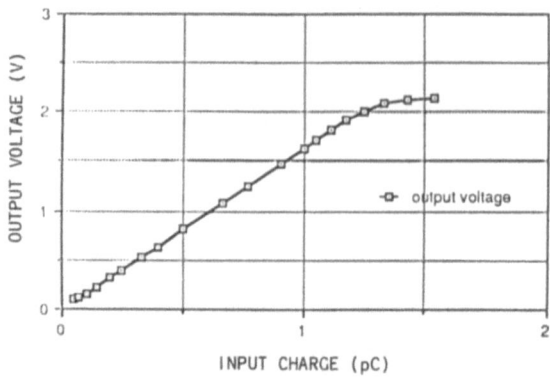

Fig. 14   Sensitivity of the SAPE 3 current sampling input configuration.

3.3 Design of SAPE 1.5

The sampling speed of SAPE 3 reached 20 Mhz which is not sufficient for the event rate expected at supercollider experiments. The speed limitation came from a rather complicated logic controlling the transfers of data between the first and the second analog pipeline elements. Therefore, we have developed a single pipeline simultaneous read/write circuit in the MIETEC N.V. 1.5μm double-poly, double-metal process. The block diagram in fig. 15 shows the pipeline structure with 64 memory cells. The chip photograph is shown in fig. 16.

The circuit consists of 4 parallel channels with 2 voltage sampling and 2 CSI inputs. The write and read amplifiers, using differential folded cascode OTAs, are designed with p-channel transistors at the input and n-channel transistors for the cascode devices to obtain settling time. Unity gain bandwidth is 220 Mhz with 0.8 pF load. Power consumption is 5 mW. The value of the storage capacitor of 0.4 pF is a trade off between accuracy and noise on one hand and speed and silicon area on the other hand. The shift registers WSR and RSR act as pointers to the storage capacitors to be written or read. The size of each AME cell is 100 μm × 38 μm. The total length of each channel is 3150 μm.

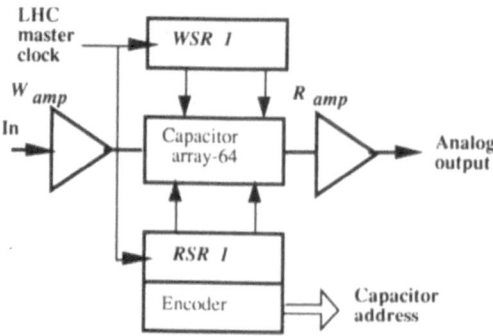

Fig. 15   Block diagram of the SAPE 1.5 circuit.

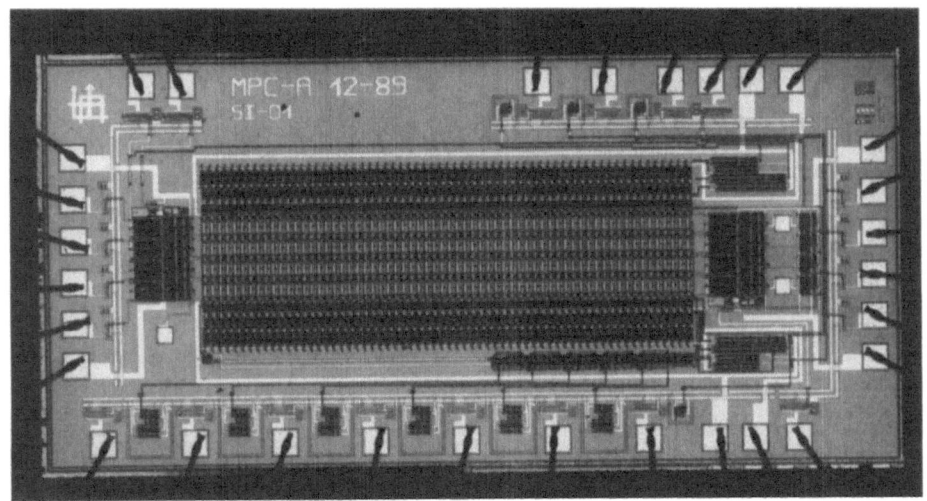

Fig. 16   SAPE 1.5 view.

3.4   Experimental results of SAPE 1.5

The operation of the circuit as a simple delay line in voltage sampling mode is shown in fig. 17 and the maximum write/read clock frequency is found to be 70 MHz.

The operation of the circuit as a waveform analyser is shown in figs 18 and 19. In this test, the analog pipeline is firstly filled at a write frequency of 70 Mhz , and read out at a read frequency of 1 Mhz. Figure 18 illustrates the response of the voltage sampling channel while fig. 19 shows the CSI output. The results indicate a precision of at least 11 bits as seen with SAPE3 which is difficult to obtain with flash ADC.

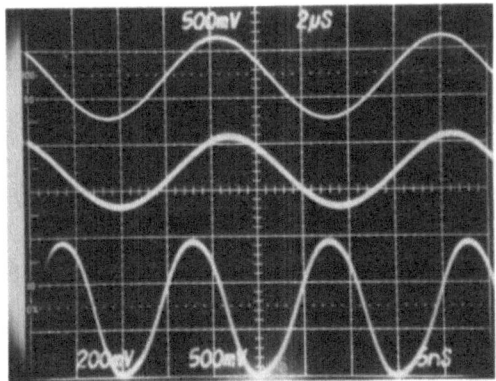

Fig. 17 Oscilloscope picture of the operation of the SAPE 1.5 circuit as a delay line, operating at 66.6 MHz. The top trace is a 100 kHz sinusoidal input signal which is reproduced at the output, middle trace. The bottom trace is the clock frequency.

The sensitivity of the CSI circuit channel (fig. 20) is 2.29 mV/fC and the linearity extends from -0.7 pC to +1 pC. A 2 MIP signal equivalent in Si detector corresponding to 8 fC is visible in one of the cells in fig 19. This proves that the circuit can operate even without an external front-end amplifier. The pedestal distribution of the 64 memory elements is plotted in fig. 21. The mean value is -1.28 V and is related to the offset voltage of the source follower amplifier used to measure the output of the read amplifier. The variance of the pedestal non-uniformity is 1.4 mV and also the variance of the output signals for a single memory element is in this range (1.6 mV) with a sampling clock of 67 Mhz and a readout clock of 1 Mhz. At present we are studying which are the limiting factors and it would be premature to regard these fluctuations as the noise performance of the circuit.

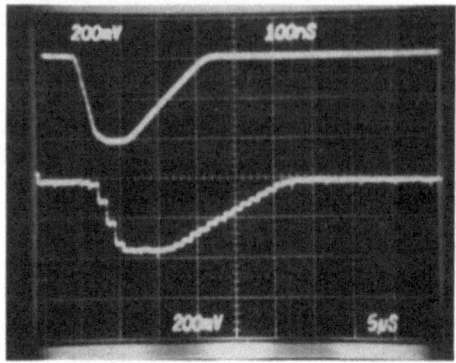

Fig. 18 Oscilloscope picture of the SAPE circuit as waveform recorder in voltage sampling mode. The input pulse (upper trace), is stored in the SAPE at the write frequency (sampling) of 67 MHz and at a lower read frequency of 1 MHz (lower trace).

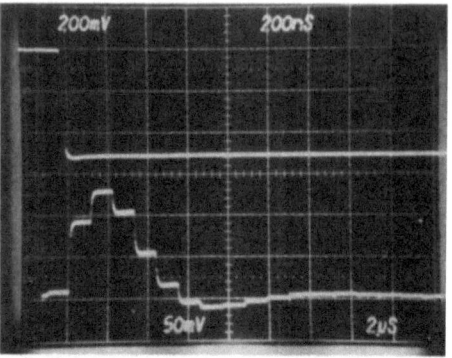

Fig. 19 Oscilloscope picture of the operation of the SAPE circuit as waveform recorder in current sampling integrator mode. The input step voltage (upper trace), injects a charge at the input of the circuit via a test capacitor of 4.8 pF. The input charge is sampled at 67 MHz.The impulse response is readout at a frequency of 1 MHz (lower trace). The sum of the samples read at the output is the image of the input charge. The waveform of the output response is determined by the write amplifier impulse response.

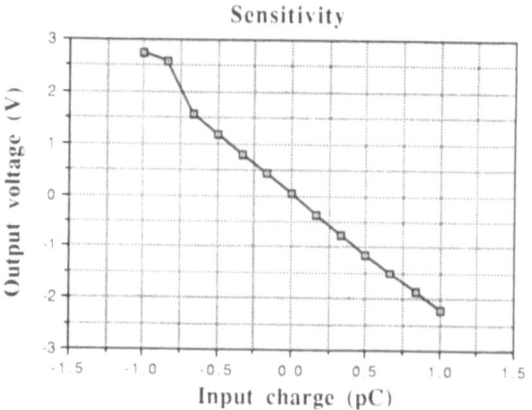

Fig. 20 Graph of the output voltage as a function of the input charge for SAPE 1.5, measured at 1 MHz.

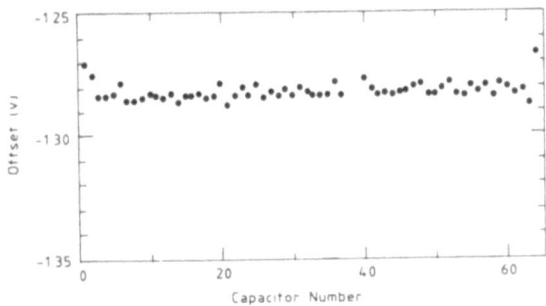

Fig. 21 Plot of the pedestal valves of the 64 memory elements of SAPE 1.5 in the charge sampling mode. The average valve of 1.28 V comes from the offset of the source follower circuit used as output buffer.

4. READOUT SYSTEM WITH DIGITAL BINARY OUTPUT

The HARP architecture is suitable for readout systems, where pulse height analysis of the input signals with a large dynamic range is to be done, e.g. multi-hit detection in a preshower detector, energy deposition in a calorimeter. For tracking applications digital information given by a discriminator is often sufficient.

With this in mind, we developed a readout architecture called FASTPLEX, of which the functional diagram is shown in fig. 22. The basic building blocks in the circuit are the front-end amplifier/shaper, the comparator and the digital FIFO with some additional logic.

The chip is designed to run at clock speeds between 20 Mhz and 50 MHz in 3 μm CMOS technology.

4.1 BiCMOS current sensitive preamplifier

Regarding speed one would prefer silicon bipolar devices over CMOS devices [7]. In spite of this, the better accessibility of CMOS and the need for integration of complicated logic circuits led us to study alternatives in CMOS technology. Although a true BiCMOS technology may ultimately prove to be the best solution, we have for the moment, as a cheaper alternative, restricted ourselves to the use of the lateral and vertical bipolar transistor, which are available in a CMOS process. The characteristics of these bipolar devices are quite acceptable in the Faselec 3 μm SACMOS process, that has been used.

The CMOS compatible vertical bipolar npn utilizes the substrate as the collector, a pwell as the base and an n-type diffusion inside the pwell as the emitter. Due to the fact, that the collector is connected to the substrate, this device can only be used in common collector configurations.

A more useful device is the CMOS compatible lateral bipolar npn. For this device a second n-type diffusion inside the pwell, closely surrounding the emitter, to get a short base width, is used as the collector. Characteristics like current gain, transconductance and noise are quite good compared to bipolar devices in a regular bipolar process. The transition frequency F_t is limited by basewidth of the lateral bipolar transistor which is equal to the minimum physical gate length L_{gmin} of the MOS structure. In 3 μm SACMOS $L_{gmin} =$ 2.4 μm , $F_t = 140$ MHz. One of the drawbacks of using the lateral bipolar is that approximately half the emitter current flows through the parallel vertical bipolar transistor directly into the substrate. To avoid this one would have to use a true BiCMOS technology.

The schematic diagram of the input amplifier is shown in fig. 23. The amplifier is based on the previously developed amplifier 'bipoltest' [8] LAA review. The input transistor T1 in the common emitter configuration is of the lateral bipolar type. The current gain of the

117

first stage is 170 for a standing collector current of 150 µA. A second stage with unity gain is formed by the n-channel transistor T2. Together, T1 and T2 form a current feedback pair. To overcome the large Miller capacitance of the lateral bipolar T1, which limits high-frequency operation, a neutralization capacitance Cnf is used. The sharp frequency cut-off filtering inherent to this circuit design produces a nearly Gaussian shaping of the output signal.

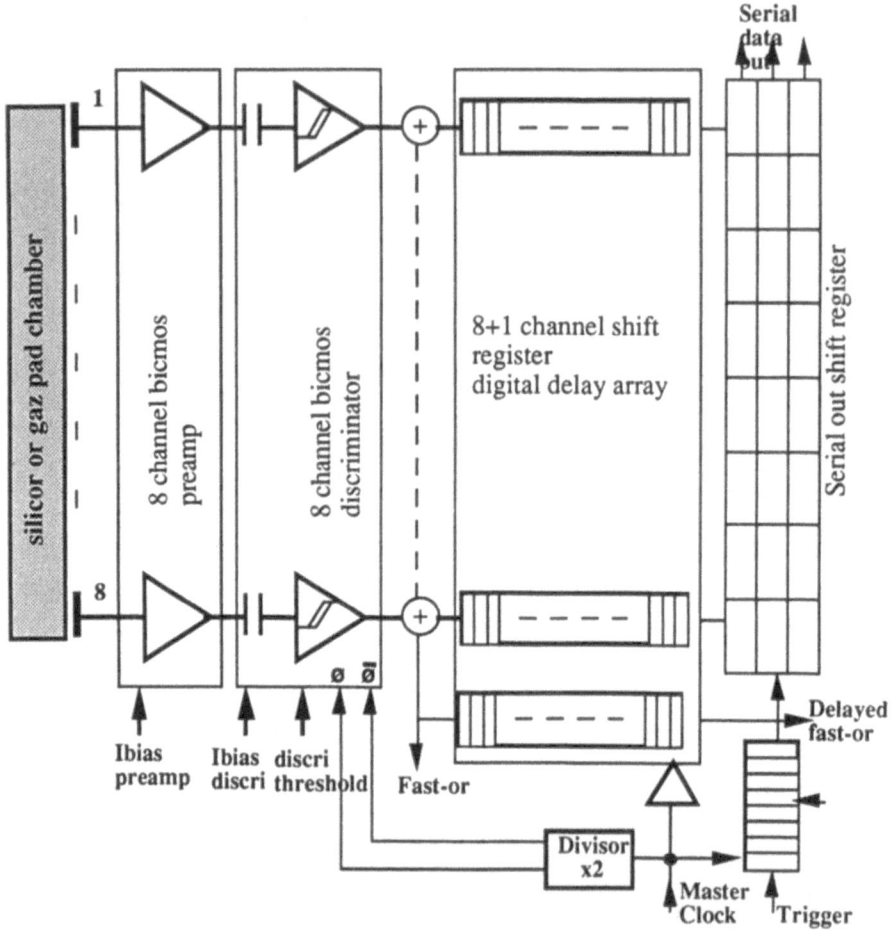

Fig. 22 Fastplex block diagram.

To increase the output level of the amplifier, the resistive emitter load in the source follower T3 of the original amplifier is replaced by an active load, formed by the p-channel transistor T4 and the resistor R = 10 KΩ. This increases the voltage level of the output signal with a factor gm * R compared to the original circuit 'bipoltest' as shown in fig. 24. In our case gm * R is approximately 10. The conversion gain of the amplifier is 5 mV per fC or 20 mV per MIP. The peaking time, measured in fig. 25 is 15 ns. The amplifier is designed to operate at 3 V and dissipates then 2 mW. The Equivalent Noise Charge (ENC) is 800 e-r.m.s. with a slope of 60 e$^-$ per pF.

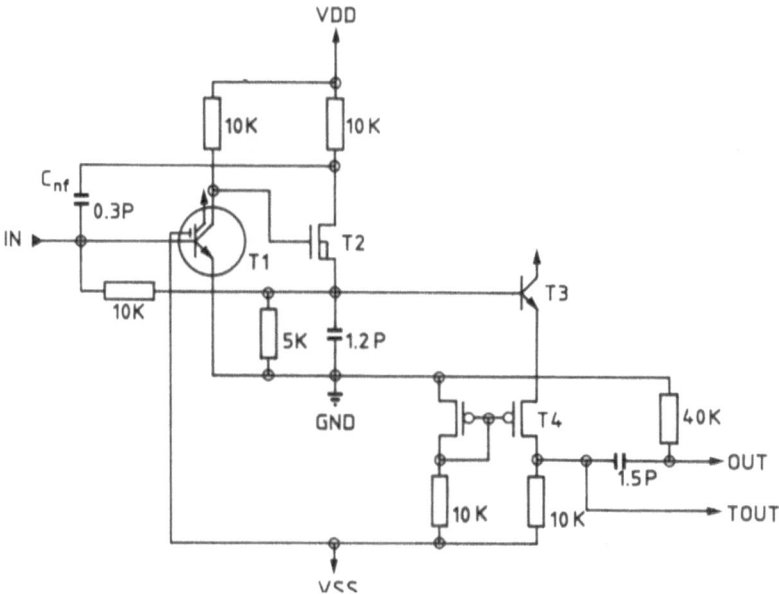

Figure 23 Schematic diagram of the current sensitive amplifier, using a lateral bipolar transistor T1 and a vertical bipolar transistor T3.

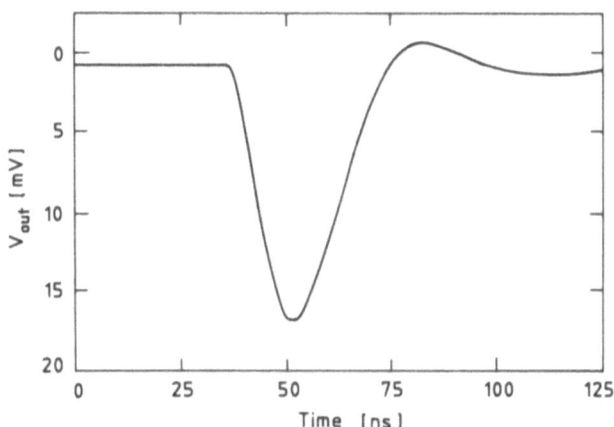

Figure 24 Simulated output signal of the bipolar preamp: peaking time is 15 ns, and gain is 20 mV/MIP.

119

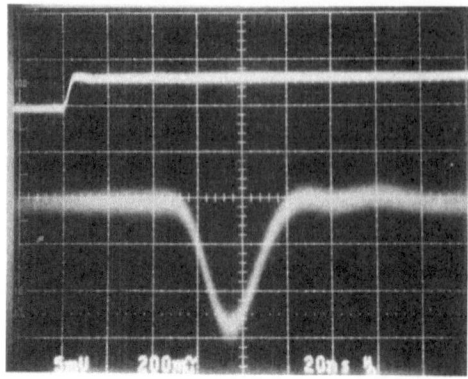

Figure 25   Experimental output response of the bipoltest amplifier. The input charge generated by the input voltage of 5 mV and a test capacitor of 1.6 pF, is 2MIP Si equivalent

4.2   Fast low power CMOS comparator

The comparator is based on a comparator developed by Krummenacher [9].The principle of the comparator is shown in fig. 26. The comparator consists of a differential transconductance input stage, which converts the differential input voltage dVe to a differential output current $dI = I+ - I- = gm * dVe$. The differential input stage uses lateral bipolar transistors as input devices to reduce the offset. The bistable output stage is controlled by the clock signal CLK.

During the initialisation phase, when CLK is high (1), transistor T3 is conducting, and transistors T1 and T2 are connected in parallel, with gate and drain short circuited. When CLK goes low (0), transistor T3 is switched off, and the difference in the currents I+ and I- will rapidly switch the output stage to one of its stable states depending on the sign of the differential input voltage dVe. A shortcoming of this circuit is, that the comparator is only sensitive to an input voltage, when CLK is low.

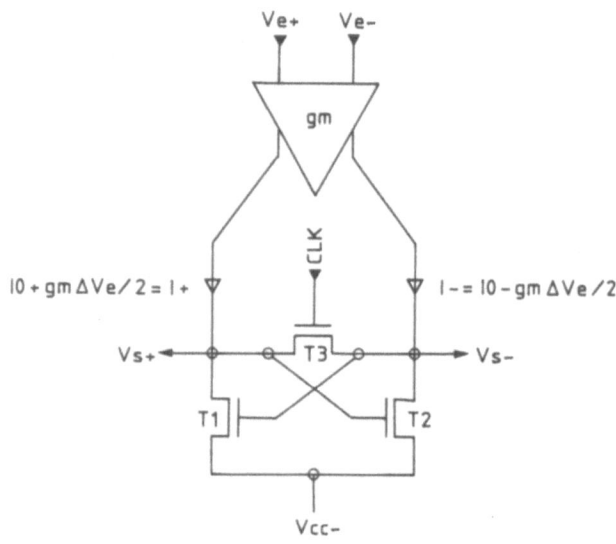

Figure 26   Diagram, showing the principle of Krummenacher's comparator.

To avoid the dead time of the comparator, the circuit shown in fig. 27 is proposed. The input stage is the same as in Krummenacher's scheme, whereas the output stage is changed. Instead of one output stage, two output stages are used, in which for each output stage two transistors T4 and T5 (T4' and T5') are added to redirect the currents I+ and I- to the currently active output stage. Deadtime is avoided at the cost of a slight increase in area and power. The maximum clock speed of the comparator is 25 MHz, which means, every half clock period a comparison can be done this corresponds to a true clock frequency of 50 MHz. The power dissipation is < 1mW. Again the power- dissipation can be reduced by almost a factor 2, by using true BiCMOS transistors instead of the CMOS compatible lateral bipolar transistors in the input stage.

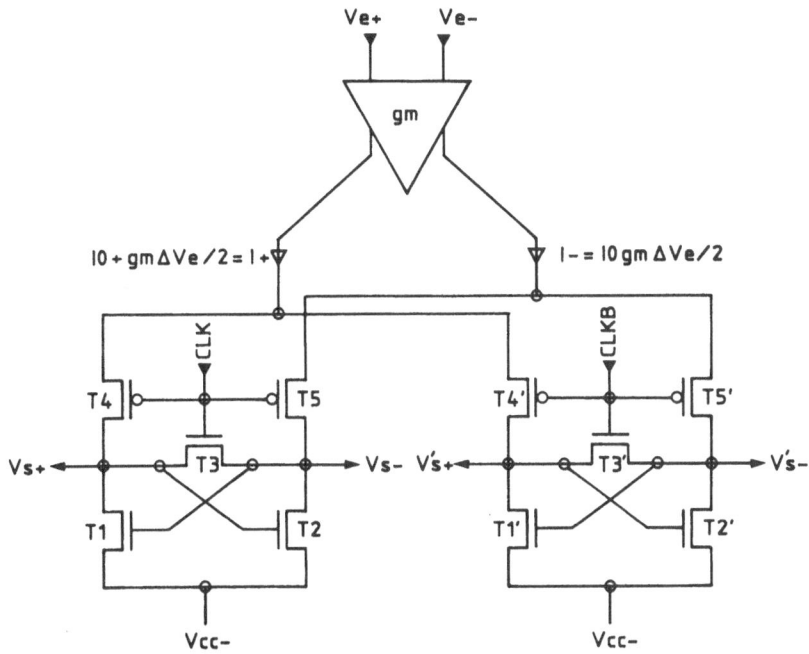

Fig. 27 Diagram, showing the principle of the comparator using two output stages.

4.3 The digital shift register (FIFO)

A digital shift register (FIFO) is used to store the information coming from the comparator. The length of the FIFO is 56 cells, which for a 50 MHz clock results in a maximum delay of 1.1 µs to take a trigger decision. A FIFO built with 112 latches, implemented as a fully serial architecture would be too costly in power dissipation and area.

To reduce power and area the FIFO (DMUX) structure shown in fig. 28 has been used. It consists of an 8-bit wide input demultiplexer, an 8-bit long, 8-bit parallel FIFO and an 8-bit wide output multiplexer MUX. The shift register uses 8 non-overlapping clock phases as shown in fig. 29. Suppose, that initially column 7 of the 8×8-FIFO is empty, in the first clockperiod following, column 7 will be filled with the content of column 6, and column 6 will become empty. In the second clock period column 6 will be filled with the content of column 5, and column 5 will become empty. This continues until in the eighth clock period the outputs of the input demultiplexer are shifted into the first column of the 8×8-FIFO. The process then starts again at column 7.

Every clock period 1 bit is shifted into the 8-bit wide input demultiplexer and saved until it can be shifted into column 1 of the 8×8-FIFO. Also every clock period 1 bit of column 7 of the 8×8-FIFO is shifted out at the 8-bit wide output multiplexer. The area needed is less than 70% of the fully serial FIFO, and the power dissipation is less than 10% of the fully serial FIFO. The basic cell, which is used in the FIFO is shown in fig. 30. To

121

reduce the size of the FIFO-cell and reduce the number of control signals for the FIFO-cell, a single n-pass transistor is used as the input transistor, a pullup- transistor is used to pull the input of the invertor to Vdd. The output of the FIFO is shifted into a serial output register. The serial output register consists of 3 cells per FIFO (fig. 22). As long as no trigger signal is present, bits are shifted from left to right in the serial output register, when a trigger occurs the bits in the output register are shifted out serial (bottom to top).

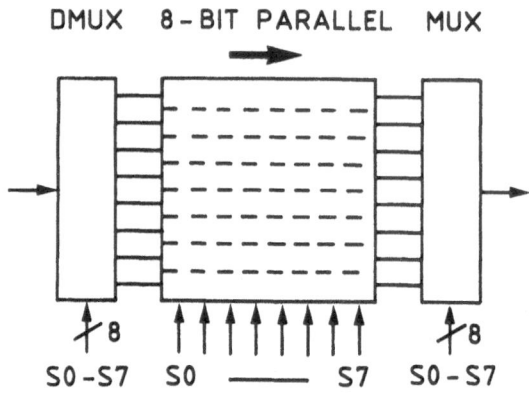

Fig. 28 Functional description of the FIFO.

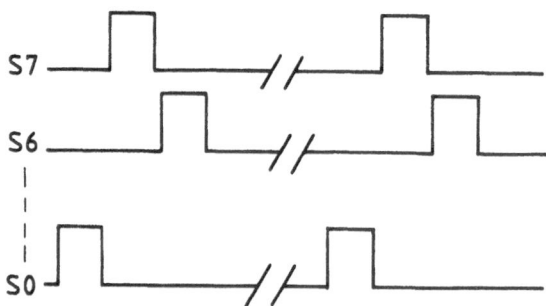

Fig. 29 Timing diagram, showing the different clockphases for the FIFO

There is some additional logic present in the diagram of fig. 22. The Fast-or performs an OR on the output signals of the comparator. This signal is directly output and also put in an additional FIFO, this delayed-or is output after 56 clockperiods. To tune the triggering a delay-tuning is added. The delay, which has to occur between the trigger signal and the start of the serial shiftout, can be set between 1 and 8 clock periods.

For testing purposes, some additional logic and bondpads are added, which give the following possibilities for testing:

- The output signal of the amplifiers can be read directly through one common output pad via an analog multiplexer.
- One input of the comparator is used to set the threshold voltage, the other input is connected to the opamp
- The output of the comparators can be read through the fast-or.
- The input of the FIFO can be directly set.

- The FIFO, together with the serial output register, can be set as one long shift register, where the serial output pads are used as outputs.
- The FIFO element to be tested individually can be selected.
- The serial inputs of the first serial output registers are connected to a common inputpad.

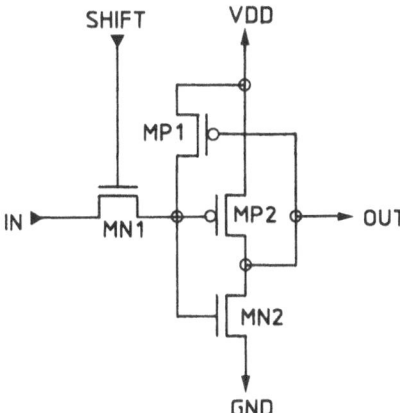

Fig. 30 Basic cell used in the FIFO.

CONCLUSION

We have built demonstrator chips employing available industrial microelectronics processes which show great promise for readout systems with pipelined architectures.

This novel approach shows the road towards meeting the challenge of constructing particle detector systems for future supercollider experiments with millions of channels running at around 100 MHz clock rate.

REFERENCES

[1] A. Zichichi et al., The LAA project, Nuovo Cimento, 13 (1990) 5.

[2] CERN/LAA SD 9018.

[3] CERN/ECP 9013.

[4] H. Anders et al., Feasibility study for a charge coupled device (CCD) as a multiplexing signal processor for microstrip and other particle detectors, CERN/EF 85–10.

[5] K. Matsui et al., CMOS video filters using switched capacitor 14 MHz circuits, IEEE Journal of solid state circuits, 20/6 (1985) 1096.

[6] V. Radeka, Trapezoidal filtering of signals from large germanium detectors at high rates IEEE trans. Nucl. Sci. NS–16 (1972).

[7] P. Jarron, Fast silicon detector systems for high luminosity hadron collider experiments, ECFA proceedings Barcelona, (1989) 287.

[8] F. Anghinolfi, P. Aspell, M. Campbell, E.H.M. Heijne, P. Jarron, G. Meddeller, Ch. Cenz, F. Krummenacher, L. Moult, P. Sharp and A. Olsen, Progress report 1988–1989, Development of integrated CMOS circuits and silicon pixel detectors in the CERN–LAA project, CERN/LAA SD 89–11.

[9] Private communication, (1989) EPFL Lausanne–CH.

A STUDY OF A TOROIDAL MUON SPECTROMETER WITH A DISCRETE AIR COIL FOR A MULTI-TeV HADRON-HADRON COLLIDER

F. Bergsma, G. Laurenti, G. Levi, Q. Lin, G. Susinno

Large Area Devices group, CERN, Geneva
Switzerland

1. INTRODUCTION

The behaviour of a muon spectrometer with toroidal magnetic field for a future multi-TeV hadron-hadron collider was investigated by computer simulation, using the finite element program TOSCA[5].

A toroidal magnetic field configuration in the forward region is a natural choice. For the central region two options are possible: a solenoidal[2] or a toroidal[1] field. The solenoidal field deflects muons in a plane perpendicular to the beam axis. This allows the precise vertex position to be used as a constraint in the high p_T muon trigger. However there is a transition region between the barrel solenoid and the end cap toroids, that creates limitations for the muon momentum measurement accuracy in the correspondent pseudorapidity range. The choice of a toroidal magnetic configuration in the barrel part gives a uniform field geometry, with good momentum resolution up to high pseudorapidity regions. The toroidal magnetic field was generated by a set of separated super conducting coils, each contained in its own cryostat, see fig. 1. A set of 16 large coils was covering the central region, two sets of 16 smaller coils were used as end caps to cover the forward regions. This layout was chosen for ease of construction and accessibility. No iron return yoke is necessary, giving a light, compact detector. The momentum resolution is high, due to the reduced multiple scattering, compared with the momentum resolution in an iron spectrometer which is restricted to about 10%[6].

A three dimensional field calculation shows that a good field uniformity is achievable in the proposed modular design. A more homogeneous coil would require more complicated cryostats and reduce the accessability to the inner muon chambers and other inner detector parts, without introducing benefits on the mo-

New Technologies for Supercolliders, Edited by L. Cifarelli and
T. Ypsilantis, Plenum Press, New York, 1991

mentum measurement accuracy. The effect of field inhomogeneity on the momentum resolution has been investigated and found negligible as discussed in the following. In addition other important aspects as current densities, forces on the conductors and stored magnetic energy are presented.

The performances of our proposed design were simulated assuming the information coming from three layers of drift chambers, with a space resolution of 100 μm.

The present chosen parameters are to be considered as a reasonable starting point for further optimization.

2. SPECTROMETER GEOMETRY

It was assumed that the part left free to house the inner detectors, was contained in a cylinder with diameter of 6 m and length of 8 m. A region of 50 cm around the coils was kept free for the cryostats. A view on the coils, looking along the beam direction, is given in fig. 2. The set of forward coils is rotated by 11.25° degree along the beam axis relative to the set of barrel coils in order to reduce the effect of the inhomogeneity of the magnetic field on the momentum resolution. The coils consisted of straight sections, connected by quarter parts of a cylinder. In Table 1 coil caracteristics are given.

Table 1

	Central detector	Forward detector
# of coils	16	16 (rotated by 11.25 degrees rel. to central coils)
width	1.0 m	0.3 m
largest straight section	8.50 m	2.00 m
smallest straight section	2.60 m	
radius of arcs	1.00 m	1.00 m
distance centres of two opposite coils	12.00 m	4.00 m
current per coil	2700 kA	1200 kA
central field	2 Tesla	2.5 Tesla
maximum field	2.7 Tesla	3.7 Tesla
outer diameter	16.60 m	6.00 m
length	19.00 m	4.00 m

With the dimensions given in Table 1 the spectrometer fits in a cylinder of 19.00 m length and of 16.60 m diameter.

3. MAGNETIC FIELD STRENGTH AND DIRECTION

At first glance one might expect that a magnetic field produced by an ensemble

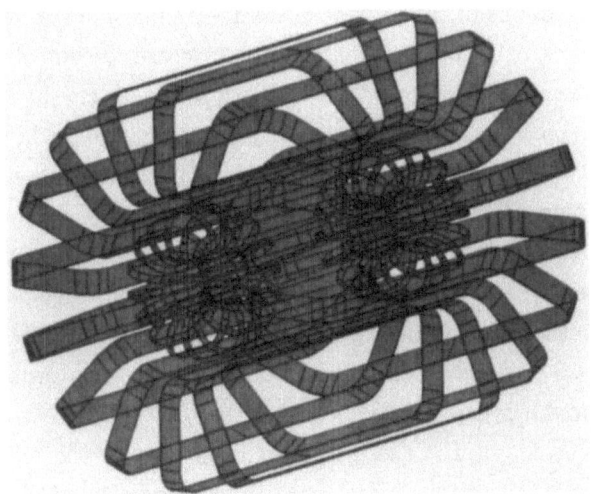

Fig. 1. Detector concept.

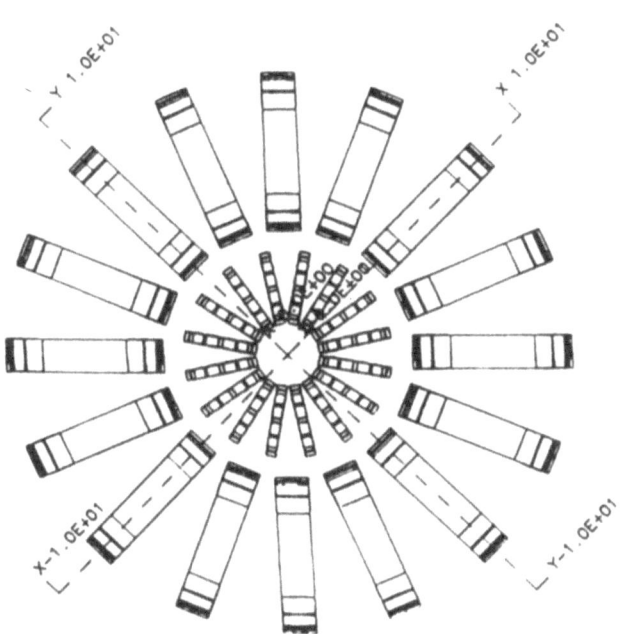

Fig. 2. Front view of the detector. The forward coils are rotated by 11.15° relative to the central coil.

of coils as given in fig. 3 heavily deviates from a toroidal field. In fig.4 the field direction is indicated for a 45° section in a plane perpendicular to the beam axis in the middle of one of the forward coil-sets. As one can see, the field direction is perpendicular to the beam axis almost everywhere, except at the very edge and outside of the coils, where the field strength is low, as can be seen in fig. 4 which is a contour plot of the absolute value of the magnetic field. From this figure it follows that, for a given radius, the field strength is the lowest in between the coils.

For a more detailed evaluation the field strength was scanned along lines radial to the beam axis. In fig. 5 the result of such a scan is shown for a radial line passing inside a barrel coil and through a gap between the forward coils. The line was taken in the plane perpendicular to the beam axis at the centre of the forward coil. In fig. 6 field values are given along a radial line passing inside a forward coil and through a gap between two barrel coils. The effect on the momentum resolution will be given in the next paragraph. Due to the 11.25° rotation of the forward coils relative to the barrel ones a high energetic particle can never traverse both minima of field strength.

In fig. 7 the absolute value of the field strength is given along an arc of 45° for different radii in a plane perpendicular to the beam axis and crossing the centre of the spectrometer. In fig. 8 a similar field scan is shown for the forward coils. The scanplane crosses here the centres of the forward coils. Thus fig. 5 and fig. 6 give the field strength in the maxima and minima of fig. 7 and fig. 8.

To investigate the longitudinal variation of the magnetic field, scans were made along lines parallel to the beam axis inside the coils. For the barrel coil the result is shown in fig. 9. Scans were made at five different radii: just outside the coil, just inside and at the centre of the coil. As one can see from fig. 9, the field strength is to a large extent constant at a certain radius, moving along the beam direction. For a forward coil the result is shown in fig. 10. Also here the variations in field strength are small, although due to the difference in length between forward and central coils, the influence of the curved coil edges is here more important.

4. MOMENTUM RESOLUTION

As an example of the behaviour of above described spectrometer, the tracks of 5 GeV/c muons are given in fig. 11, emitted from the interaction centre at angles with the beam axis increasing in steps of 10°. One sees the rotational symmetry around the beam axis, and the forward-backward asymmetry.

To reconstruct muon tracks, detection surfaces were simulated consisting of cylinders coaxial with the beam axis in the barrel region and disks perpendicular to and centred on the beam axis in the forward and backward regions. The positions and dimensions of cylinders and disks are given in fig. 12 and Table 2.

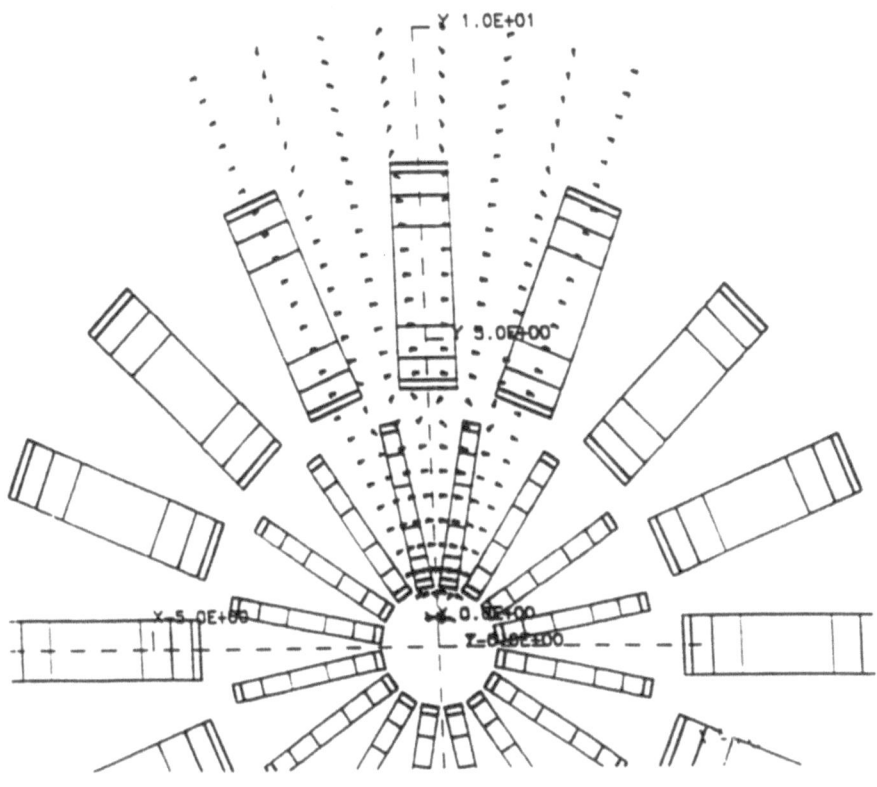

Fig. 3. Direction of the magnetic field. (The spectrometer in this figure has slightly different dimensions as the one described in the text).

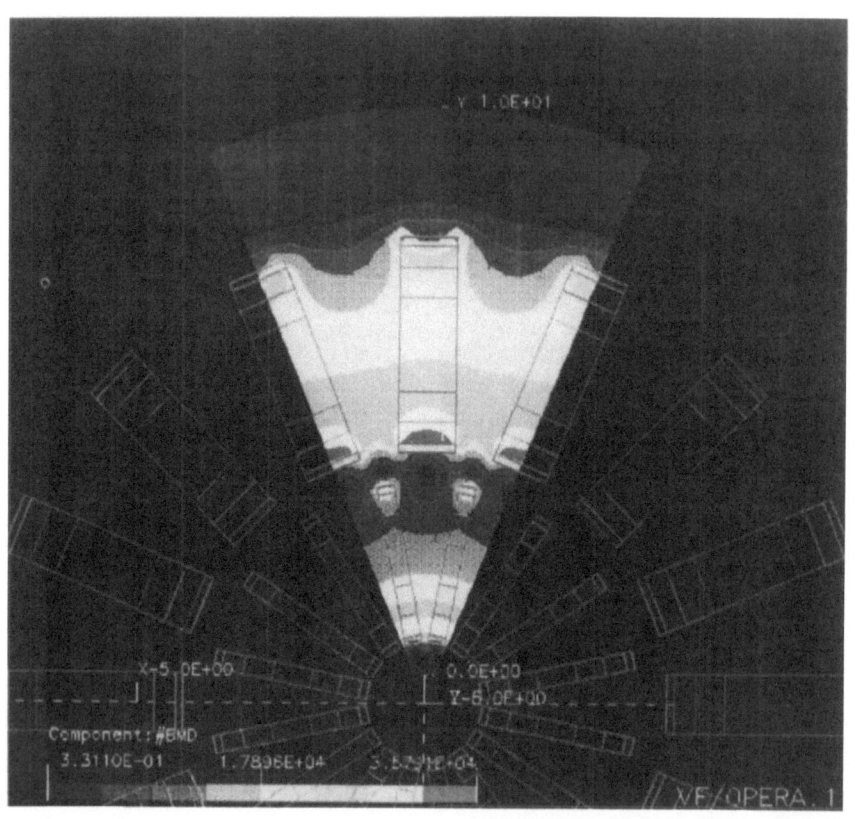

Fig. 4. *Contour plot of the field strength. Minimum = 3.3 10^{-1} gauss, maximum = 3.6 10^4 gauss. (The spectrometer in this figure has slightly different dimensions as the one described in the text).*

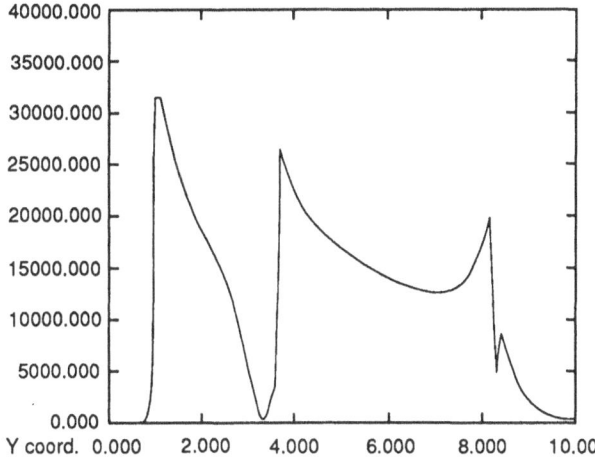

Fig. 5. Radial scan of field strength along a line passing through a barrel coil.
The scan line lies in a plane perpendicular to the beam axis through
the centre of the forward coils. At the left $(y = 0)$ the scan line crosses
the beam axis.

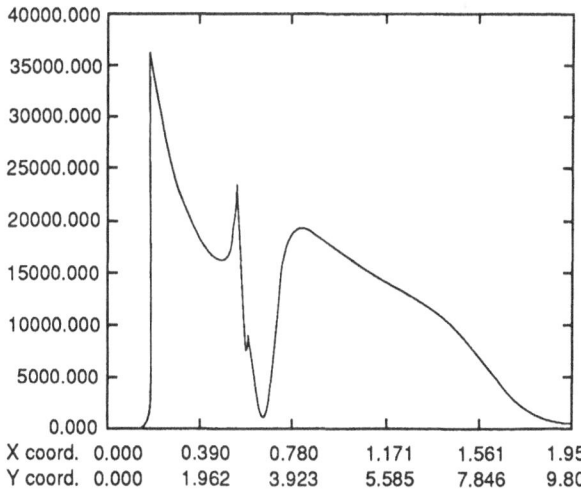

Fig. 6. Radial scan of field strength along a line passing through a forward coil.
The scan line lies in a plane perpendicular to the beam axis through
the centre of the forward coils. At the left $(y = 0)$ the scan line crosses
the beam axis.

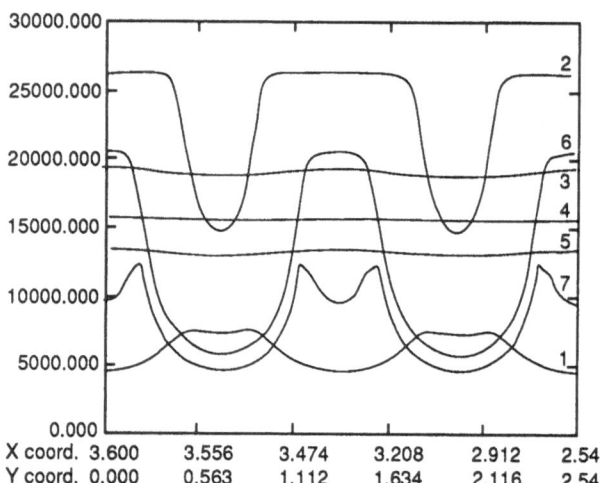

Fig. 7.   Scan of field strength for the central coils along an arc of 45° and at
          different radii. The arc lies in a plane perpendicular to the beam axis
          through the centre of the spectrometer. 1: R = 3.60 m; 2: R = 3.80
          m; 3: R = 4.50 m; 4: R = 5.50 m; 5: R = 6.50 m; 6: R = 8.20 m;
          7: R = 8.40 m.

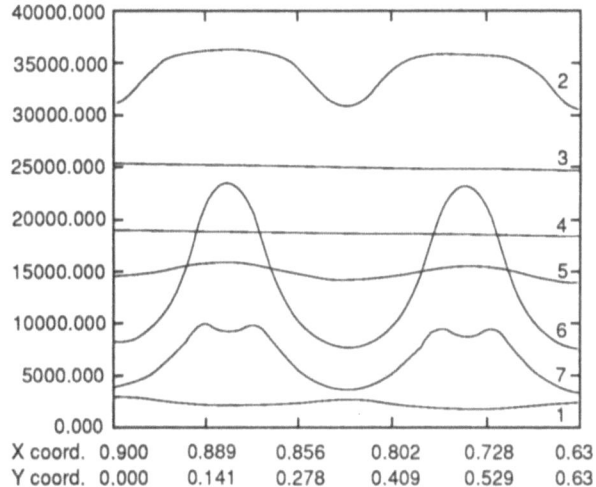

Fig. 8. Scan of field strength for the forward coils along an arc of 45° at different radii. The arc lies in a plane perpendicular to the beam axis through the centre of the forward coils. 1: R = 0.90 m; 2: R = 1.10 m; 3: R = 1.50 m; 4: R = 2.00 m; 5: R = 2.50 m; 6: R = 2.90 m; 7: R = 3.10 m.

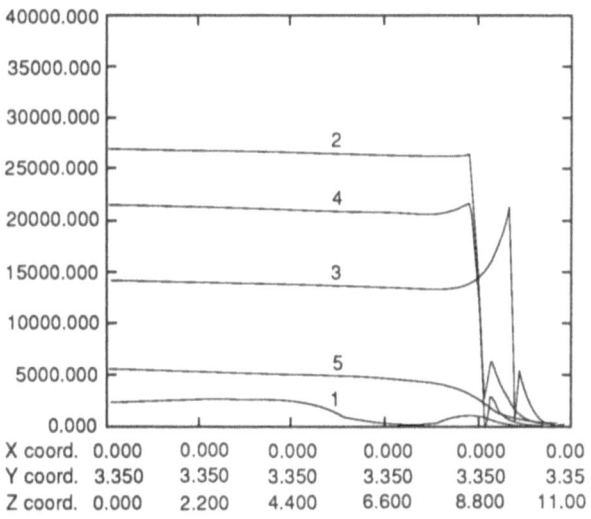

Fig. 9. Longitudinal scan of field strength in central coil 1: R = 3.35 m; 2: R = 3.75 m; 3: R = 6.00 m; 4: R = 8.25 m; 5: R = 8.65 m.

133

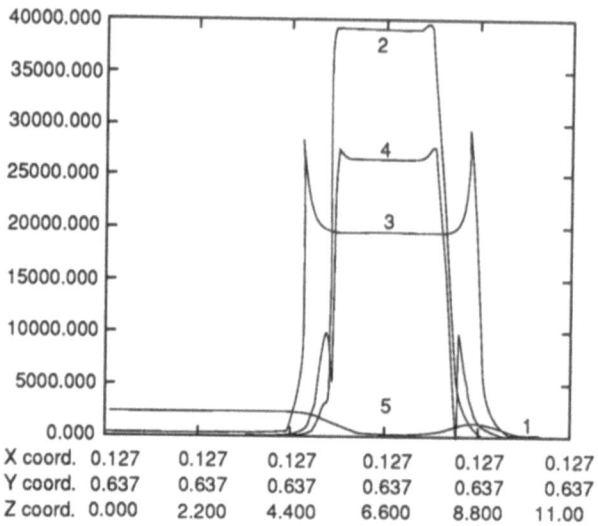

X coord.	0.127	0.127	0.127	0.127	0.127	0.127
Y coord.	0.637	0.637	0.637	0.637	0.637	0.637
Z coord.	0.000	2.200	4.400	6.600	8.800	11.00

Fig. 10. *Longitudinal scan of field strength in forward coil1: R = 0.65 m;
2: R = 1.07 m; 3: R = 2.00 m; 4: R = 2.93 m; 5: R = 3.35 m.*

The cylindrical detection surfaces can be approached in practice by planes of
flat drift chambers. For the disk-shaped detection surfaces a new type of detector
is under development, using thin, bended blades in stead of wires[3].

The space resolution of the detection elements was assumed to be 100 μm.

The momentum resolution for 500 GeV/c muon tracks coming from the central
interaction region was determined in the following way. First a 500 GeV/c track
was generated starting from the detector centre at a certain angle with the beam
axis. The coordinates of the crossing points of this track with the detection surfaces
were stored.

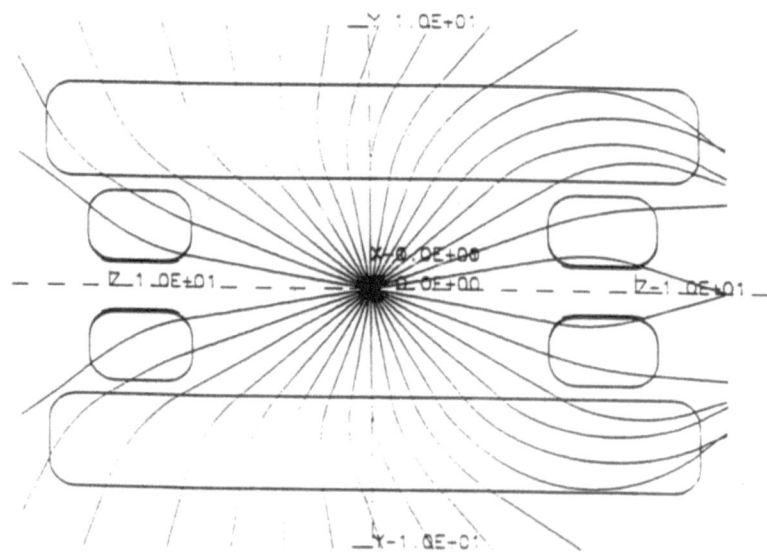

Fig. 11. *Tracks of 5 GeV/c, step in angle is 10°. The detector shown is not the
same as described in the text.*

Table 2. Detector positions (all dimensions in metres)

CYLINDERS CO-AXIAL WITH BEAM AXIS			
name	radius	Z1	Z2
A	3.20	-4.30	+4.30
B	6.00	-8.78	+8.78
C	8.80	-11.00	+11.00

DISKS PERPENDICULAR TO BEAM AXIS			
name	distance from centre	Rmin	Rmax
D	4.30	0.55	3.20
E	5.52	1.62	2.59
F	8.08	1.41	2.38
G	8.78	4.11	6.00
H	9.10	1.17	1.59
I	11.00	1.41	8.8

Next tracks of higher and lower momentum at the same starting point and angle were generated, then displaced and rotated until a minimum was reached for:

$$\chi^2 = \sum \{(x_{0n} - x_n)/\sigma_n\}^2 \tag{1}$$

in which x_{0n} = coordinate of crossing point of original 500 GeV/c track with detection surface n, x_n = coordinate of crossing point of probe track with detection surface n, σ_n = space resolution of surface n. The momenta at which (1) became equal to the number of degrees of freedom were taken as 68% probability limits. Since the deviation from a straight line for $\simeq$ 500 GeV/c tracks is small (of the order of 1 mm over the traversed magnetic volume), the above mentioned displacements and rotations were extremely small (of the order of 10^{-4} m for displacement and 10^{-4} rad for rotation) and thus the method is reliable. The resolution found is a lower limit, since it neglects multiple scattering. The effect of multiple scattering can only be studied when more design parameters are determined.

In fig. 12 the momentum resolution of 500 GeV/c muons is indicated for tracks originating around the centre and lying in a plane through the middle of a central coil. It has to be stressed that no knowledge from inner detection elements was used, it was only assumed that the tracks cross the beamline somewhere. As can be seen, the resolution is almost everywhere better than $\Delta p/p = 4\%$, except below

7°, where the forward coil ends, and around 25° at the change-over from barrel to forward magnet. In this region the resolution can be improved by optimizing the detector positions and resolutions. Around 40°. the resolution becomes smaller than 1%. A more uniform distribution of $\Delta p/p$ could be obtained by rounding of the coils at the outer edges, thus creating D-shaped coils.

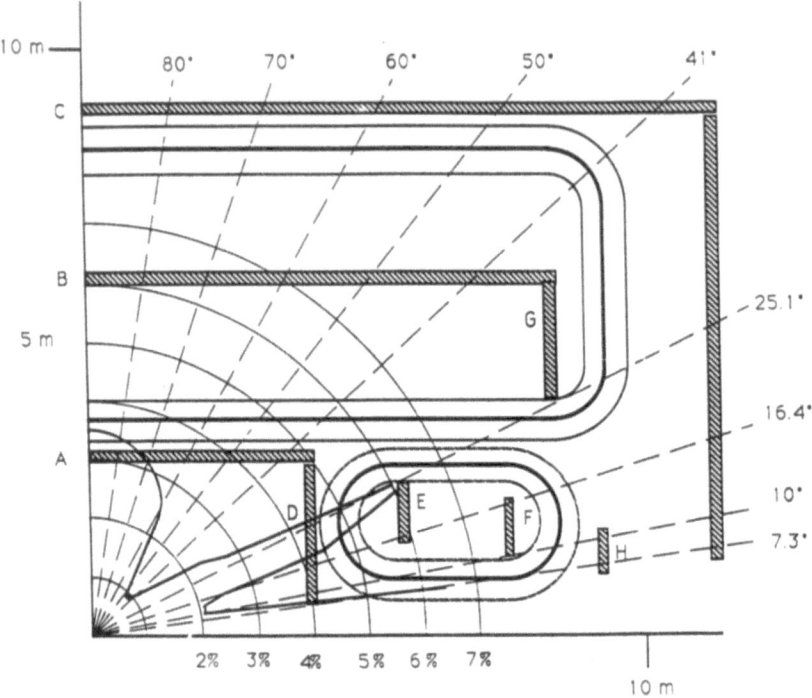

Fig. 12. *Position of detection surfaces. Also the momentum resolution as function of angle is indicated for 500 GeV/c tracks and chamber resolution of 100 μm. For more explanation see text.*

In fig. 13 the resolutions are given in a cartesian $\Delta p/p$ vs. Θ plot. For the full curve the tracks were lying in a plane through the beam axis and crossing the centre of barrel coil, thus passing through a gap of the forward coils. The dotted curve gives the resolution for tracks in a plane through the beam axis rotated by 11.25° with respect to the former scan plane. Thus the tracks of the full curve traversed the highest field region of the central coil and the lowest field region of the forward coil, for the tracks of the dotted curve the situation was reversed. The difference between the two curves is smaller than 0.5% from which one can conclude that the chosen layout with discrete coils is satisfactory.

136

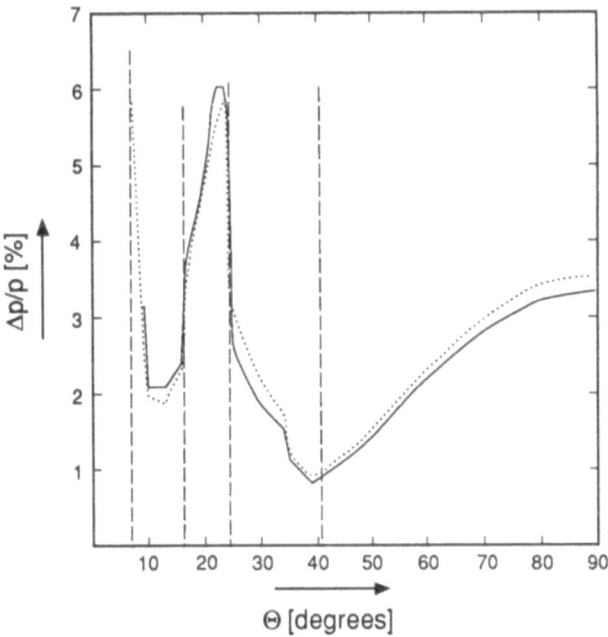

Fig. 13. Momentum resolution as function of angle with beam axis. The momentum resolution is given for 500 GeV/c tracks and chamber resolution

5. TRIGGER

For a fast low momentum cut trigger in the forward region special round chambers are under design[3] These chambers give, after a largest drift-time of 120 ns, directly the distance to the beam axis of a traversing track.

For a fast low momentum cut trigger in the central region one can take advantage of the fact that the interaction point along the beam line is well determined: due to the small distance between bunches the two beams have to cross each other under a small angle in order to avoid collisions of particles belonging to one bunch with particles from more than one other bunch. For current LHC-designs[4] the interaction region is contained within ± 15 cm of the beam crossing point. In fig. 14 the trajectories of muons with different momenta are given. One sees that muons which leave the inner detector with a momentum smaller than 1 GeV/c do not reach the outer muon chambers, which gives a very efficient trigger cut, making negligible any trigger contribution from calorimeter leakage. A 50 GeV/c muon, emitted perpendicular to the beam axis, would at the place of the middle muon chamber (R = 6 m) have a deviation of 2.63 cm from a straight line and at the position of the outer muon chamber (at 8.80 m) a deviation of 12.11 cm. The line through the two hits in middle and outer chamber would cross the beam axis at a distance of 18 cm from the interaction point, which shows that a fast trigger on muons with momentum higher than about 50 GeV/c is feasible, using only the middle and outer muon chambers.

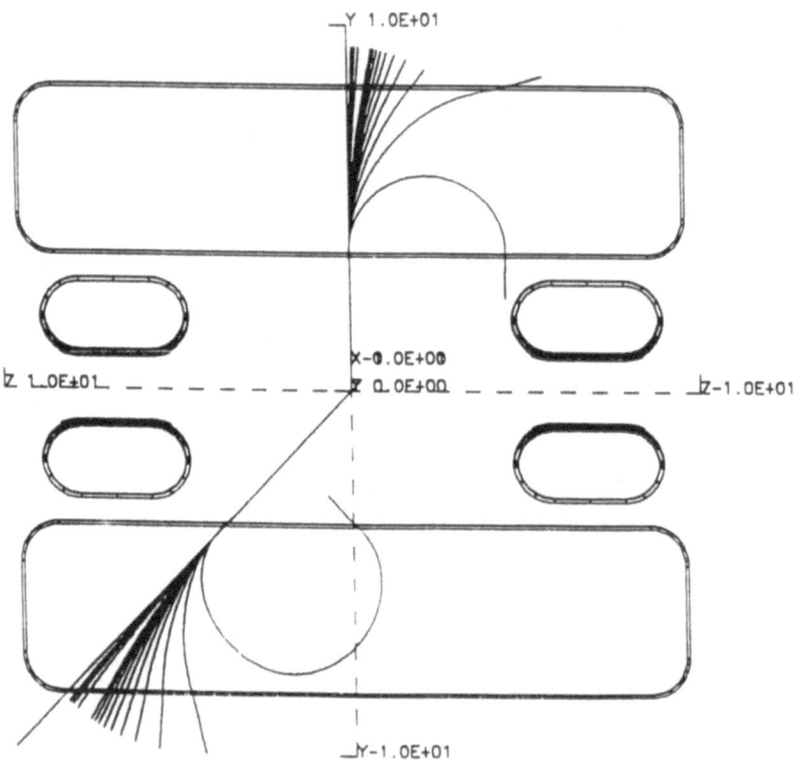

Fig. 14. Muon tracks at 90° and 45° angle with beam axis. The momenta are 1,2,3,4,5,6,7,8,9,10,20,30,40,50, 10000 GeV/c.

6. DESIGN

6.1 Forces and energy

The forces on the different coil segments are given in fig. 15. In this figure the coil segments are in between two dashed lines.

The forces on the conductor elements tend to make the coil surface bigger in order to reduce the magnetic energy. The coils as a whole are pressed inwards towards the beam axis: the force on the outer part is smaller than on the inner part. The

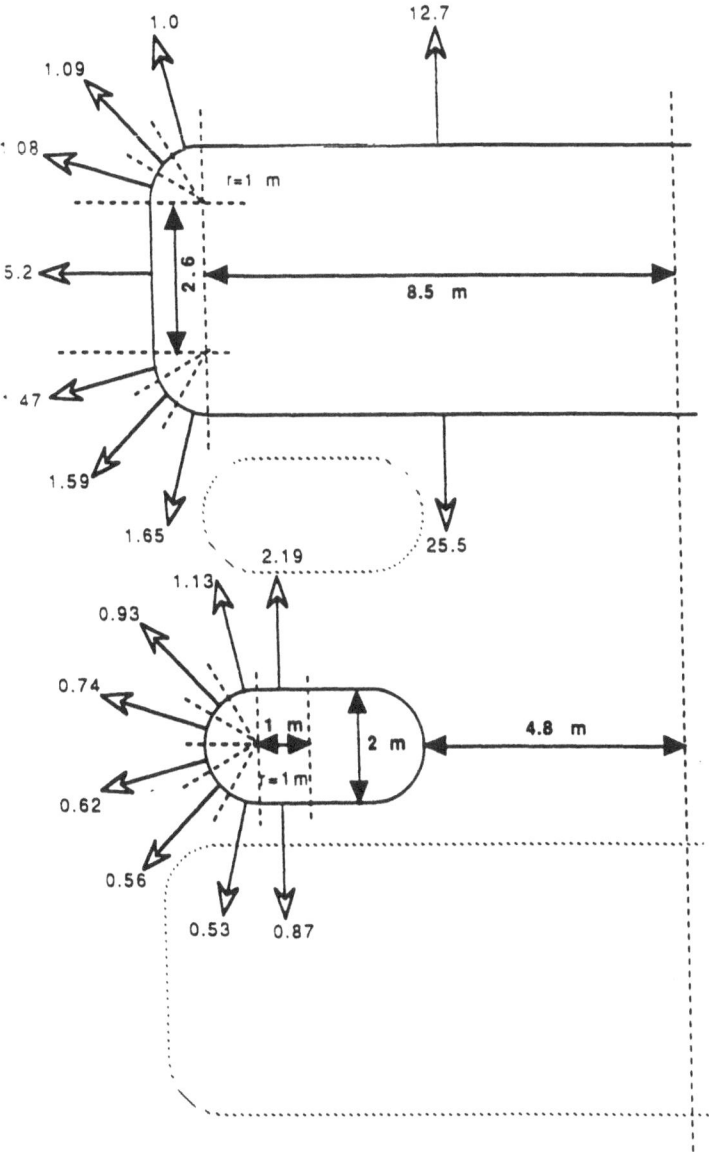

Fig. 15.   Forces on coil segments. Each segment is delimited by two dashed lines.
Forces are given in 10° N.

maximum pressure on the inner sections amounts for the central coils up to 30 atm. and for the forward coils up to 73 atm. The total force on a central coil was 28 MN, radial towards the beam axis, for a forward coil this number was 4.4 MN.

The magnetic energy stored in the coils is:

$$\begin{aligned}
\text{energy central magnets} &= 2.89 \ 10^9 \text{ joule} \\
\text{energy forward magnets} &= 2 \times 0.124 \ 10^9 \text{ joule} \\
\text{total energy stored} &= 3.14 \ 10^9 \text{ joule}
\end{aligned}$$

The large amount of stored energy and the high forces on the coils will cause the main difficulties in this design. A strong support for the conductors will be necessary to absorb the high forces, which increases the amount of material in the magnetic volume and will increase the heat transfer. To reduce the effect of multiple scattering low-Z and composite materials will be used where possible.

6.2. Superconducting cables

The choice of the magnet current is the essential parameter which drives the rest of the design of the coil. It was selected to be 54 or 27 kA for the central coils and 40 kA for the forward coils, after considering a number of factors including: the number of turns and coil layers, the length and cross-section of the conductor, the flexibility of the conductor in the coil windings, the inductance and resistance of the coil, the characteristics of the current leads and power supply, etc. A large current value has the advantage of simplifying the mechanical structure by reducing the number of turns and layers. In addition, a low number of turns reduces the coils inductance and resistance, making the quench protection simpler. On the other hand, a large current requires a thick conductor, large current leads and a large power supply. The main parameters of the magnets are listed in Table 3.

To keep the Helium mass flow at a low level and to obtain a low cable thickness, inner direct cooling has been chosen. The inner direct cooling is achieved by forced-flow liquid helium circulating in pipes with a diameter of 6 mm inside the conductor. The conductors are aluminium-stabilized NbTi/Cu superconductors, insulated with 0.5 mm thick Kapton foil, as shown in fig. 16. At the rated currents of 54, 27 and 40 kA, the current densities in the NbTi superconductor are 1.68, 1.59 and 1.42 kA/mm^2. These values are less than half of the critical current at the temperature of 4.5 K and at 2.7 and 3.7 Tesla for commercially available superconducting wires. The magnetic field inside the conductor for the central magnets is shown in fig. 17 at several places. To share the magnetic pressure with the stainless steel hoops, an aluminum alloy sheet of 2 mm thickness is used. The main parameters of the superconducting cables are summarized in Table 4.

Table 3. Main parameters of superconducting magnets

	central detector		forward detector	
# of coils	16		2×16	
cross-section of coil	$4.6 \times 19 \ m^2$		$2 \times 4 \ m^2$	
current per coil (kA)	2700		4000	
total ampere turns	4.32×10^7		2×10^7	
maximum field on the cable (T)	2.7		3.7	
stored energy toroid (MJ)	2890		2×124	
	version A	version B	version A	version B
Average current density (kA/cm^2)	4.5	3	4	3.2
rated current (kA)	54	27	40	40
self inductance (H)	1.98	7.93	0.16	16
cross-section of coil (cm^2)	6×100	9×100	10×30	12.5×30
# of layers per coil	1	2	2	2
# of turns per layer	50	50	15	15
# of turns per coil	50	100	30	30
total weight of all windings (Ton)	123	175	27	33

Table 4. Main parameters of superconducting cables

	central detector		forward detector	
	version A	version B	version A	version B
rated current (kA)	54	27	40	40
conductor dimensions (mm^2)	60 × 19	44 × 19	48 × 19	60 × 19
conductor unit length (m)	2 × 1137	4 × 1137	308.4	308.4
SUPERCONDUCTING INSERT				
material	NbTi/Cu		NbTi/Cu	
Cu/SC ratio	1		1	
(Al+Cu)/(S.C.C.) ratio	25.1	36.5	25.3	28.8
# of strands	32	16	16	28
strand diameter (mm)	1.6	1.6	2	1.6
filament diameter (μm)	50		50	
twist pitch (mm)	50		50	
STABILIZER				
material	Al (99.996%)		Al (99.996%)	
dimensions (mm^2)	56 × 15	40 × 15	44 × 15	56 × 15
spec.resist. @ 4.3K B=0T (Ωm)	< 4.0 10^{-11}		< 4.0 10^{-11}	
RRR	> 1000		> 1000	
thermal conductivity @ 4.3K (W/mK)	> 1500		> 1000	
Al-ALLOY JACKET				
thickness (mm)	2		2	
$\sigma_{0.2}$ (with SC insert + matrix) (kg/mm^2)	> 15		> 15	

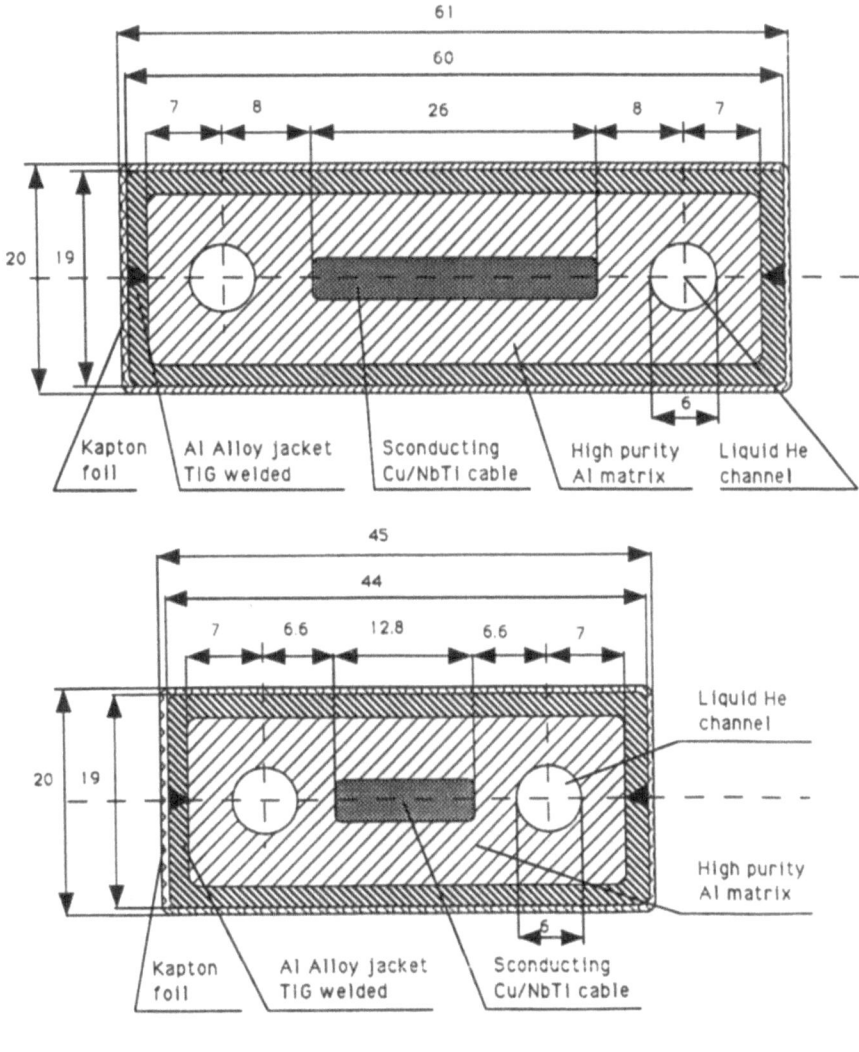

Fig. 16. Cross-section of superconducting cables for central magnets.
Above: version A, below: version B.

mm

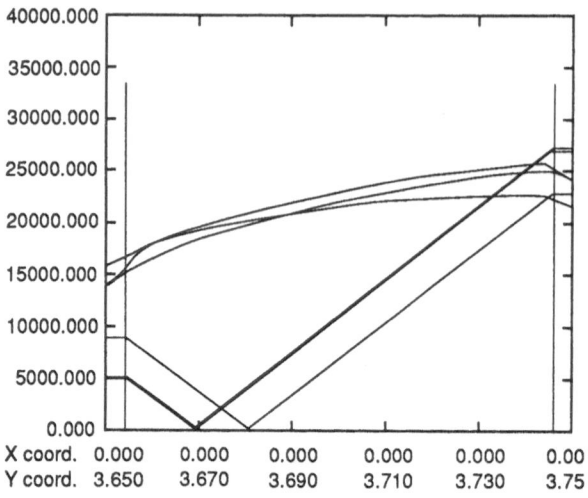

Fig. 17. *Magnetic field at several places inside the conductor for version B of the central magnet. The two lines indicate the dimension of the conductor.*

6.3. Mechanics

As was shown in paragraph 6, the forces can be decomposed in an outwards directed pressure on each individual coil, which tends to make the coil surface bigger, and an inwards force which pushes the coils towards the beam axis. To absorb the first part, which was calculated to be 15 atm., the inner and outer conductor can be connected with bars or flanges, as shown in fig. 18.

What remains is an inwards pressure on the inner conductor, which was calculated to be 15 atm. Since there is little space between the inner conductors of the toroid (see fig. 20), these forces are almost uniformly distributed over a cylindrical surface and can be absorbed by a strong inner cylinder.

A possible arangement of the muon chambers is given in fig. 19. Because of the modular structure, the chambers inside the magnets can be accessed from outside, which is very important from point of view of alignment and maintenance.

The thickness of a section of a central coil is expected to be less than 15 cm iron equivalent. The total weight of the barrel toroid could be 3500 tons, that of the forward toroids about 300 tons each.

6.4. Thermal insulation

As indicated in fig. 23, one could group the coils together in pairs per cryostat. Due to lack of space, this is the only solution for the forward toroids, in case of the central toroid one could also envisage one coil per cryostat, as shown in fig. 19. For a central coil a possible layout is given in fig. 21.

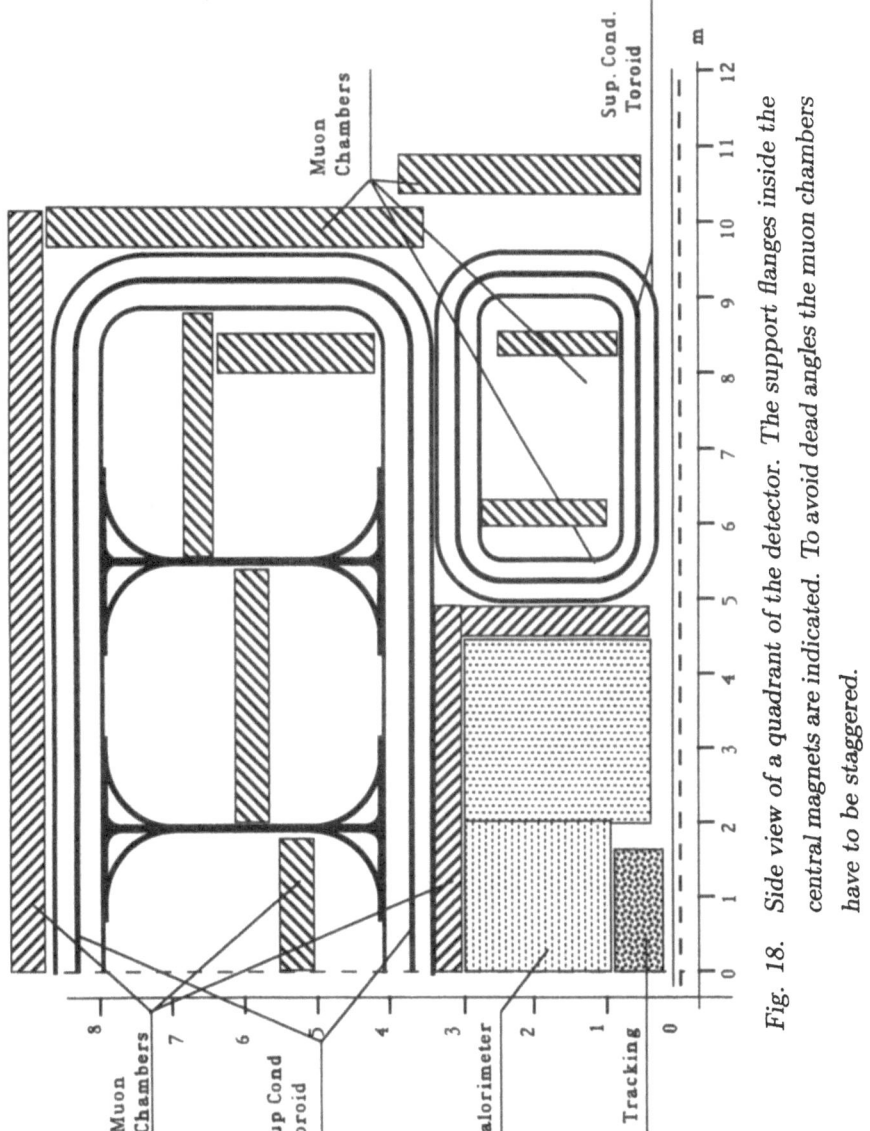

Fig. 18. Side view of a quadrant of the detector. The support flanges inside the central magnets are indicated. To avoid dead angles the muon chambers have to be staggered.

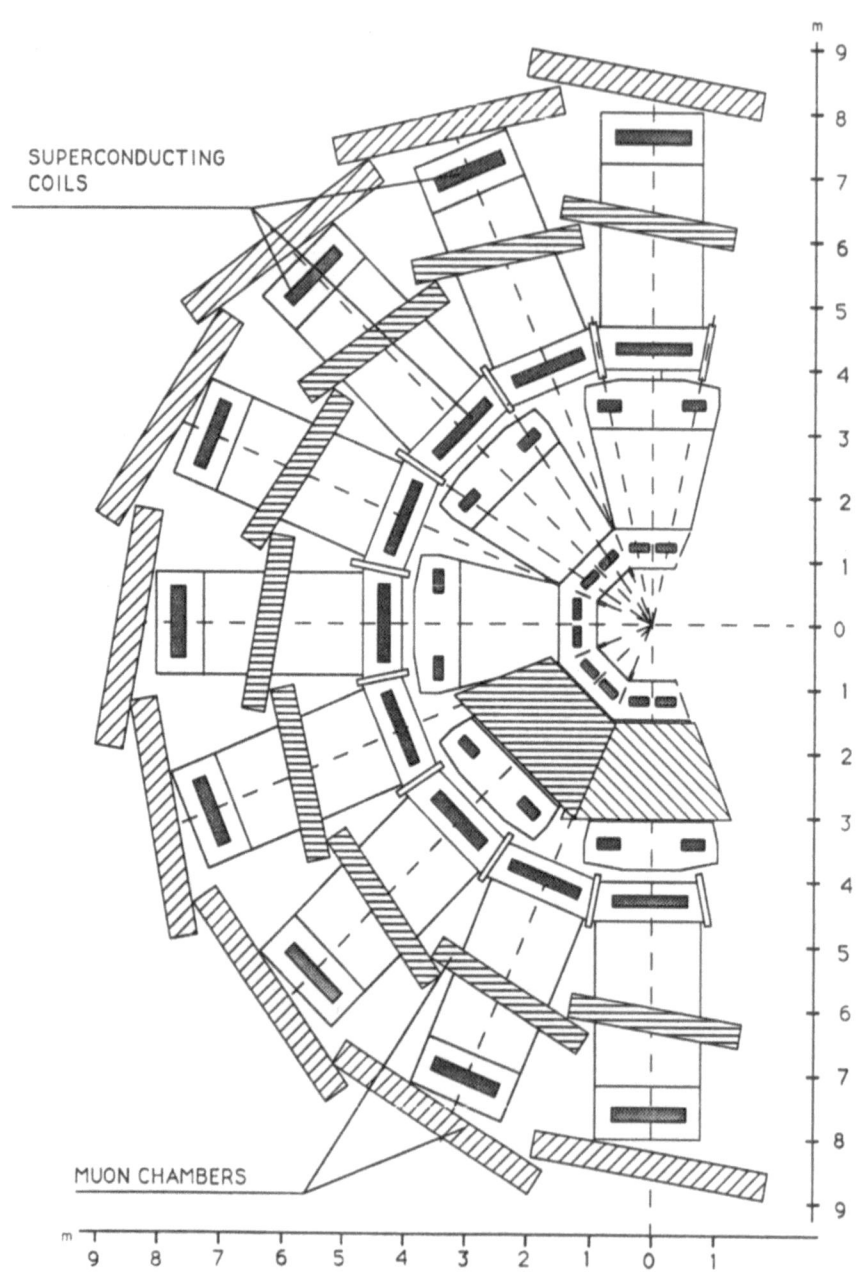

SUPERCONDUCTING
COILS

MUON CHAMBERS

Fig. 19. Transversal cross-section of the detector. Each coil is contained in a separate cryostate. The chambers inside the coils can be easily accessed.

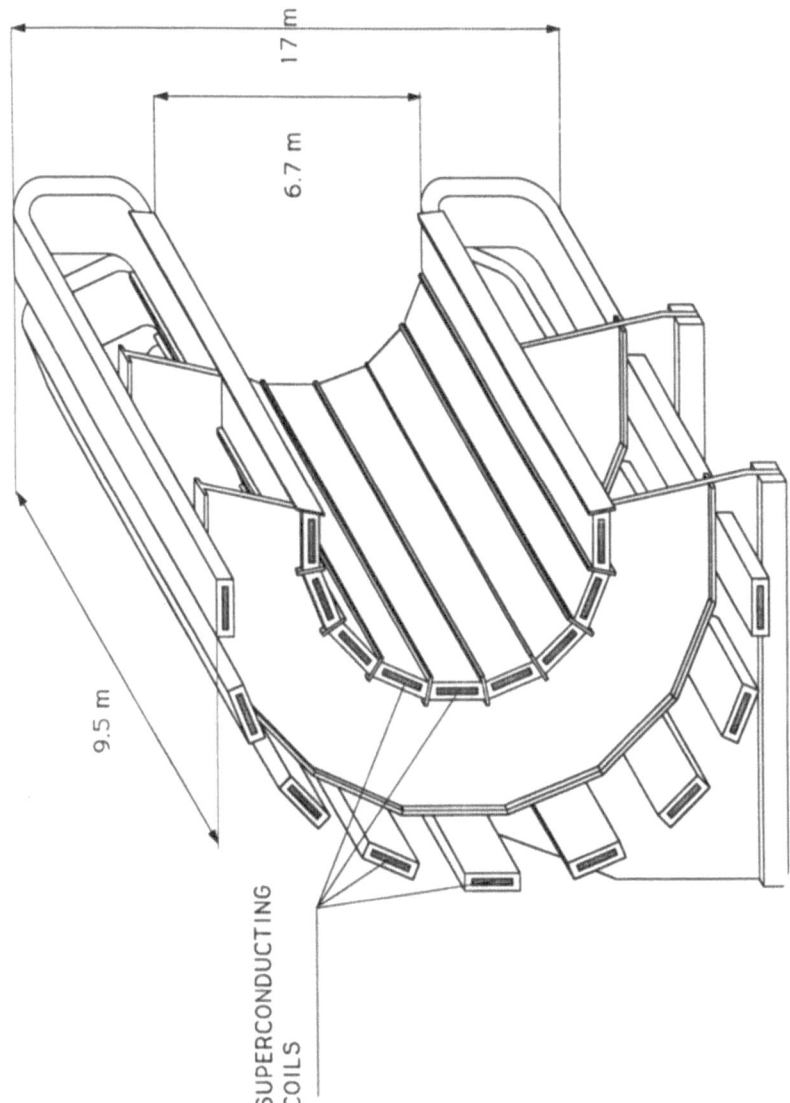

17 m

6.7 m

9.5 m

SUPERCONDUCTING COILS

Fig. 20. Perspective view of the detector.

147

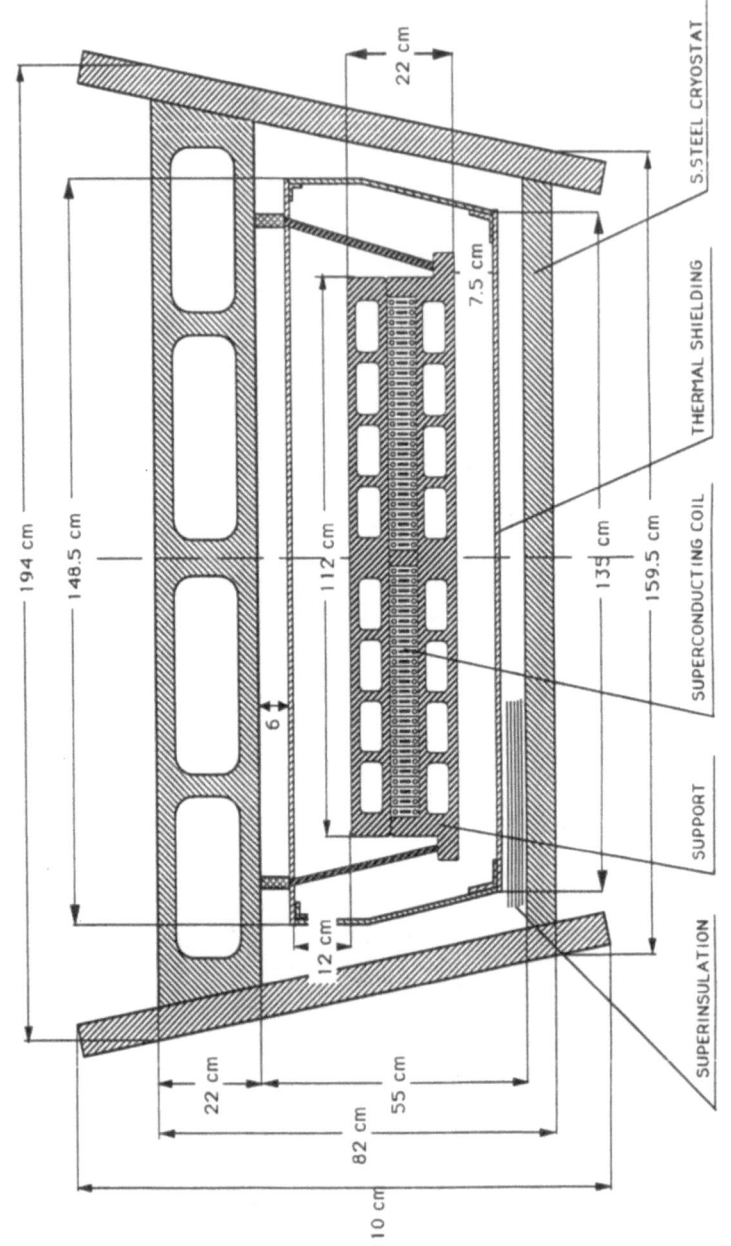

Fig. 21. Cross-section through the cryostat of the inner conductor from the central magnet. Each inner conductor has a separate cryostat. Temperature of the conductor is 4.5 K. Temperature of the thermal shielding is about 70 K.

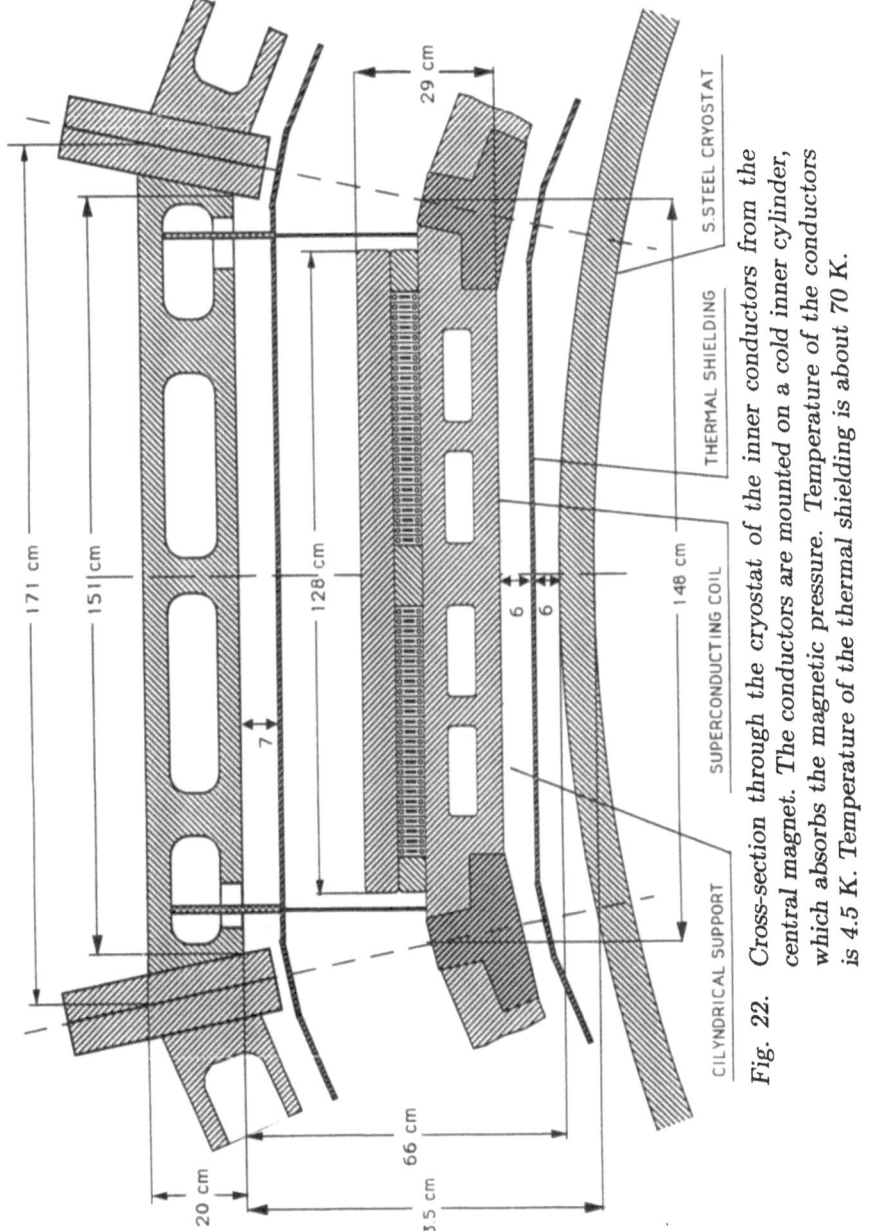

Fig. 22. Cross-section through the cryostat of the inner conductors from the central magnet. The conductors are mounted on a cold inner cylinder, which absorbs the magnetic pressure. Temperature of the conductors is 4.5 K. Temperature of the thermal shielding is about 70 K.

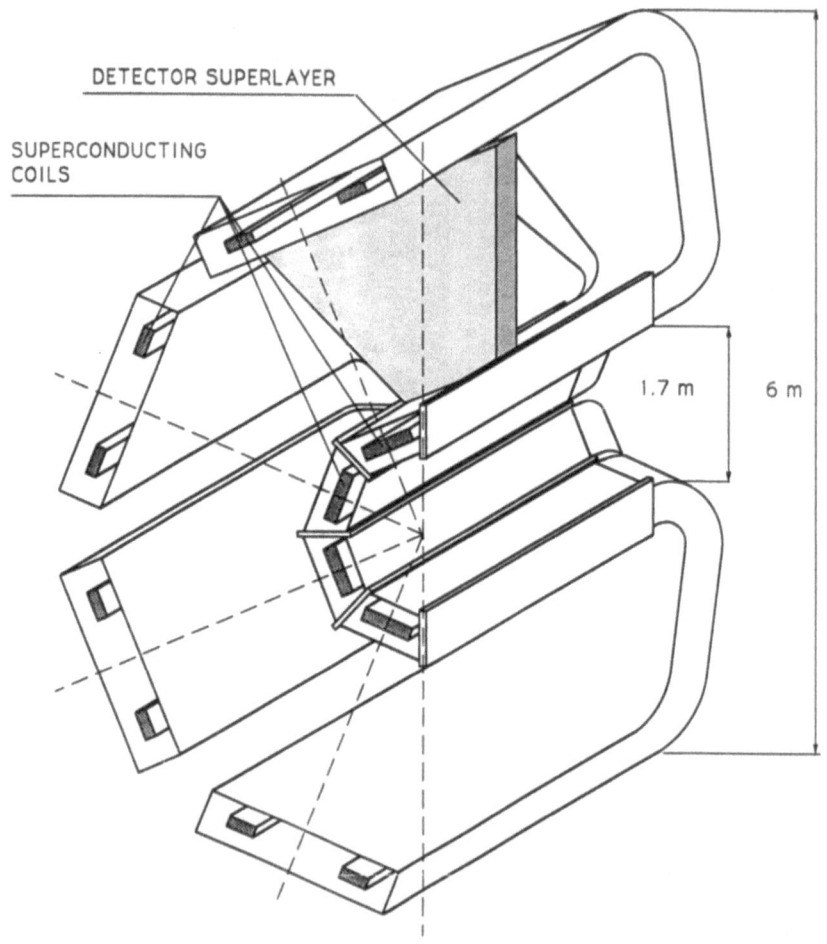

Fig. 23. Forward toroid. Two coils are contained in one cryostat.

This figure gives a cross-section through the inner conductor. The windings are squeezed inside a stainless steel cage, which is hung on the outside support by thin rods of a material with low heat conductivity. The windings and their cage are kept at a tempeature of 4.5 K. and are surrounded by a thermal shield, which is kept at 70 K. Due to the high forces, a lot of material is necessary to support the conductors. This makes difficult to keep the heat flow towards the cold parts within limits. A very promising solution to do it is to group the inner conductors together in one cryostat to form a cold cylinder which absorbs the magnetic pressure as shown in fig. 22. The support bars than have to carry only the weight of this cylinder, connecting it to the cryostat.

7. DISCUSSIONS AND CONCLUSIONS

From the results of paragraph 4, one can conclude that the here proposed design is promising: a resolution better than 4% $\Delta p/p$ for 500 GeV/c muons seems to be feasible. The modularity of the design gave no rise to unacceptabel inhomogeneities in the momentum resolution. Probably it is possible to use less than 16 coils. It should be stressed that the obtained resolution comes from the muon spectrometer only, no additional information was used from inner detection elements.

The main difficulties in this design come from the large amount of stored magnetic energy and the high forces on the superconductors which make the thermal insulation difficult.

This report is meant to intensify the discussion with experts in the field of magnet design and cryogenics, in order to come to a realistic design for magnets, cryostats, cooling system, power supplies, SC-cables, quench protection, etc. The dimensions should be optimized and construction material added. Then renewed study can be made of momentum resolution, including multiple scattering and detector details. Our general conclusion, at this stage of the work, is that an air core toroidal spectrometer is feasible within today's technology, giving the possibility of an extremely good muon momentum resolution, essential requirement for any detector at a future Supercollider.

8. BIBLIOGRAPHY

[1] EMPACT Collaboration, SSC-EOI-6 (1990).

[2] L* Collaboration, SSC-EOI-10 (1990).

[3] G. Ambrosi et al., "A new gaseous detector for tracking: the blade chamber", Nucl. Instr. and Meth. **A289**, 351-355, 1990.

[4] LHC Pink Book, in preparation.

[5] Vector Fields Limited, 24 Bankside Kidlington, Oxford, OX5 IJE, England, TOSCA reference manual VF068814 (1988).

[6] C. Zupancic, *"Physical and statistical foundations of TeV muon spectroscopy"*, CERN EP 85-03, 1985.

LATEST RESULTS ON THE DEVELOPMENT
OF THE GASEOUS PIXEL CHAMBER

D. Antreasyan[2], H. Castro[3], J. Galvez[3], Y. Liu[3], D. Mattern[1], F. Rivera[3],
A. Sharma[3], J. Solano Salinas[3], M.C.S. Williams[1] and A. Zichichi[1]

1) CERN, Geneva, Switzerland
2) INFN Bologna, Italy
3) World Laboratory, Geneva, Switzerland

ABSTRACT

The Gaseous Pixel Chamber is a new device that has been developed within the LAA Project at CERN. The excellent properties of this device have been previously described[1]. We present here the latest tests and results from further development. As this device is still being actively developed, these results are by no means complete but are so encouraging that presentation is warranted. Included are various tests that we have made to extend the range of operational voltage and extend the range of gas mixtures. We also include the first results of a device built on a ceramic substrate. Finally we will discuss some ideas we have for further development and also where this device may be used within a detector built for the Eloisatron or similar hadron collider.

1. INTRODUCTION

The gaseous pixel chamber has been previously described [1,2]. Essentially it is a foil consisting of a small anode spot surrounded at some distance by a cathode plane. Figure 1 shows a cross section of this device built on Kapton foil. The voltage is applied to the anode spot by means of a plated through hole. We mount this foil in a gas tight box, through which we flow pure isobutane. Electrons are liberated by a charged particle passing through the gas volume above the foil. These electrons undergo gas amplification as they approach the anode. Finally limited streamers are formed and therefore large signals on various electrodes. We have performed various tests to find ways of further improving this device. We have built a similar device on a ceramic substrate. We will describe this work below.

2. FOIL THICKNESS TESTS

The cross section of the gaseous pixel chamber is shown in figure 1. Although it is a surface device there are two places where the thickness of the Kapton foil may be important. The first effect is the surface of the Kapton insulator just above the high voltage bus line. Initially after applying high voltage to the anode the dielectric is not charged. Thus field lines pass through the dielectric and terminate on the bus line. These field lines will transport electrons to the surface until the surface is 'fully' charged, that is when field lines no longer enter the Kapton. Obviously with a thicker substrate fewer electrons will be needed to charge the surface. Another place where the thickness of the substrate may play a role is around the anode. The anode is a 500 μm diameter pad on the surface of the Kapton. This is connected via a 200 μm diameter through-plated hole to the anode bus line on the backside of the foil. Field lines will emanate from the sides of this through-plated hole, and also from the pad and bus line on the backside of the foil. These field lines will pass through

New Technologies for Supercolliders, Edited by L. Cifarelli and
T. Ypsilantis, Plenum Press, New York, 1991

the surface of the Kapton close to the anode spot. We show these field lines in figure 2, assuming that the surface is not charged and ignoring the effect of the high voltage bus itself (but we include a 500 μm pad on the back side of the board to make the connection between the high voltage bus and through-plated hole). Again the surface will get charged by electrons. By increasing the thickness of the substrate, the pad on the backside of the board becomes less important. It is unclear exactly how charged the surface will become in the steady state for the following reasons. In the fully charged state we will have a very strong electric field parallel to the surface and in the direction of the anode, thus if an electron becomes unattached to the surface it will be transported through the gas volume to the anode. There may also be effects of surface currents and volume currents that (at a very low level) discharge the surface. The source of electrons is from the ionisation of the gas by traversing charged particles. It is these electrons, when drifted to the high electric field around the anode, that initialise the growth of limited streamers. We, therefore, do not want to lose any of these electrons to the surface.

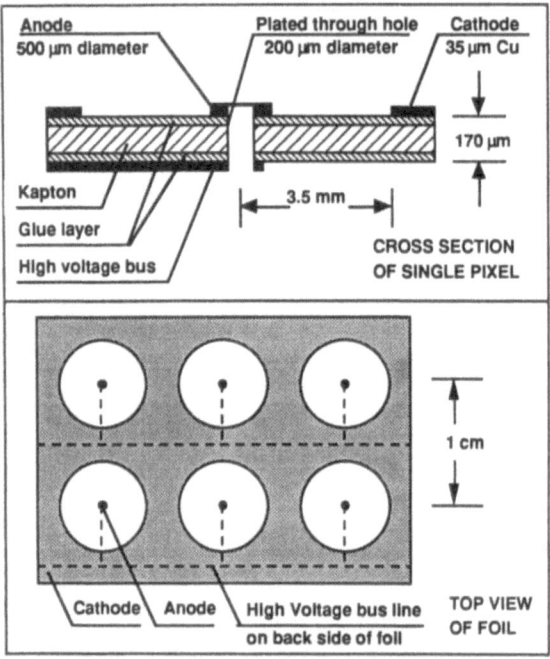

Figure 1.   Cross section and Top View of Gaseous Pixel Chamber

We have performed two tests relevant to this study. The first was to have extra dielectric printed onto the surface. This was actually 'solder mask' material[3] used in standard printed circuit manufacturing. The geometry and dimension of this mask are shown in figure 3. The second test was to make the Kapton foil thicker. The thickest Kapton foil available is 125 μm. In our case this is covered with two layers of glue to make a total substrate thickness of 175 μm. We asked the manufacturer[4] of the copper covered Kapton material to glue two layers of Kapton together before adding the copper layers. This they did and we had a laminate of two layers of 125 μm thick Kapton, which together with the glue layers made a laminate of dielectric substrate of 325 μm thickness in total.

We installed each foil in a gas tight box through which we flowed pure isobutane. There was 3 cm of gas above the foil. We irradiated the foil with a ^{55}Fe source. We measured the average pulse charge with a digital Tektronic 2440 oscilloscope. In figure 4 we show the average pulse charge versus anode voltage for 3 cases; the standard 175 μm thick foil (standard pixel chamber), the special double laminate (thick foil pixel

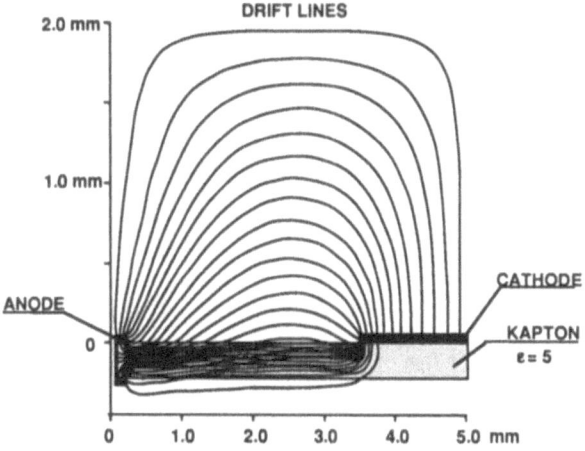

Figure 2. Drift Lines for the Gaseous Pixel Chamber with Uncharged Kapton Surface

chamber) and the foil with the extra solder mask (mask pixel chamber). We constructed 'chambers' with two of these foils (the standard pixel and mask pixel chamber) by adding a 'roof' board. This roof board has a large pad mounted opposite to the anode and a grid aligned with the foil cathode plane. The foil and the roof board are separated by 8 mm. We installed these two chambers in a test beam in the CERN East hall. In figure 5 we show the efficiency for detecting 6 GeV/c pions. In figure 6 we show the charge distribution for these two chambers with 6 kV applied across the pixel.

We find that the detection efficiency for the mask pixel chamber to be higher than the standard pixel (100% versus 98%). We think that this is due to having the dielectric thicker above the high voltage bus line, thus fewer electrons are needed to charge the surface (and also fewer needed to keep the surface charged). The higher gain measured with the [55]Fe source for the mask pixel chamber is a repeatable measurement. We are therefore surprised to find that the peak of the charge spectrum with 6 GeV/c pions is so similar, while the [55]Fe source results indicate that there is a factor 2 difference in gain. There is a small change in shape of the distribution, some events that gave small pulses close to the pedestal have shifted towards the peak for the mask pixel chamber. We are still investigating this effect. The shape of the [55]Fe source charge curve for the thick foil chamber and the standard pixel chamber are very similar in shape, except that the thick foil chamber needs about 1000 volts less. We think that this is due to the high voltage bus and pad being much further away close to the anode. We hope that this will also work at a 1000 volts less in a particle beam. This will be tested during our next beam test in October 1990.

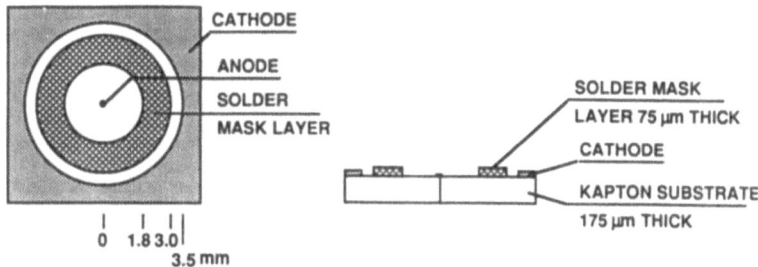

Figure 3. Solder Mask Layer Applied to Gaseous Pixel Chamber

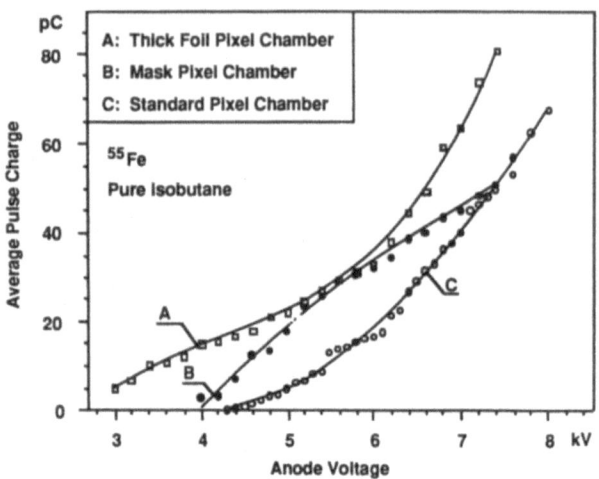

Figure 4. Average Charge of Anode Pulse when Irradiated with ^{55}Fe Source

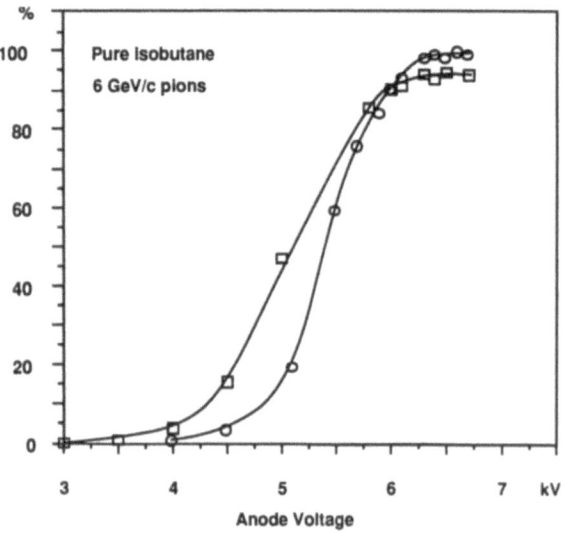

Figure 5. Detection Efficiency for 6 GeV/c pions

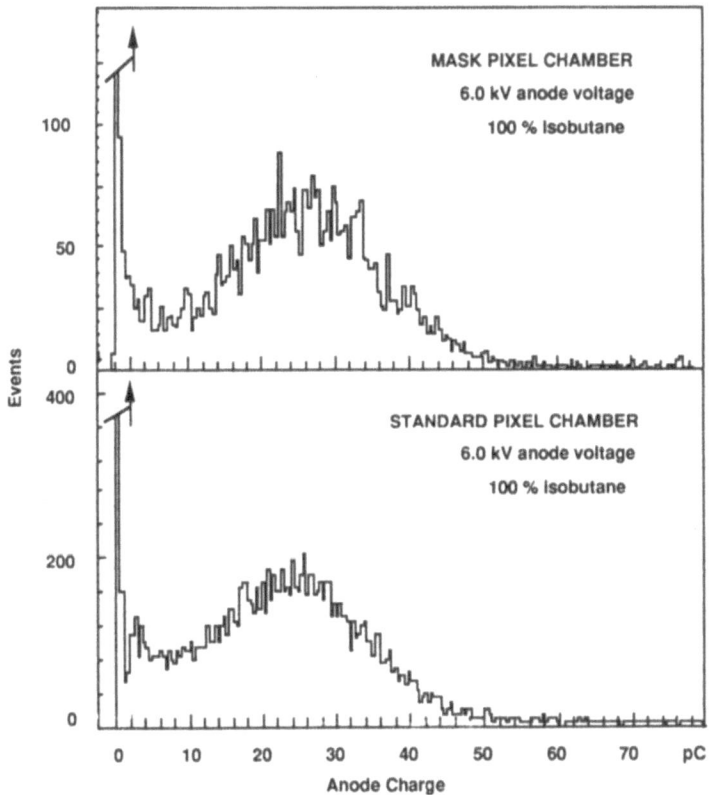

Figure 6. Charge Distribution for Standard Pixel Chamber and Mask Pixel Chamber

3. SMALL ANODE CHAMBER

One of the parameters that we believe to be important is the diameter of the plated through hole. The smallest hole that we can produce at CERN is 200 μm. This is a mechanically drilled hole. Lasers can be used to make holes (either by burning or ablation) and are already being used for mass production of circuit boards. This is still earlier days for this technology, however we have started trying to construct foils with very small laser holes. In the meantime we have tested a foil where we had a 200 μm diameter unplated hole and no anode pad. We then inserted a 50 μm gold plated tungsten wire. The connection to the high voltage bus was performed with a conductive epoxy. We tested two versions of this chamber, one where the wire stuck up above the foil by about 500 μm and another version where the wire was cut level with the surface. Both chambers worked well, however the chamber with the wire sticking out worked better especially for high concentrations of isobutane. The foil with the cut off wires was also much more noisy, we think this noise is caused by having the sharp dielectric edge close to the high field region. In some cases the cut wire disappeared down the hole. For these reasons we will only present the results from a foil with the wire sticking out by 500 μm (small anode chamber). However we think that it will not be necessary to have anodes that stick out from the surface by 500 μm to get the good results that we obtained with this device.

We constructed a chamber using a roof board in a similar manner as described above. All the anodes were bussed together. In figure 7 we show the efficiency for detecting 6 GeV/c pions for a variety of Argon and isobutane mixtures. For pure isobutane the knee of the plateau is at 4 kV. This should be compared with 6 kV for a standard pixel chamber. The voltage where sparking occurs (~7 kV) should be the same (we did

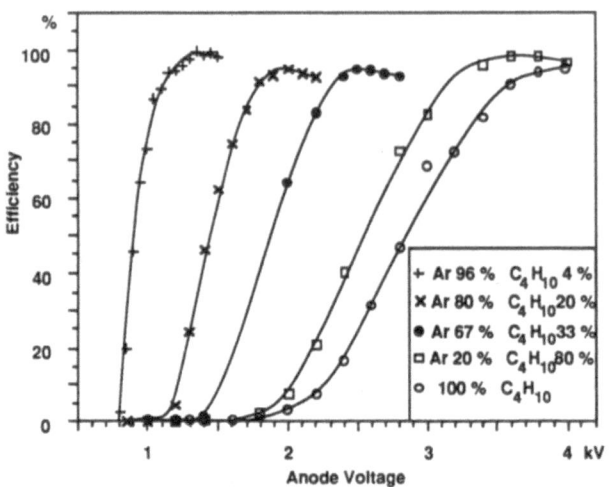

Figure 7. Detection Efficiency of 6 GeV/c pions with Various Gas Mixtures

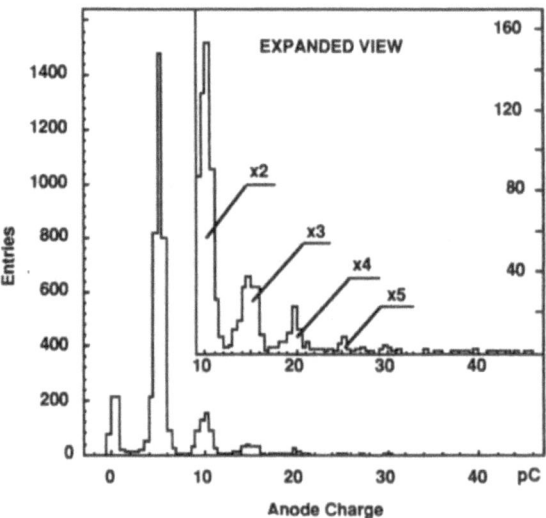

Figure 8. Charge Spectrum of Small Anode Chamber with 96% Ar and 4 % Isobutane

observe a spark at 6 kV with the small anode chamber but think that this may be due to the 'home-made'nature of the device). The remarkable thing we found was that it worked with very little quensher in the gas mixture. With the standard pixel chamber we found that with more than 20% Argon in isobutane, then the chamber had voltage breakdown before it reached full efficiency for minimum ionising particles.

In figure 8 we show the charge spectrum for the chamber with 96% Ar 4% isobutane at its operational voltage of 1000 volts. There is a very clean peak separated from the pedestal. There are also peaks at 2x, 3x, 4x and 5x the charge of the first primary peak. After some investigation we conclude that these are due to more than one pixel firing when more than one particle traverses the chamber.

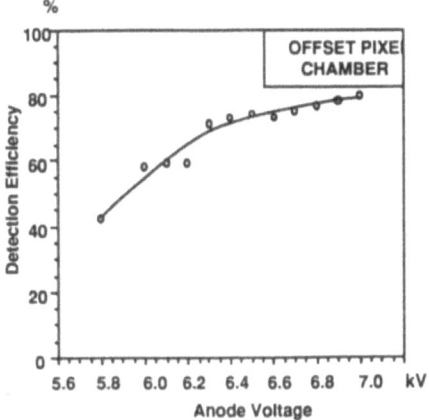

Figure 9. Detection Efficiency of 6 GeV/c pions for Offset Pixel Chamber

4. THE OFFSET PIXEL CHAMBER

There are many advantages in having a highest possible pixel density, for example faster time response for through going particles and better position resolution. However, depending on the manufacturing techniques, there will be a lower limit of pixel size due to the finite size of the anode. We have investigated the possibility of increasing this density by a factor two by having two foils facing one another (sharing the same gas volume) but staggered by half a cell width in the x and y direction. Thus the anode of one foil faces the cathode plane of the other. We constructed a chamber to test this possibility. The gap between the foils was 8 mm.

We installed this chamber in the test beam. We derived a signal from the logical 'OR' of the two foils. The efficiency for detecting 6 GeV/c pions is shown in figure 9. In figure 10 we show the drift time of this chamber, for comparison we show the drift time spectrum of the standard pixel chamber. A scatter plot of the position of the particles that do not fire either foil is shown in figure 11. It is easy to see the structure of inefficient hits follows the boundaries between the pixel cells. We expect that we can reduce this inefficiency by changing the gap beween the foils. We are also investigating whether the shape of the cathode plane plays a role. Further work is in progress.

5. POSITION RESOLUTION

We have taken some data with the standard pixel chamber to determine the position resolution. The chamber was mounted in a test beam in the CERN East hall. The position of the incoming particle was measured with 2 sets of drift chambers. In figure 12 we show a scatter plot for radial distance of the particle from the anode versus the drift time recorded from the pixel chamber. A clear band is seen. We fitted a curve

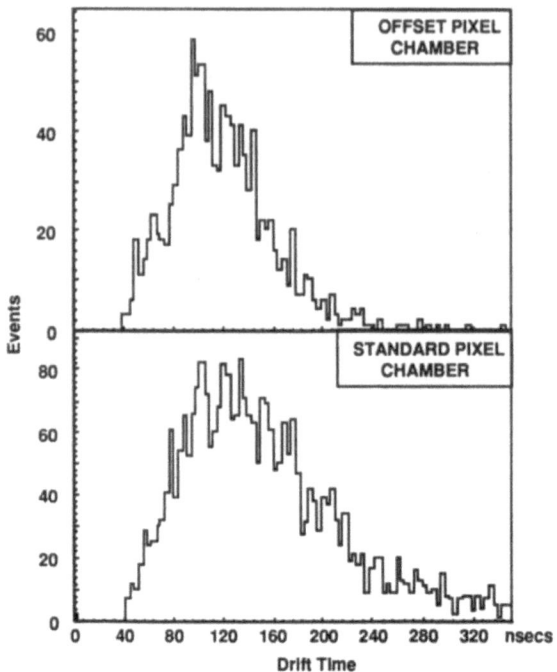

Figure 10. Drift Time Spectrum for Offset Pixel Chamber compared with Standard Pixel Chamber

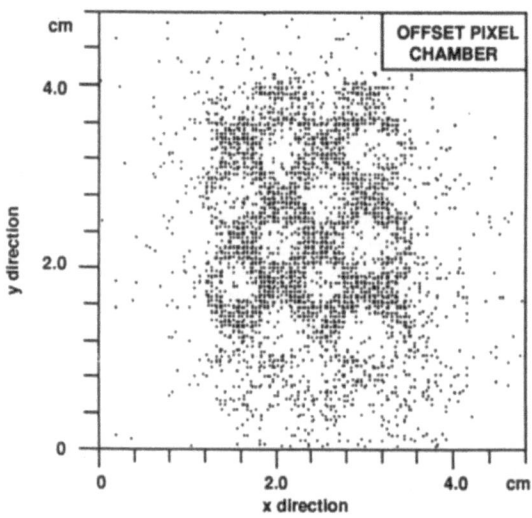

Figure 11. Scatterplot of Inefficient Hits showing Dead Region between Pixels

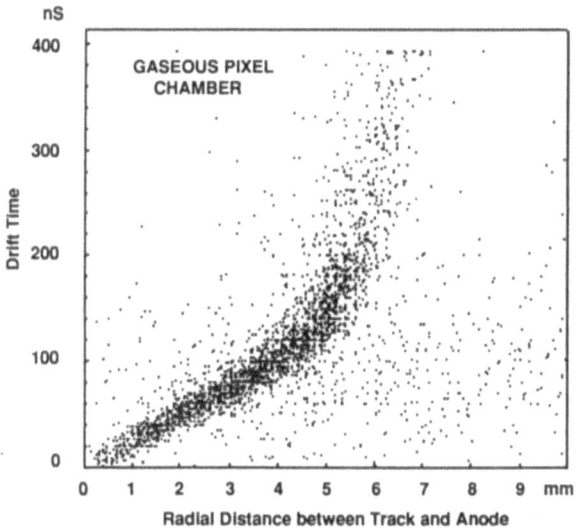

Figure 12. Scatterplot of Drifttime measured by the Gaseous Pixel Chamber and the Radial Distance from the Anode (measured with external drift chambers)

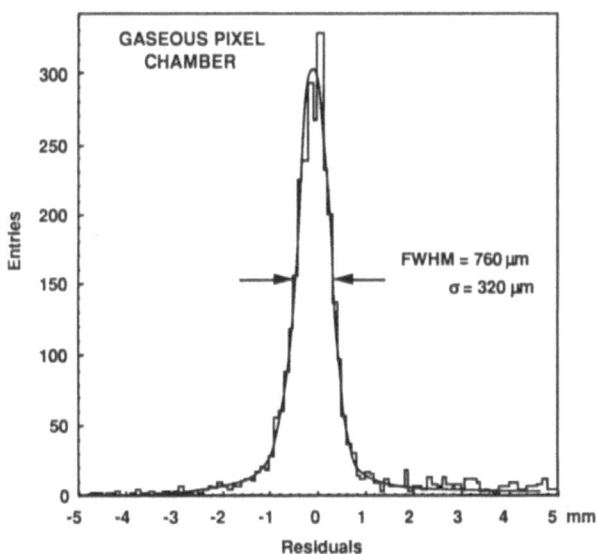

Figure 13. Distribution of Residuals of Radial Distance from Anode

to this band and used this to calculate a radial distance from the drift time measurement. A histogram of the residuals is shown in figure 13. The FWHH is 760 μm, this corresponds to a sigma of 320 μm. We think that the contribution to this width from the measurement error made by the external drift chambers to be small. We hope to improve this result when we start working with smaller hole, smaller anode chambers.

6. CERAMIC PIXEL DEVICE

We have built a pixel device using a ceramic substrate[2]. Although lasers are being used to create small holes in ceramic, it is still in a development stage. We have, therefore, chosen a different technique to connect the voltage to the anode pad. Ceramic circuits are made by 'printing' layers of conducting or insulating paste onto a ceramic plate. The plate is then 'fired' at about 900°C. Then another layer of paste can be printed which can be subsequentially fired. With this technique one can fabricate a multilayer circuit board. The printing is by serigraphic techniques using a stainless steel mesh. The ceramic plate has to be flat, otherwise the screen has to stretch to accommodate the different levels, and this introduces distortions. Lines of 100 μm width can be successfully printed. To generate an anode spot we first print a narrow ridge of dielectric material and then print a 100 μm conductive line at right angles to this ridge. Thus the conductive line goes up one side of the ridge and down the other. We then fill each side of this ridge with dielectric material. This results in having a thin anode line buried under dielectric material except at certain spots where it comes to the surface. The size of the exposed anode spot is defined by the width of the conductive line and by the width of the top of the ridge. In our case the anode is a square spot of 100 by 100 μm. We need to have the layer of dielectric covering the anode bus line as thick as possible. We thus make the dielectric ridge by printing three layers of widths 350, 275 and 100 μm successively. The anode lines can be successfully printed over this ridge. At some distance we need the cathode plane. Since the anode bus line has to run beneath it we need a certain thickness of dielectric to hold off the applied high voltage. The figure supplied by the manufacturer is > 28 volts/μm. Thus 15 layers of dielectric should hold off more than 5460 volts. At a radius of 2 mm from the anode spot we start building these layers of dielectric. The cross section of a pixel is shown in fig. 14.

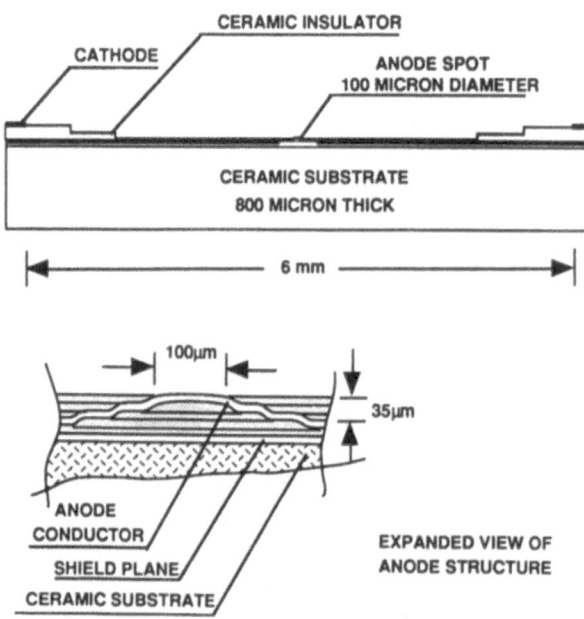

Figure 14. Cross Section of Gaseous Pixel Chamber built on Ceramic Substrate

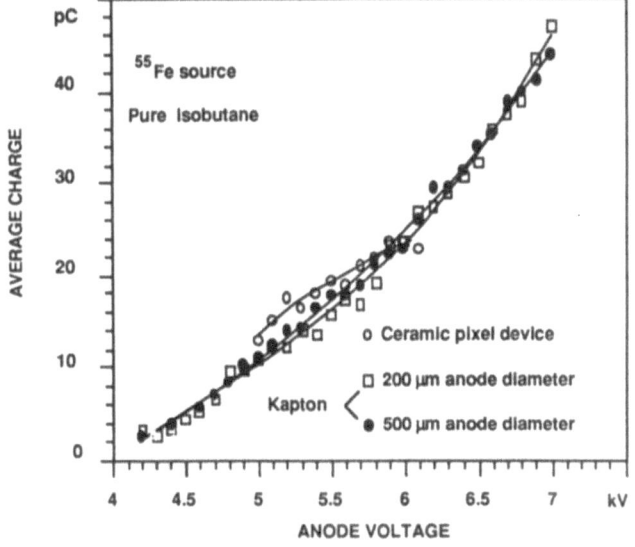

Figure 15. Average Charge on Anode when Chamber Irradiated with ^{55}Fe source

We performed similar tests with ^{55}Fe source as with the other pixel chambers. The average charge versus anode voltage is shown in fig. 15. Shown for comparison on this figure are measurements performed on the Kapton foil chamber with similar geometry. The anode spot size for the Kapton foil chamber was 500 μm and 200 μm, while the ceramic chamber anode spot is 100 μm. It is remarkable that the gain of the ceramic chamber is so similar to the Kapton foil. The geometry of the anode is somewhat different in the two cases, as well as the thickness and type of dielectric used.

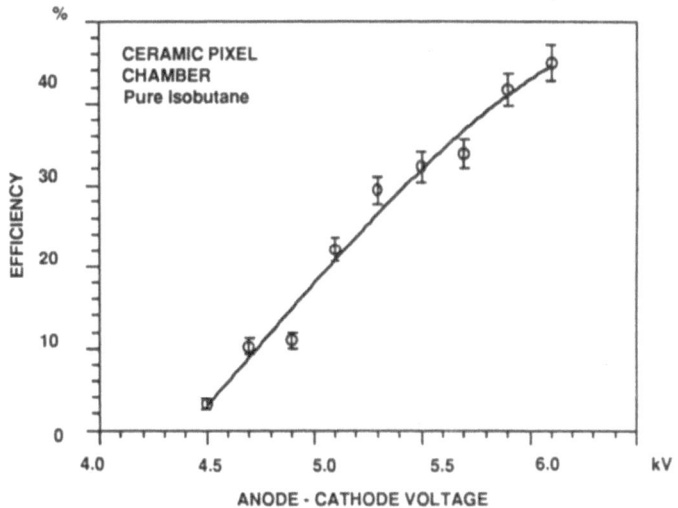

Figure 16. Detection Efficiency for 6 GeV/c pions for Ceramic Pixel Chamber

We constructed a chamber using a 'roof' board of similar design to that described above. We installed this chamber in a 6 GeV/c pion beam in the CERN East hall. We flowed pure isobutane through the chamber (some argon was added to the mixture later). We applied 500 volts to the central roof spot. Figure 16 shows the increase of efficiency for detecting through going particles as a function of the voltage across the pixel. Unfortunately we had some voltage breakdowns for voltages above 6.2 kV so that we could not reach full efficiency. Figure 17 shows the charge distribution of the cathode pulse for 3 different gas mixtures but with the same applied high voltage. It is interesting to note that there is a slow increase in pulse height for increasing argon concentration, but a much more dramatic change in the shape of the charge distribution. The distribution of large pulses on the right of the peak is due to two (or more) pixels firing when more than one particle traverses the chamber. The efficiency for detecting particles also increases slightly with increasing argon concentration.

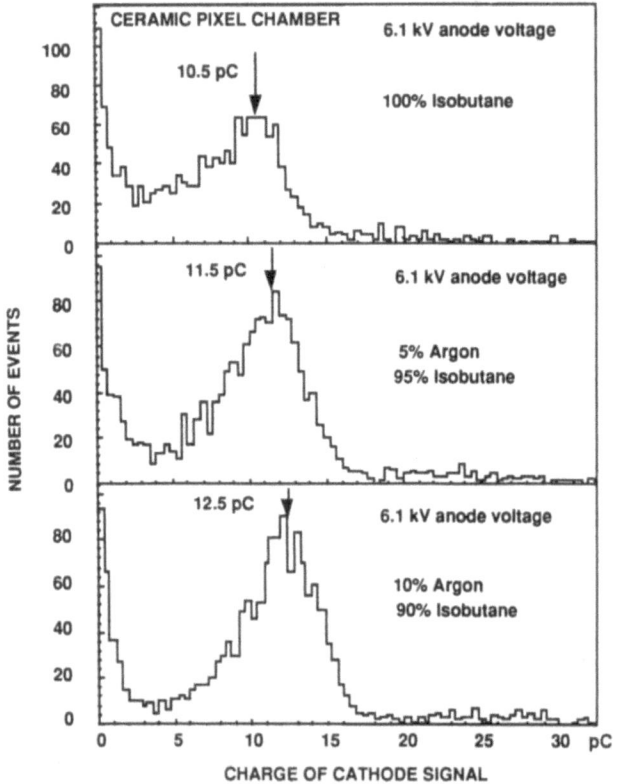

Figue 17. Charge distribution of Anode Pulse for various Gas Mixtures

We believe that the inefficiency for detecting minimum ionising particles is caused by having the anode bus line shielded by only 35 μm of insulator. We are fabricating a new ceramic board with a thicker layer of dielectric. We also expect that the gain will increase by having thicker dielectric around the anode spot. Using this technique one has an anode spot of 100 μm in size compared to 500 μm for the standard pixel device. Maybe this defines more precisely the position of the high electric field needed for gas amplification, and thus could lead to better spatial resolution.

7. APPLICATIONS WITHIN AN ELOISATRON EXPERIMENT

There are many advantages of using detectors with gas amplification. They can be made relatively radiation hard. Gas amplification is intrinsically a high-gain, low-noise process. Wire chambers have been in use in particle physics for the last 25 years, therefore they are known devices. However there are disadvantages related to constraints imposed by having straight stretched wires. By having a device that keeps the advantages of gas amplification without having this constraint opens up many possible areas of use.

For the barrel muon detection one needs to cover a large area with cheap, easy to build and reliable detector. One also needs good position resolution and x and y readout for triggering purposes. The gaseous pixel detector may be the ideal device for this application. Most designs use toroidal magnets in the forward direction. Therefore for muon detection it would be ideal to have a detector that has circular geometry. It is very easy to arrange to have the pixels to follow a curve and thus have a device with r,ϕ readout. We have already shown[1] that the gaseous pixel chamber has no problems with rates up to 2.10^4 / cm^2 / sec (it is highly likely that we can work at much higher rates than this when the device is fully developed). Thus again the gaseous pixel chamber appears to be an ideal device.

Various detector designs call for the instrumentation of the iron between the hadron calorimeter and the muon detection system. This system is to measure the hadronic shower leakage, detect catastrophic energy loss of muons and also to track muons. The detectors used for this purpose have to fit in many awkward places. The gaseous pixel device is ideal for this application as it can easier be made in any geometry to fit into a certain space. More importantly the gas gain is defined by the anode-cathode spacing (which is defined by the surface of the foil) rather than by a distance between two surfaces. Thus the chamber will operate even if it is severely distorted.

There are various schemes to have a pre-shower device in front of the electromagnetic calorimeter. It is possible to consider layers of lead followed by gaseous pixel chambers. The pixel readout is ideal for this purpose. It is also possible to imagine using a gaseous pixel chamber with a photocathode and use it to read out BaF crystals. Again the pixel nature is highly desirable. Again the gain is set by the anode-cathode spacing so mechanically it would be an easy device to build. It should be pointed out that there may be problems with poisoning of the chamber with these photosensitive gases, so it is not clear that the gaseous pixel chamber can be used for this. For both these devices it would be necessary to work in proportional mode. As yet we have only observed limited streamer mode of operation, however we expect to proportional mode when we have devices with smaller anode structures.

For the leading particle detector one needs a high rate and radiation hard device. The size of the pixel of our current chamber is too large for this. However there are some interesting developments[5] in the field of vacuum microelectronics where they are building structures similar to vacuum valves, except that they are scaled down by many orders of magnitude. The thermionic cathode is replaced by field emission cathodes to provide a source of electrons. A cross section of such a device is shown in figure 18. One can see that the geometry is similar to the gaseous pixel chamber, however their cathode (the tips) are equivalent to our anodes. Thus by reversing the voltage applied to this device and by adding a gas, maybe it would work as a particle detector. We are about to study the feasibility of using such devices as gaseous detectors.

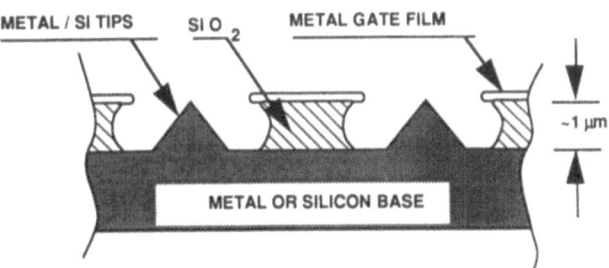

Figure 18 . Cross section of Typical Field Emission Diode structure

For tracking there have been many elegant solutions suggested. However it may be possible to supplement this central tracking either for trigger purposes or maybe to provide extra z information. Since this device is pixel in conception it does not suffer from the occupancy problems usually found with gaseous detectors in the central region. For example a simple jet trigger could be built that just demands a certain track density in given ranges of ϕ and psuedo-rapidity. The clear separate peaks shown in figure 8 for different number of pixels firing indicates that this would be a very simple device to build and instrument.

8. FUTURE DEVELOPEMENTS

Some of development that we have in mind has already been covered above. Maybe one of the most critical things is to find a way to reduce the anode size. This would also allow us to reduce the anode cathode spacing and thus make big gains in resolution and time response. At present we are limited by the minimum drill size of 200 μm used to drill the plated through hole. However lasers are becoming used more and more in various production facilities. For example it is possible to use a CO_2 laser to punch a 75 μm hole through 300 μm glass fibre board[6]. Also this laser can be pulsed at 50 Hz. 50 holes a second is a much higher rate than with mechanical drilling. Thus it seems that we will want to use lasers both to decrease the hole size and to increase the ease of fabrication.

The technique of printing used for the ceramic construction is also very interesting as it is basically a low technology fabrication process. Line widths in the order of 50 μm are possible today. We will continue with these studies.

ACKNOWLEDGEMENTS

As before we have enjoyed the excellent work of the printed circuit workshop here at CERN run by Mr. Gandi. We are also very much aware that Mr. Donato who runs the 'Thick Layer' section is a master of creating ceramic structures.

REFERENCES

1. D.Mattern, M.C.S. Williams and A.Zichichi, A New Approach for Constructing Sensitive surfaces: The Gaseous Pixel Chamber. CERN EF 90-4 and to be published in Nucl. Instr. and Meth.

2. D.Mattern, M.C.S.Williams and A.Zichichi, First Tests of the Gaseous Pixel Chamber Fabricated on a Ceramic Substrate. Submitted to 2nd London Conference on Position-Sensitive Detectors, September 1990, to be published in Nucl. Inst. and Meth.

3. The material we used was LAMINAR DM Dry Film Solder Mask manufactured by Morton Thiokol, Sa, Division Dynachem, 2, rue Ampère, B.P. n° 36, Z.I. 91430 Igny, France.

4. GTS FRANCE, GTS Materiaux Flexibles SARL, ZA de Courtaboeuf, Batiment Auvidulis, Avenue d'Oceanie, BP 90-91943, Les Ulis Cedex, France.

5. N.A. Cade and R.A. Lee, Vacuum Microelectronics, The GEC Journal of Research, Vol.7, No. 3 (129-138) 1990.

6. A.N. Pargellis, D.T.W. Au, T.V. Lake, A. Kestenbaum, Optics and Laser Technology, Vol 22,No 3 (205-207) 1990.

THE MUON SPECTROMETER: WHAT IS NEEDED?*

M.C.S. Williams

LAA Project, CERN, Geneva, Switzerland

1. INTRODUCTION

This report considers the requirements for a muon spectrometer for an experiment at some future hadron collider. Whatever system is built to identify muons, the cost will be a substantial part of the experiment. Does this large investment in manpower and capital allow us to observe new physics processes? What is really needed in terms of angular coverage and momentum resolution?

There are two basic reasons for a muon spectrometer, the first is to make the detector hermetic and the second is that muons can be used to signal the presence of new particles. We will consider each of these points in turn.

2. HERMETICITY

The muon carries energy that is not deposited in the calorimeter. If we do not detect the muon we create an event with missing p_t. The muon also signals the possible presence of neutrinos, thus by detecting the muon we can flag these events. We will consider two examples from the Non-Magnetic Detector group report[1].

Example 1: The Non-Magnetic Detector Group considered the production and decay of the W (W $\rightarrow$ $\mu\nu$) with pt_W > 100 GeV/c. If the muon spectrometer starts detecting muons at θ > 5° then 16% of these decays would not have the muon detected. Of these 40% would have a missing p_t > 100 GeV/c (assuming perfect calorimetry to η of 5). Since this number of events is small compared to ineradicable contribution from W $\rightarrow$ $\tau\nu$ and Z $\rightarrow$ $\nu\nu$, they point out that it is not worthwhile extending the angular coverage below 5°.

Example 2: The same group also considered events that contained a Z and a jet with pt_Z > 100 GeV/c. They generated 135,000 of these events with ISAJET with the top mass in the range 30-85 GeV (the results were independent of the value of the top mass). Using perfect calorimetry to η of 5 then events with a missing p_t > 50 GeV/c were almost exclusively due to semi-leptonic decays of heavy quarks. The identification of extra leptons in these events would flag these decays. If the muon spectrometer could only identify muons

---

* Presented in a preliminary form at the 9th Workshop of the INFN ELOISATRON Project "Perspectives for New Detectors in Future Supercolliders" Erice, Italy. 17th- 24th October 1989.

with $p_t > 10$ GeV/c and $\theta > 5^O$ then 25% of these events would be flagged. This efficiency improved to 28% if the muon coverage was extended to 2^O. If the muon momentum selection was improved to 5 GeV/c then 32% of these events would be flagged. A much more dramatic improvement is obtained if both electrons and muons are used to flag these events (53%) which increases to 69% if the lower momentum constraint is used.

Both these examples show that muon coverage at $\theta > 5^O$ is sufficient using hermeticity arguments. It is also clear from these examples that the muon coverage should match the electron coverage.

3. SIGNAL OF NEW PHYSICS

Many major discoveries have been made via the lepton channels. Future hadron colliders will be dominated by extremely high event rates. Leptonic and semi-leptonic decays of new heavy particles will produce isolated leptons. Events with isolated leptons may well be the only way to select events of interest. We will consider two examples of particle searches to see what it implies for the muon spectrometer. The first will be the asymmetry measurement of lepton pairs, and the second will be the search for a neutral Higgs.

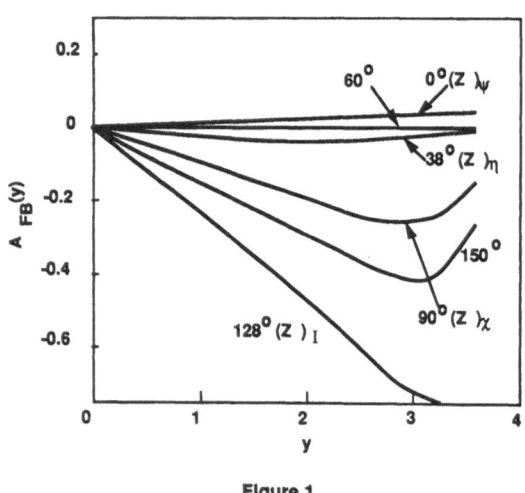

Figure 1

3.1 Asymmetry measurement

If another heavy gauge boson exists it will decay into two leptons, $Z' \rightarrow l\,l$. The energy resolution of an electron calorimeter will be superior to that of a muon spectrometer. Thus the direct observation of this resonance will be far easier to observe in the electron channel. However if such an object exists it will be very interesting to measure its couplings. The forward-backward asymmetries test the vector and axial-vector couplings of the new boson to quarks and leptons[2,3]. A class of bosons have been discussed of the form $Z(\theta) = Z_\chi\cos\theta + Z_\psi\sin\theta$, where Z_χ and Z_ψ are two bosons arising from E_6 GUTs. The leptonic forward-backward asymmetries are odd functions of rapidity y, and depend strongly on θ. Rosner[4] has calculated the forward-backward asymmetry on the resonance peak as a function of y and for various mixing angles θ. This is shown in figure 1. They are small for a wide range of θ.

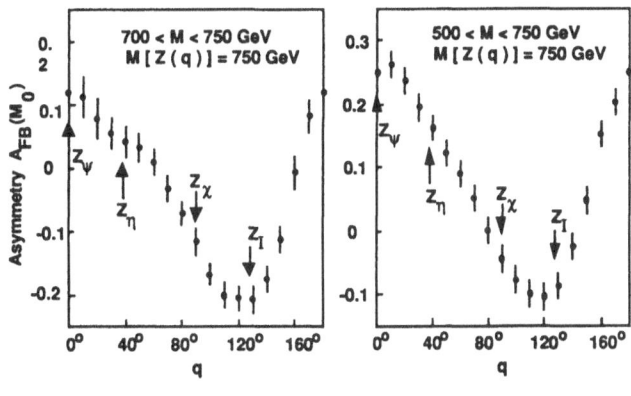

Figure 2

Off-peak asymmetry measurements tell us more more about the couplings from interference of their resonance amplitudes with those of the standard Z and the photon. In figure 2 we show Rossner's calculation of the forward backward asymmetry for two mass intervals. The error bars correspond to an integrated luminosity of 10^4 pb^{-1}. The asymmetries are quite pronounced for a wide range of angle θ, and it appears possible to measure an asymmetry even for the Z_ψ, which displays no asymmetry on the resonance peak.

What does this imply for the muon spectrometer? First the asymmetry is biggest for large y, so we need coverage in the forward direction. In figure 3 we show the acceptance for detecting the leptons from the decay $Z_\chi \to l\,l$ for various masses of Z_χ as a function of minimum detection angle. Also shown is the effect of a 1TeV/c cut on the lepton momentum. To make an asymmetry measurement it is necessary to measure the sign of the lepton (without systematic errors - delta rays tend to follow μ^- but separate from μ^+). There are various proposals for the central detector. Either they have no magnetic field, or a solenoidal magnetic field. Even with a solenoidal magnetic field the sign of forward going electrons is very difficult to determine. If we are going to measure the lepton forward-backward asymmetry we need to have a muon spectrometer covering the high η range capable of measuring the sign of the muon a momenta in excess of 1 TeV/c. Also as the off peak asymmetry varies as a function of the mass of the muon pair, we need good momentum resolution.

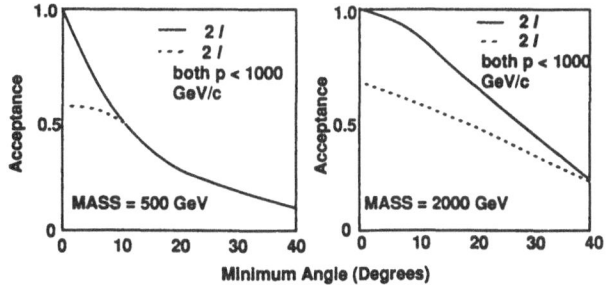

Figure 3

3.2   Higgs Search

A possible channel to search for the Higgs particle is via its decay into two Z^0s and then its subsequent decay into four leptons. $H^0 \to ZZ \to llll$. The acceptance for detecting all four leptons as a function of minimum detection angle has been computed by Carlsmith et al.[5] This is shown in figure 4. A 5^0 minimum angle cut loses 15% of the Higgs events if the mass is 800 GeV and 23% if the mass is 300 GeV. Clearly forward angle coverage is important. At high masses the Higgs resonance is expected to be broad, however it is still necessary to have good momentum resolution in order to reduce the background. This is illustrated by the following example as calculated by the Non Magnetic Group at Snowmass[1].

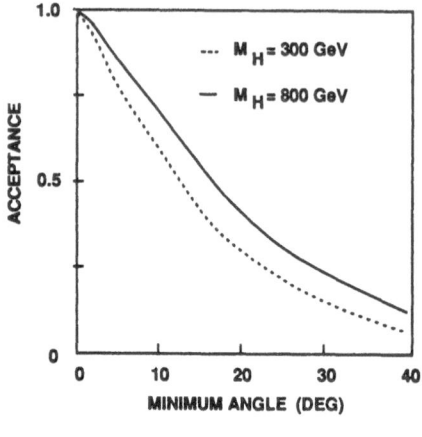

Figure 4

The predominant source of di-muon background is due to the decay of heavy quarks. A heavy top decaying into W's and b's would be a great source of high p_t muons, however when isolation cuts are imposed then the background is reduced to negligible levels as only a fraction of the b decays produce an isolated muon. The ability to impose these isolation cuts is very important and a finely segmented calorimeter is crucial. For the subsequent discussion it is assumed that the calorimeter is segmented into cells of 0.1 in $\Delta\phi$ and 0.1 in η. The only process that the above group could find to produce isolated muons in large enough quantities is from the decay of a fourth generation quark (≈ 200 GeV) into a heavy top (≈ 100 GeV). They generated events for a 400 GeV Higgs and a 200 GeV fourth generation b quark using ISAJET. Events were retained if they had 4 isolated muons with a $p_t > 30$ GeV. For an integrated luminosity of 10,000 pb^{-1} they were left with 3000 background and 80 signal events. Using a 10% momentum resolution for the muon spectrometer then selecting on the Z mass peak they were left with 60 signal and 100 background. The invariant mass of the ZZ pair from the background peaks at low mass. If one looks in the mass range 350 to 450 GeV then one is left with 50 signal and 15 background. They make the point that if the momentum resolution is 20% then the signal to background goes from 10:3 to 5:6. At a higher Higgs mass then the Higgs will appear as an enhancment at the high mass range. It therefore becomes even more critical to suppress the background when selecting the Z. From the work of G.Herten [6] we show in figure 5 the Z peak from the decay of a 1 TeV Higgs with a background from heavy top decay for the two cases of momentum resolution 2% and 10%.

4.   CONCLUSION

From hermeticity arguments it appears that a moderate muon spectrometer with a 10% momentum resolution and an angular coverage of $\theta > 5^0$ is sufficient. However if we want to use the leptons in order to search for evidence of new particles, then it is clear that the angle coverage should be extended as far as possible

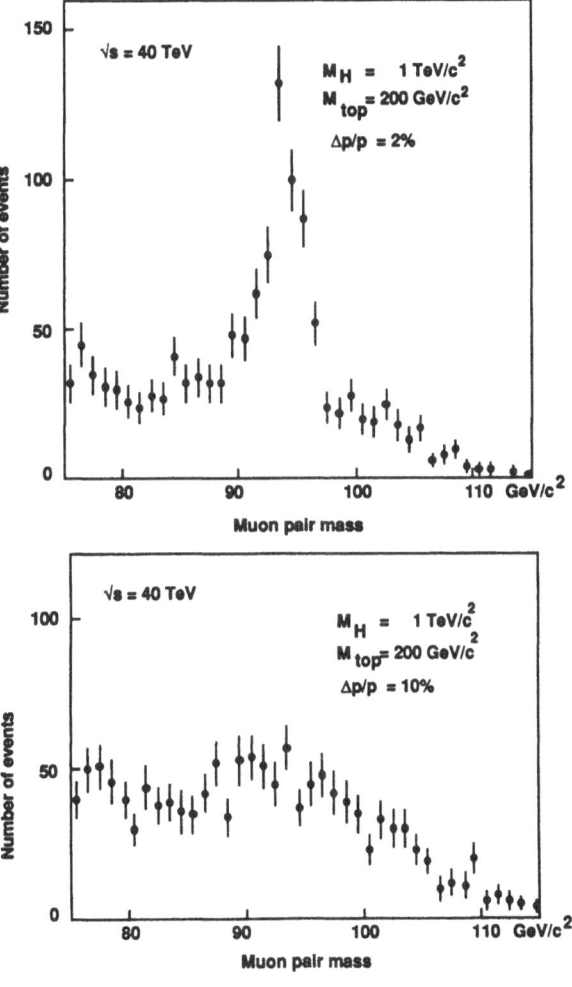

Figure 5

in the forward direction. Also in the forward direction the muon momentum is highest so that the analysing power of the spectrometer should be good above 1 TeV/c. As it will be very difficult to measure the sign of the electron in the forward direction, asymmetry measurements can only be done with the muons.

If a central magnetic field exists we can measure the muon momenta for isolated muons in the central region. In this case the barrel part of the muon spectrometer need not have such a good momentum resolution.

Not only does the muon spectrometer have to identify and measure muons off-line but it has to be capable of triggering on the muons. This in itself is a challenge to the designer of the muon system.

REFERENCES

1) Report of the Non-Magnetic Detector group, T. Akesson et al. SLAC-PUB-4457 (1987) and in Proceedings of the Workshop on Experiments, Detectors and Experimental Areas for the Supercollider, Berkeley, California (1987).

2) V.Barger, N.Deshpande, J.Rosner, and K.Whisnant, Univ. of Wiscousin Preprint MAD/PH/299 (1986).

3) D.London and J.Rosner, Phys. Rev. D34, p. 1530 (1986).

4) J.L. Rosner, "New Contributions to Forward Backward asymmetries in Hadronic Lepton Pair Production", Summer Study on Physics of the Superconducting Supercollider, Snowmass (1986).

5) D.Carlsmith et al. "Leptonic Angular Acceptance", Summer Study on Physics of the Superconducting Supercollider, Snowmass (1986).

6) G.Herten, "Heavy Top Quark Production as a Background in the Search for the Higgs Boson", Proceedings of the Workshop on Experiments, Detectors and Experimental Areas for the Supercollider, Berkeley, California (1987).

SCINTILLATING FIBRES FOR CENTRAL TRACKING

OF CHARGED PARTICLES(*)

C. D'Ambrosio[1], P. Destruel[4], U. Gensch[2], H. Güsten[3], H. Leutz[1] (speaker),
D. Puertolas[4], S. Schlenstedt[2], T. Shimizu[5], O. Shinji[5], S. Tailhardat[1] and
M. Taufer[1]

1 CERN, European Organization for Nuclear Research, Geneva, Switzerland
2 Institut für Hochenergiephysik, Zeuthen, Germany
3 Kernforschungszentrum Karlsruhe, Institut für Radiochemie, Karlsruhe, Germany
4 Laboratoire de Génie Electrique, assoc. au CNRS (UA304), Toulouse, France
5 Kuraray Co. Ltd., Tokyo, Japan

ABSTRACT

A scenario for central tracking at future hadron colliders based on small diameter scintillating fibres is outlined. Assemblies of these fibres within fused multibundles were produced and test results with a small prototype of a tracker shell are reported: spatial resolution of 35 μm and two-track resolution of 83 μm were achieved. The importance of local light emission in these fibres is discussed together with the results obtained from newly synthesized scintillators of large Stokes shifts. Monte Carlo simulations of top, Higgs, and minimum bias events show that tracking with fine-grain scintillating fibres is well feasible and occupancy at 10^{34}cm^{-2}s^{-1} luminosity does not present problems.

INTRODUCTION

In comprehensive reviews [1, 2] on central particle tracking within the free barrel-space of a calorimeter, it was pointed out that general-purpose detectors at future hadron colliders would be incomplete without tracking. The main issues, why central tracking is vital, are:

- improvement of momentum resolution for high p_t tracks including muon tracking in front of the calorimeter;

- location of primary vertices to reject uninteresting beam collisions, which provide minimum bias events only;

- capability of secondary vertex tagging by linking reconstructed tracks with the coordinates provided by a vertex detector;

- provision of topological trigger decisions to reject minimum bias events.

(*) The work reported here is part of the LAA project.

Mac/PPE/8005R/CA/sj

To achieve these goals, a central particle tracker must provide:

- fine granularity to achieve enough spatial precision and two-track resolutions, as well as to avoid occupancy problems;
- long track lengths for sufficient momentum resolution and high angular resolution for vertex location;
- low-gamma conversion to avoid disturbing high p_t electron production;
- fast response time to meet the drastic timing constraints of future colliders.

There exist three basic techniques to realize central particle tracking: gaseous chambers, silicon detectors, and scintillating fibres. Some parameters of them are compared in table 1. We have decided to apply scintillating fibres for central tracking, which have inherent advantages, provided we can meet the following conditions:

- production of small diameter fibres in fused coherent multibundles;
- reduction of light losses due to bulk absorption and multiple reflections at the core-cladding interface;
- prevention of cross talk to neighbouring fibres;
- delay of track patterns to enable coincidences with the first-level trigger.

Table 1 Comparison of relevant parameters for the three-tracking techniques

Parameters	Silicon	Fibres	Gas
Price kSF m^{-2} (hit thickness)$^{-1}$	150–200	8	~ 8
Number of hits $(0.1\ X_0)^{-1}$	15–20[a]	42–85[b]	> 400[c]
Occupancy	No problem	No problem	R = 50 cm 12% R = 20 cm ⩾ 100%
Signal resp. time ns	5–10	2.5–5	50–150
Spatial precision µm	5–10[d]	35[d]	60–120
2 track resolution µm	25[d]	80[d]	500–1000
Number of readout channels m^{-2} (hit thickness)$^{-1}$	10^6–10^{7}[e]	10–20[f]	300–500
Heat prod. kW m^{-2} (hit thickness)$^{-1}$	0.6–1[g]	0.01–0.02	SMALL

(a) X_0^{Si} = 9.36 cm; for 1 hit 0.3 mm (0.2) detector + 0.3 mm electronics = 0.6 (0.5) mm Si(*).

(b) X_0^{PS} = 42.4 cm; for 1 hit 1 mm (0.5)(*).

(c) Depending on gas fill: Xe at 2 bar X_0 = 700 cm; other gases negligible; cell size 4 mm with 50% Xe ~ 380 hits/0.1 X_0(*).

(d) Strip width 25 µm; fibre Ø 30 µm.

(e) One silicon layer; pitch 25 µm; strip length 4 mm or 10 mm.

(f) PS-layer thickness 1 mm; CCD units 10 mm × 10 mm; 2.5 × 10^5 pixels/unit.

(g) Assuming 0.6 mW × mm^{-2} for pads (3 mm × 3 mm) and 1 mW × mm^{-2} for pixels (200 µm × 200 µm).

(*) X_0 means radiation length.

DETECTOR LAY-OUT

Several scenarios for central tracking with scintillating fibres are imaginable, one of which is presented in fig. 1. The beam pipe of the collider is surrounded by four concentric cylindrical shells. Each shell is composed of four coherent scintillating fibre layers arranged in z–u–v–z directions to provide the radius R from the beam and the polar angle ϕ (via the z fibres) and the z coordinates via the stereo angle α between the u–v pair. Table 2 shows the relevant dimensions of the four shells and the precision of the z coordinate achievable with the indicated stereo angle α.

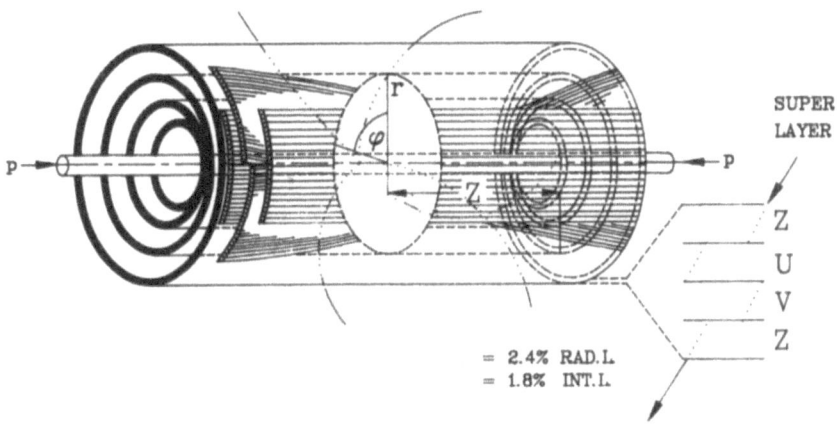

Fig. 1 Scenario for central tracking with scintillating fibres. Each of the four concentric cylindrical shells is composed of four coherent scintillating fibre layers. The two z layers provide radius R and polar angle ϕ. The z coordinates are derived from the u–v pair with its stereo angle α.

Table 2 Detector parameters, with α meaning the stereo angle, Φ_f the fibre diameter and δ_z the z-precision

Radius in centimetres	α	Φ_f, µm	δ_z, µm
25	8	30	216
50	8	60	431
75	8	90	647
100	8	120	862

Since fibres with diameters below ~ 200 µm cannot be manipulated individually (diameter of human hair is between 40 µm and 100 µm), we arrange them as square multibundles of edge dimensions between 1 mm and 2.5 mm. Depending on its edge dimensions and the individual fibre diameter, such a bundle contains hundreds of hexagonal fibres separated from each other by 5 µm of cladding material (fig. 2). The digital readout of the scintillating signal is provided by an opto-electronic tube, which is presented in the contribution of Gys et al. [3].

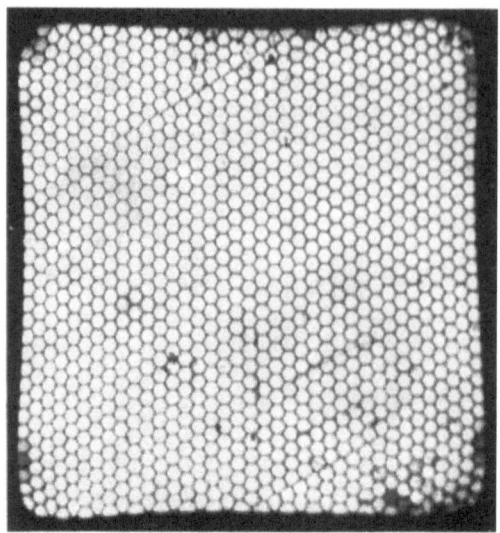

Fig. 2 Cross section of a fused square multibundle with 1 mm edge length. It contains ~ 900 individual fibres of 30 μm diameter.

NEW DOPANTS WITH LARGE STOKES SHIFTS

In order to achieve enough hit densities for the tracker and to avoid cross talk to neighbouring fibres we developed new scintillators. The difference to the generally-adopted dopants is illustrated in fig. 3: the energy loss of an ionizing particle is transferred to the basic matrix of the fibre core, in most cases polystyrene (PS). Since the quantum yield of PS is rather poor, it must be enhanced by adding a scintillator, in most cases p-terphenyl. The energy transfer between the basic matrix and the added scintillator occurs, depending on the scintillator concentration, in two competitive ways: non-radiative (Förster transitions) or radiative via photon exchange (low concentration). For both mechanisms an overlap of the respective absorption and emission bands is necessary (fig. 3).

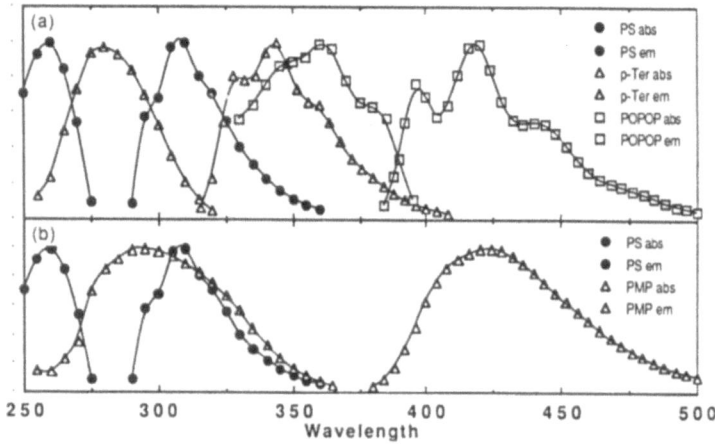

Fig. 3 Absorption and emission bands of polystyrene doped with: (a) p-terphenyl and POPOP (wavelength shifter), and (b) PMP. Absorption coefficients and emission intensities are not to scale.

Since p-terphenyl emits at around 340 nm, normally a second dopant must be added, whose emission matches the transparent region of the basic matrix and the sensitivity of suitable photocathodes. Since absorption and emission bands of such wavelength shifters overlap considerably (fig. 3(a)), their concentrations must be kept low to achieve a reasonable light transmission through the scintillating fibre. We have therefore radiative energy transfer between scintillator (p-terphenyl) and wave shifter (POPOP in fig. 3(a)). The absorption length of the wave shifter is of the order of 300–500 μm.

In applying PMP as a scintillator (fig. 3(b)), the addition of a wave shifter is not necessary. Due to its large Stokes shift, the PMP absorption band overlaps conveniently with the PS emission band and the PMP emission peaks at 420 nm. According to the small overlap between absorption and emission, we can apply PMP concentrations high enough to ensure non-radiative (Förster)-transitions between PS and PMP. Therefore, no primary light escapes even from very small diameter fibres. This can be seen from fig. 4, which shows the light emissions of two fibre bundles, one with PMP and the other with traditional doping; both bundles are excited in the same way by a Nd: YAG laser at 265 nm. Whereas the PMP bundle emits only from the first excited fibre layer, the bundle with a wave length shifter shows clearly the cross talk of primary light over several fibre layers.

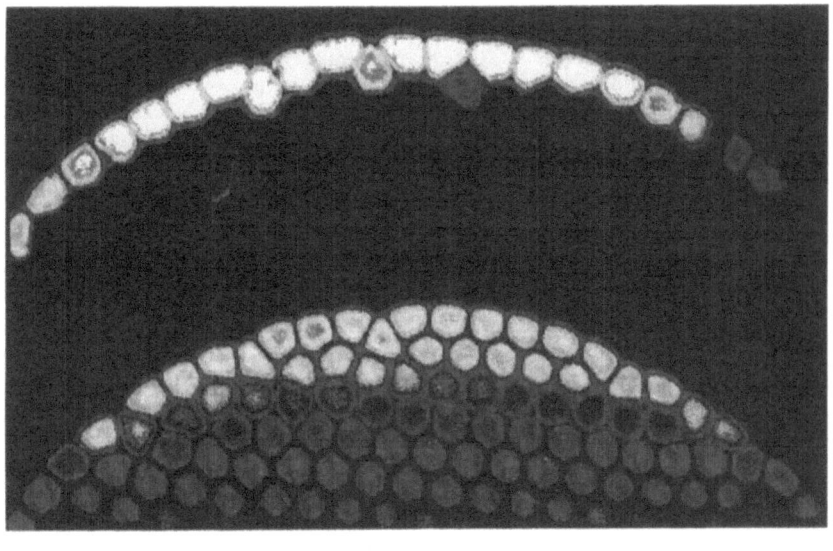

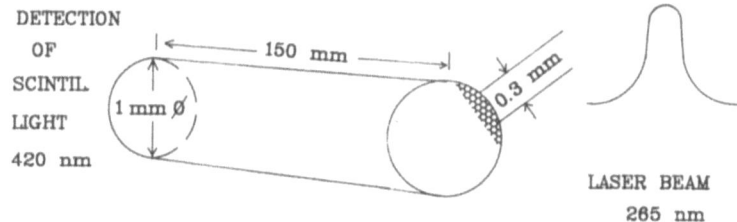

Fig. 4 Excitation of differently-doped multibundles by a Nd: YAG laser at 265 nm (PS absorption). The PMP-doped bundle (up) emits only from the first directly excited layer. The bundle with a wave shifter (down) shows cross talk over several fibre layers.

Looking for further improvements, like reduction of self-absorption, we synthesized four new scintillators on the basis of PMP, which we call now PMP 420: 1-phenyl–3-mesityl–2-pyrazoline. The new compounds were characterized by their fluorescence spectra: 1-p-tolyl–3-mesityl–2-pyrazoline, mp 105–106°C (PMP 430); 1-p-tolyl–3-(2', 6'-dimethoxy-phenyl)–2-pyrazoline, mp 182–183°C (PMP 440); 1-p-anisyl–3-mesityl–2-pyrazoline, mp 88–89°C (PMP 450); 1-p-anisyl–3-(2', 6'-dimethoxyphenyl)–2-pyrazoline, mp 187–188°C (PMP 460).

Polystyrene bulk samples doped with the different PMP compounds at 0.025 mole ℓ^{-1} concentration were produced at Toulouse[*]. For comparison, a sample containing 3-HF (3-hydroxyflavone) was also manufactured under equivalent conditions. With these samples we measured the absorption and emission spectra, the scintillating yield and the decadic extinction coefficients. The results are given in table 3 and in figs 5 and 6.

We also measured the self-absorption by exciting the scintillations at a wavelength of 270 nm (within the PS-absorption band) at the surfaces of the samples. Introduction of equivalent samples with different lengths between the surface emission and the spectral analyzer of a spectro-fluorimeter resulted in different emission spectra depending on the lengths of the inserted sample filters. The results of this procedure are shown in fig. 7. Clearly, the self-absorption of POPOP dissolved at 2.5×10^{-3} mole ℓ^{-1} in toluene is considerably higher (fig. 7(a)) than that of PMP 420 (fig. 7(b)) and in particular that of PMP 450 (fig. 7(c)).

Table 3 Spectroscopic properties of the investigated scintillators at 0.025 mole ℓ^{-1} concentration in PS.

Scintillator 0.025 m/ℓ	Absorption maximum	Emission maximum	Stokes shift		Extinction coefficient maximum	Scintillation yield
in PS	[nm]	[nm]	[cm^{-1}]	[eV]	ℓ/mole.cm	(%)
PMP 420	302	414	8960	1.11	10100	100
PMP 430	302	420	9300	1.15	9630	102
PMP 440	307	428	9210	1.14	10200	95.8
PMP 450	304	436	9960	1.24	10150	83.5
PMP 460	311	440	9430	1.17	10140	78.7
3 HF	343	528	10215	1.27	13200	27.2

This favourable difference was achieved, although the POPOP concentration was 10 times lower than that of PMP 420 and PMP 450. Moreover, POPOP was dissolved in toluene, and liquid solutions exhibit smaller light extinction than solid PS samples (no Rayleigh-scattering contribution). In view of the low self-absorption of PMP 450 we have now launched a project to produce scintillating fibre bundles with this dopant.

(*) Laboratoire de Génie Electrique, assoc. au CNRS, 31062 France.

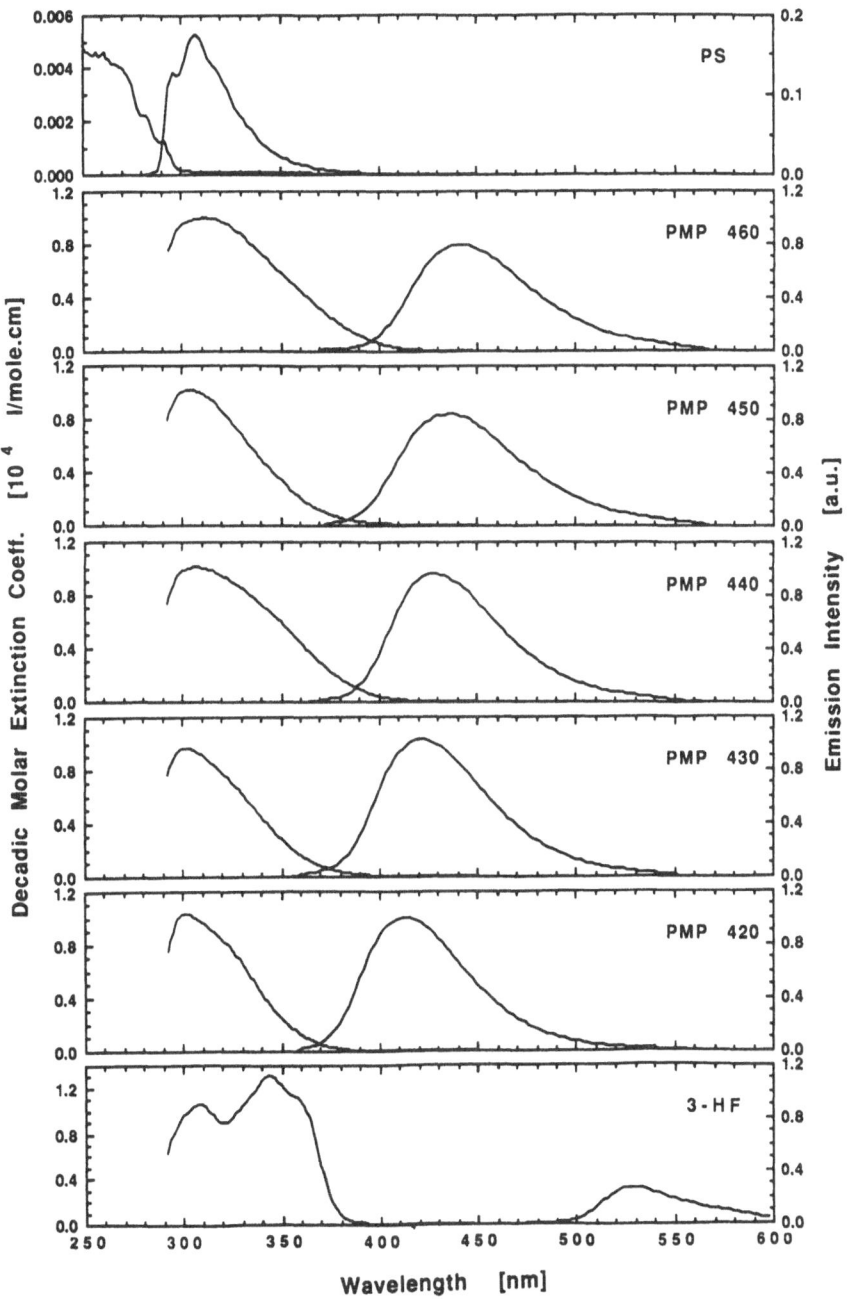

Fig. 5 Measured absorption and emission bands of pure polystyrene (PS), the five PMPs and 3-hydroxyflavone (3-HF).

179

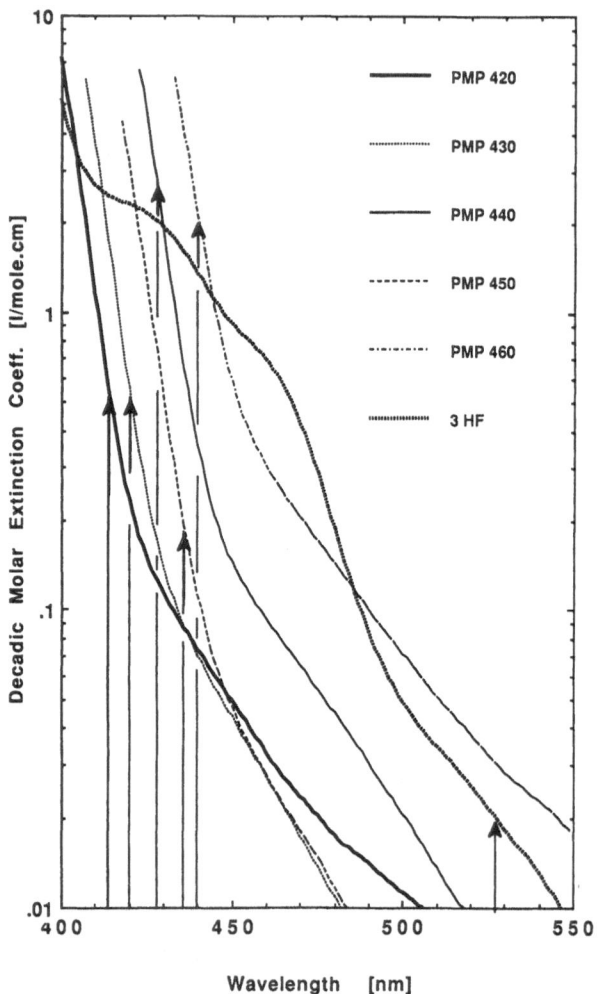

Fig. 6 Decadic molar extinction coefficients within the overlap regions of the absorption and emission bands of the different dopants. The maxima of the emission wavelengths are indicated for the scintillators by vertical arrows.

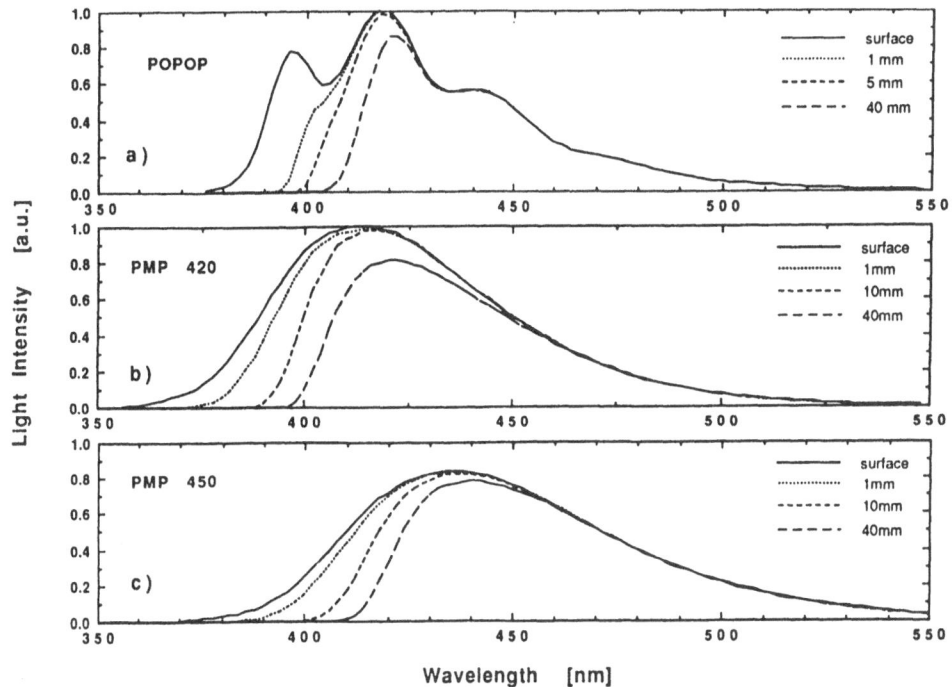

Fig. 7 Distortions of the emission spectra caused by different absorption paths of the sample:
(a) POPOP solution of 2.5×10^{-3} mole ℓ^{-1} in toluene,
(b) PMP 420 at 2.5×10^{-2} mole ℓ^{-1} concentration in PS,
(c) PMP 450 at 2.5×10^{-2} mole ℓ^{-1} concentration in PS.

MONTE CARLO SIMULATIONS FOR TRACKING

These simulations were performed for a central tracker consisting of four shells (fig. 1). Shell radii, fibre diameters and stereo angles are indicated in table 2. The tracker covers the central part of the kinematic range of the pp interactions and its acceptance is confined to $|\eta| < 1.7$ at $\sqrt{s} = 16$ TeV, with η meaning the pseudo-rapidity. Each tracker shell is composed of four superlayers with 2.5 mm thickness. The presence of two spaced z layers in each shell, which determine the polar angle ϕ, eases the pattern recognition. The z layers are interspaced by the u and v layers, which are mounted with a stereo angle α. These layers provide symmetric hits with respect to the z hits. The distances β at the readout end of the fibre shells provide the z coordinates of the tracker

$$Z = \beta R / \tan \alpha,$$

where α is the stereo angle and R the radius of the shell, supposing that the readouts are mounted at the centre of each shell, i.e. two readouts per shell.

The actual size of the stereo angle α must be chosen with care. The precision Δz of the z measurement ($\Delta z \sim (\sin \alpha)^{-1}$) has to be balanced against the ambiguities of u–z–v stereo triplets.

The detector performance was studied with pp interactions at $\sqrt{s} = 16$ TeV generated with PYTHIA 5.3[(*)]. Three event classes were considered: top production ($m_{top} = 90$ GeV/c), Higgs production ($m_{Higgs} = 200$ GeV/c), and minimum bias interactions to enable comparisons with the always present background. These generated events were tracked through the four shells with GEANT [4]. Secondary interactions, gamma conversions, δ rays, multiple scattering, bremsstrahlung and energy loss were taken into account.

[(*)] The event samples used were generated by F. Anselmo, member of the LAA-SMC group

In table 4 we present for the three-event classes the total charged multiplicities ($\langle n \rangle$), the multiplicities within the detector acceptance ($\langle n_{acc} \rangle$) and those remaining after cuts at 1 GeV/c and 2 GeV/c transverse momentum. We have also listed the fibre occupancies for shell 1 (R = 25 cm) and shell 4 (R = 100 cm).

Table 4 Events characteristics $\langle n \rangle$ = number of charged particles. Acceptance: 1.7 rapidity.

Events	$\langle n \rangle$	$\langle n \rangle_{acc}$	$\langle n(p_t) 1\ GeV \rangle\rangle_{acc}$	$\langle n(p_t) 2\ GeV \rangle\rangle_{acc}$
Higgs	176	62	24.4	13.4
Top	182	68	27.5	15.8
Min. bias	75	15	1.4	0.3

These occupancies have been computed for pure top events, top events overlayed with 10 or 20 minimum-bias events, and one minimum-bias event only. To study the influence of looping low p_t-momentum tracks, the occupancies have been calculated for zero, 1T and 3T magnetic induction parallel to the beam tube. On the average, we expect 15 minimum-bias events per bunch crossing at $10^{34} cm^{-2} s^{-1}$ luminosity. The occupancies are rather small and will therefore not present a problem for a fibre detector of the given dimensions and granularities. From table 5 it can be clearly seen that the loopers are compressed to the inner shells with increasing magnetic induction.

Table 5 Fibre occupancy (%) for zero, 1 or 3 T magnetic induction B.

B, T	Shell	Top	Top + 10 min. bias	Top = 20 min. bias	Min. bias
0	1	0.20	0.61	1.01	0.049
	4	0.45	1.38	2.46	0.099
1	1	0.18	0.55	0.98	0.055
	4	0.15	0.46	0.80	0.057
3	1	0.22	0.79	1.47	0.065
	4	0.10	0.26	0.42	0.039

In future, a simple algorithm for pattern recognition will be developed on the basis of track simulations. We also investigate the possibility, to use the information of a fibre tracker for a fast trigger decision. From table 4 we learn that interesting events (top or Higgs) of $p_t > 2$ GeV/c produce considerably more charged particles (13.4; 15.8) than 15 minimum-bias events (4.5). Accepting only interactions with a certain threshold of high p_t events, rejects the vast majority of uninteresting minimum bias events. With a fast readout scenario, as discussed in the contribution of Gys et al. [3], it might be possible to select interesting events efficiently from the minimum-bias background.

BEAM TESTS WITH PROTOTYPES

The performance of a fibre tracker has been demonstrated with fibre-shell prototypes. For these tests, fibre bundles of 1 mm × 1 mm cross section and with 30 µm diameter of individual fibres have been assembled (fig. 8(a)) to simulate a 10 mm thick tracker shell over a width of 5 mm and a length of 200 mm. They were exposed to a 20 GeV π^- beam at the CERN PS. Figure 8(b) shows a resulting track at a distance of ~ 150 mm from the bundle end facing the first photocathode of the opto-electronic chain. Figure 9 shows a histogram of the residuals of 272 tracks with 83 mm FWHM (two-track resolution) and a sigma (precision) of 35 mm.

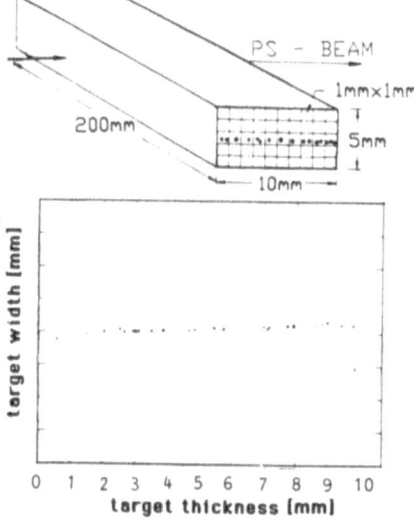

Fig. 8 Beam exposure of a small prototype for a tracker shell (a) consisting of 50 fused multibundles which contain 30 μm diameter fibres. The track (b) shows 24 hits along 10 mm.

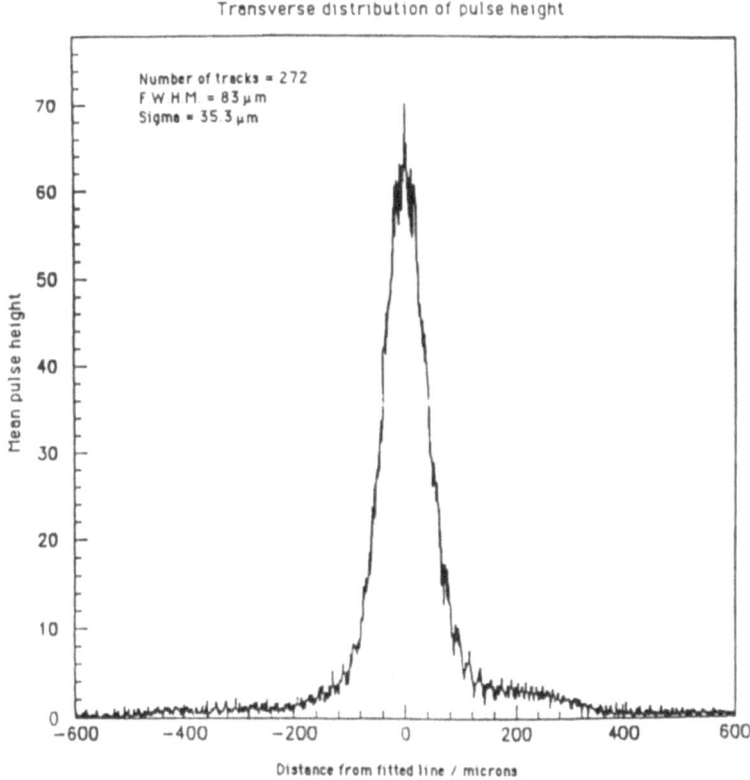

Transverse distribution of pulse height

Number of tracks = 272
F.W.H.M. = 83 μm
Sigma = 35.3 μm

Mean pulse height

Distance from fitted line / microns

Fig. 9 Histogramme of track residuals obtained with the exposure described in fig. 8.

CONCLUDING REMARKS

In terms of price, signal response time, spatial precision and number of readout channels, scintillating fibres provide a reasonable compromise to other tracking techniques, i.e. gas chambers and silicon detectors. Due to the technique of fused multibundles, the handling of very small fibre diameters became feasable. According to our tests they are coherent and a precise mounting of such bundles should be manageable. With 30 μm diameter of individual fibres in multibundles arranged as a small prototype of a tracker shell, we achieved in a π^- beam a spatial precision of 35 μm and a two-track resolution of 83 μm. With polystyrene as basic matrix and PMP 420 or PMP 450 as scintillator, we obtained fibres of local light emission, low cross-talk and high light-output.

Monte Carlo simulations showed that scintillating fibres are well-suited for central tracking within a multi-purpose detector at a LHC experiment. The majority of high p_t-tracks can be measured and reconstructed by applying a simple algorithm for pattern recognition. Occupancy will cause no problems, because of the fine granularity of the proposed fibre tracker. In addition, the information provided by the tracker can be used for a fast trigger decision to reject minimum-bias events.

To further improve the basic features of scintillating fibres, we started a programme to reduce the refractive index of the fibre cladding. This can be achieved by applying fluorinated PMMA. In this way we doubled already the fraction of trapped light, which increases the hit density accordingly. However, this first encouraging result was counterbalanced by a quality decrease of the core-cladding interface, which reduces the attenuation length for light transmission. In collaboration with Kuraray Ltd, Tokyo (Japan) we will try to improve this quality. The light transmission will be further enhanced by new PMP derivatives which we synthesize at present.

Acknowledgement

We would like to thank Prof. A. Zichichi for his continuous interest and support for our work.

REFERENCES

[1] D.H. Saxon, The tasks of tracking and vertex location at future hadron colliders, ECFA Study Week, Barcelona, Spain and Yellow report CERN 89–10.

[2] G.G. Hanson, M.C. Cundy and A.P.T. Palounek, Nucl. Instr. and Meth. A289 (1990) 365.

[3] T. Gys, H. Leutz and D. Piedigrossi, Opto-electronic delay for the readout of particle tracks from scintillating fibres (these proceedings).

[4] R. Brun et al., GEANT 3, CERN Program Library (1990).

OPTO–ELECTRONIC DELAY FOR THE READOUT OF PARTICLE TRACKS FROM SCINTILLATING FIBRES[(*)]

T. Gys (speaker), H. Leutz and D. Piedigrossi

CERN, Geneva, Switzerland

ABSTRACT

Due to the high-event rate envisaged in future hadron colliders, the opto-electronic readout of a scintillating fibre tracker will need a device, able to pipeline optical images during the first-level trigger decision time. We describe in this paper the principle of a gatable opto-electronic delay line. Two prototypes are also presented, and preliminary results confirm the expected space resolution and the capability of this device to delay optical images and select them by external trigger signals. Finally, different readout scenarios are discussed.

INTRODUCTION

Within the framework of the LAA Project [1], we are developing a compact scintillating-fibre tracking detector for future hadron colliders. It will be operated in a uniform magnetic induction parallel to the beam axis. The tracker is made up of narrow-diameter scintillating fibres arranged in concentric, cylindrical shells surrounding the beam pipe. Table 1 summarizes, as an example, the detector lay-out described in the contribution of D'Ambrosio et al. [2].

The tracker covers, in this case, a total surface of 64.7 m^2 while the total fibre cross section to be readout is only 0.16 m^2. The total fibre number is 3.4 10^7. From the surface-to-X section ratio, we recognize the advantage of such a fibre arrangement in comparison to silicon strip (pixel) detectors, where the readout channel number is proportional to the surface.

(*) The work reported here is part of the LAA project.

PPE/8006R/TG/sj

Table 1

Shell radius (cm)	Fibre diameter (μm)	Packing[a] fraction (%)	Shell[b] X section (cm^2)	Fibre[c] number $\times 10^6$	Shell[d] length (cm)	Shell[e] surface (m^2)	Surface-to-[f] X section ratio $\times 10^2$
25	30	73.5	157.1	14.8	137.4	2.2	1.4
50	60	85.2	314.2	8.6	274.7	8.6	2.7
75	90	89.8	471.2	6.0	412.1	19.4	4.1
100	120	92.2	628.3	4.6	549.5	34.5	5.5

(a) Packing fraction = fibre core surface/fibre (core + cladding) surface

$$= \frac{[\text{fibre diameter}]^2}{[\text{fibre diameter} + 5 \ \mu\text{m}]^2}$$

(for hexagonal fibres separated by 5 μm cladding).

(b) Shell X section = $2\pi \times$ [shell radius] $\times$ [shell thickness]
(the shell thickness is assumed to be 10 mm).

(c) Fibre number = $\dfrac{\text{shell X section}}{\text{fibre X section}}$.

Fibre X section = $[\sqrt{3}/2] \times$ [fibre diameter + 5 μm]^2 .

(d) Shell length = $2 \times$ [shell radius]/tan 20° (20° = zenith angle ~ 1.75 rapidity).

(e) Shell surface = $2\pi \times$ shell radius $\times$ shell length = $4\pi \times$ [shell radius]^2/tan 20°.

(f) For each shell.

THE ROLE OF THE DELAY TUBE IN THE TRIGGER SEQUENCE

In future hadron colliders, the expected time needed to process and transmit the first-level trigger signals to the gates of the opto-electronic chains will be ~ 1 μs and therefore much longer than the 15 ns between two successive bunch crossings. Consequently, optical images produced by our tracker must be pipelined. For that purpose, we designed an opto-electronic device able to delay in a controlled way those images, and to select and amplify a small fraction of them [3]. Assuming a rejection capability of 10^3 for the first-level trigger, the rest of the opto-electronic chain is fed at an image rate between 10^4 and 10^5 Hz.

MAIN FEATURES OF THE DELAY TUBE

The image delay should have the following characteristics:

- a good detection efficiency,
- a delay capability to wait for the first-level trigger decision,
- a space resolution which is comparable with that of the fibres (i.e. ≤ 30 μm),
- an intrinsic time resolution (≈ 10 ns) which minimizes the mixing of successive images,
- gating facilities to select triggered events (data reduction ≈ 10^{-3}). .

THE BASIC SCHEME OF THE DELAY TUBE

The proposed delay tube is a 600 mm long, electromagnetically focused image intensifier. The tube is divided in 6 sections separated by 5 grids labelled G1 to G5 (fig. 1).

The requirements stated in the previous paragraph are fulfilled in the following way:

- the blue light coming from the scintillating fibres is detected by a bialkali-like photocathode (QE ≈ 15%, including the input fibre optic window transparency);
- the delay is ensured by *to* and *fro* drifts of very low-energy (a few electron volts) photo-electrons in the first four sections of the tube;
- the focusing of the images is ensured by the detector magnetic field, parallel to the photo-electron trajectories;
- the required time resolution is achieved by compensating the inevitable velocity spread of the photo-electrons with the corresponding drift lengths explained in the following paragraph;
- the gating is realized by blocking the photocathode before applying an ultra-short negative pulse on grid G1 (see below).

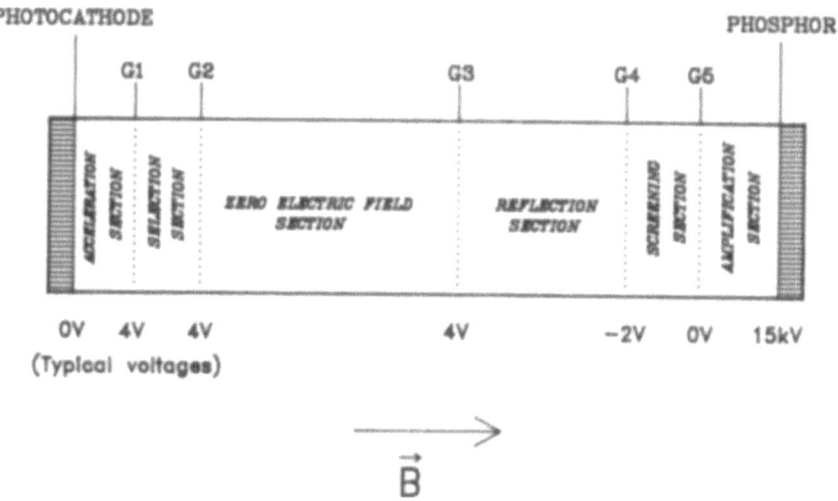

Fig. 1 Basic scheme of the delay tube.

THE BUNCHING PRINCIPLE

Electrons are emitted from a photocathode with complex angular and energetic distributions. In case of blue incident light and with bialkali photocathodes, the maximum energy of an emitted photo-electron can be as high as 1.8 eV. Without additional bunching of the electronic images, fast photo-electrons of the actual image will catch up slow ones of the previous images.

In the delay tube, the bunching is ensured by the reflex klystron principle (sect. 17 of ref. [4]). After their emission, electrons are accelerated by G1 and drift at constant velocity between G1 and G3 (G1, G2 and G3 being at the same potential of a few volts above the cathode potential). A reflector grid G4, operated below cathode potential, makes the electrons return to G1. Since the faster electrons penetrate farther into the reflection region and therefore take longer to return, all the electrons belonging to the same image finally form in bunches. The condition for two electrons of emission energy $e \times V_{01}$ and $e \times V_{02}$ (with $V_{01} < V_{02}$) to bunch within the selection section can be found with a mechanical analogy (fig. 2):

$$\frac{D}{d} = 2 \times \left(\frac{V_{01} + V_G}{V_{02} + V_G} \right)^{1/2} ;$$

where D (in metres) is the length of the selection and zero electric field sections, d (in metres) is the distance travelled by the fastest electron in the reflection section, and V_G (in volts) is the voltage applied at grids G1, G2 and G3. V_{01} and V_{02} are expressed in volts.

In this approach, the influence of the transit time in the acceleration section has been neglected. Starting from the above condition, we designed, assisted by computer simulations, a tube with a time resolution of 10 ns.

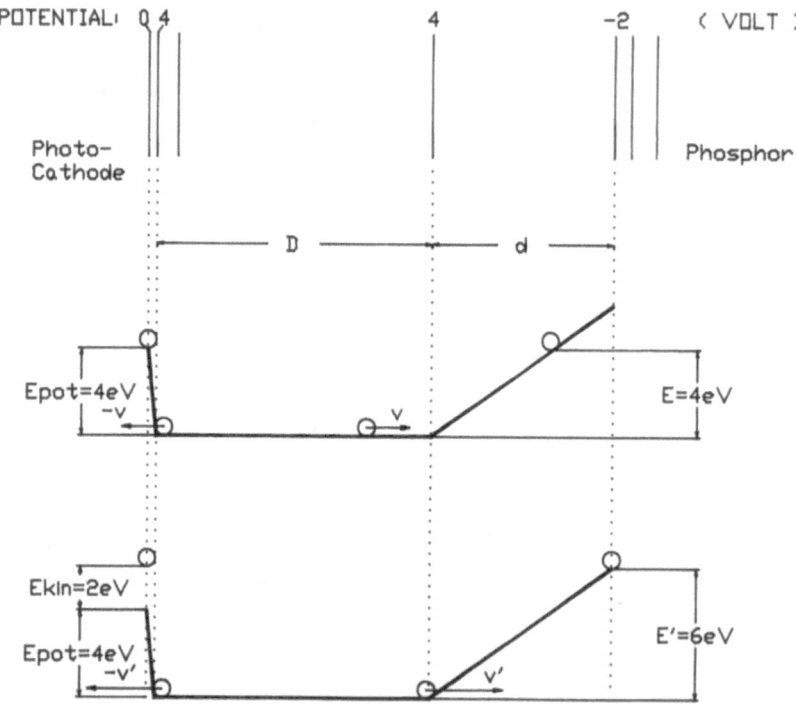

Fig. 2 The bunching principle. The particular case represented here corresponds to $V_{01} = 0V$, $V_{02} = 2V$ and $V_G = 4V$.

THE IMAGE ELIMINATION MODE

The delay tube works in "image elimination mode" as long as the first-level trigger is not active: in this case, every image arriving on the cathode must be finally discarded. After their *to* and *fro* drift, the electrons are slowed down between G1 and the cathode, and are finally absorbed by the cathode (fig. 3).

THE IMAGE SELECTION MODE

The selection process of one image, actuated by the first-level trigger, can be decomposed in three phases (figs. 4 and 5):

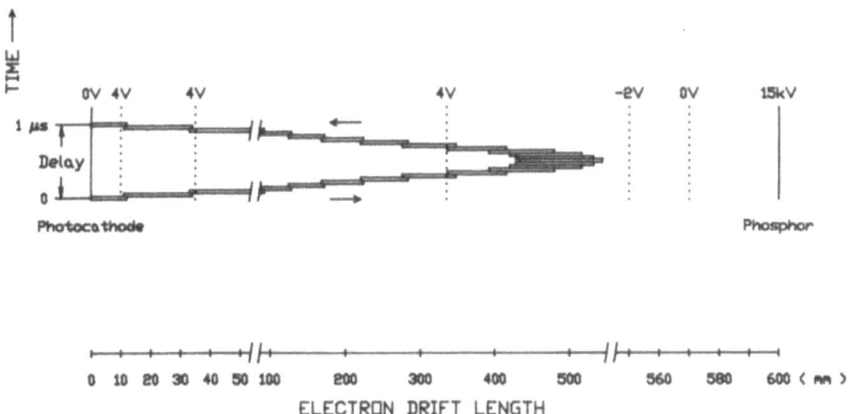

Fig. 3 Image elimination mode[(*)].

Phase 1

Just before the reflected image to be selected enters from G2 the selection zone (fig. 4), the cathode voltage is increased (from 0 V to 10 V) in order to prohibit any subsequent photo-electron from being emitted. The selection zone is thus gradually cleared from forward-drifting electrons.

Phase 2

As soon as the triggered image is completely confined between G1 and G2, a 4 ns negative (–200 V) pulse is applied on grid G1 (fig. 4). Most of the electrons inside the selection zone acquire a large velocity (20 eV) and are sent back to G3. At the same time, the photocathode is grounded, to provide again new track pictures.

Phase 3

The delay tube runs again in image elimination mode while the selected electrons due to their 20 eV energy cross the reflection and the screening zones, and are finally accelerated by the high voltage (15 kV) applied on the phosphor screen, where they produce light signals (fig. 5).

The duration of the positive voltage applied on the photocathode determines the dead time of the tube. It is of the order of 40 ns.

THE FIRST DELAY TUBE PROTOTYPE AND ITS SPACE RESOLUTION

A first, 400 mm long, simplified prototype of the delay tube (fig. 6) has been built by EEV Ltd (Chelmsford, England). It is made up of an acceleration section, a constant velocity drift space and an amplification section. The different zones are separated by two grids G1 and G2. G1, the drift space cylindrical wall (made in stainless steel) and G2 are electrically connected.

We have measured the Contrast Transfer Function (CTF) (sect. 5–6 of ref. [5]) of the tube in a 0.7 T magnetic induction. The potentials of the grids and the phosphor screen (with respect to the cathode) were 100 V and 6 kV respectively. The results of the measurements are

---

(*) The potential values given here are only typical.

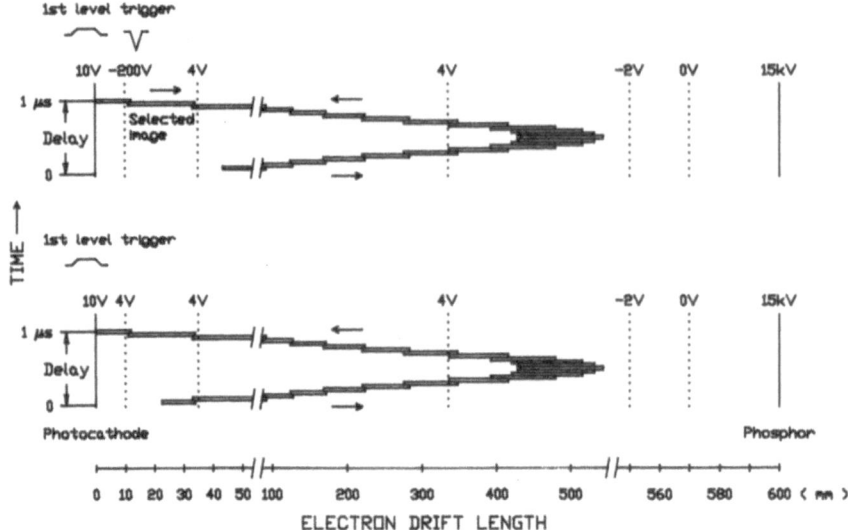

Fig. 4 Image selection mode, phases 1 and 2.

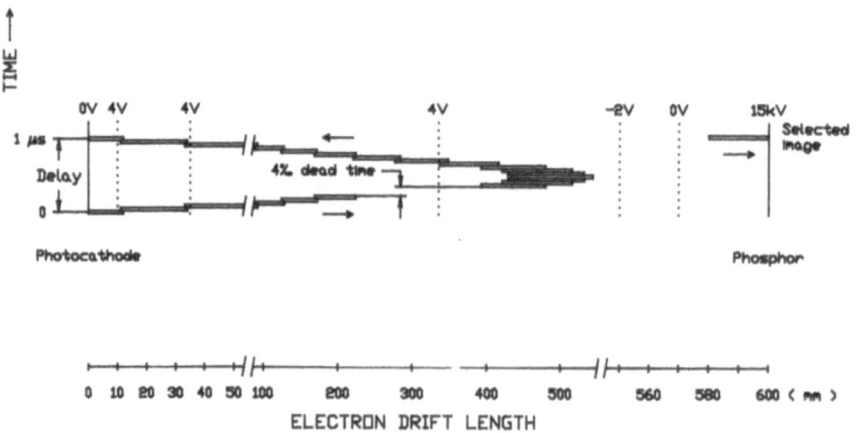

Fig. 5 Image selection mode, phase 3.Phase 3 of the image selection mode is completed while a new image elimination mode has already started.

reproduced in fig. 7. A fit with a gaussian Modulation Transfer Function (MTF) gives a limiting resolution of ~ 40 ℓp/mm (or equivalently a 25 μm FWHM point spread function). The discrepancy between the experimental points and the fit above 30 ℓp/mm is due to the fact that the resolution has been measured by a CCD camera with 20×20 μm^2 in pixel size.

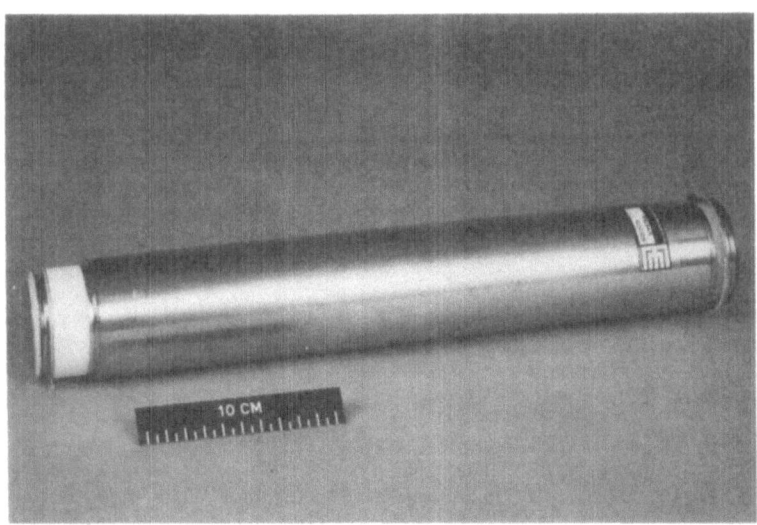

Fig. 6 A photograph of the EEV prototype.

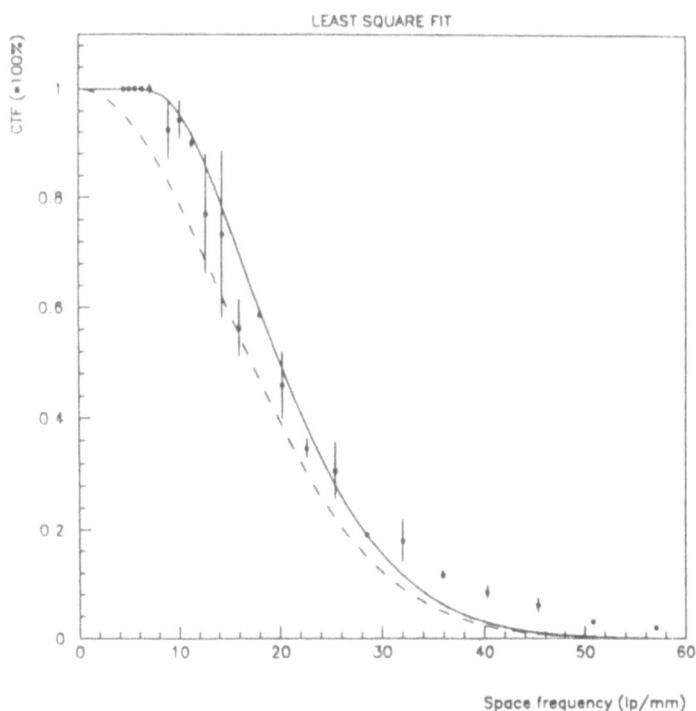

Fig. 7 Contrast Transfer Function obtained with electrons drifting 40 cm in a delay tube. The least squares fit of the measured values is indicated by the solid curve. The corresponding MTF is indicated by the dashed curve. The limiting resolution (at 0.03 contrast) is 40 line pairs/mm or 25 μm.

THE SCALE MODEL PROTOTYPE AND THE IMAGE SELECTION

A 300 mm long prototype (scale 2:1) has been manufactured by DEP B.V. (Roden, The Netherlands) and delivered at CERN in July 1990 (fig. 8). It includes all the elements schematically depicted in fig. 1. With this device, optical signals have been delayed by up to 500 ns, selected by appropriate grid pulsing, and amplified by a phosphor screen.

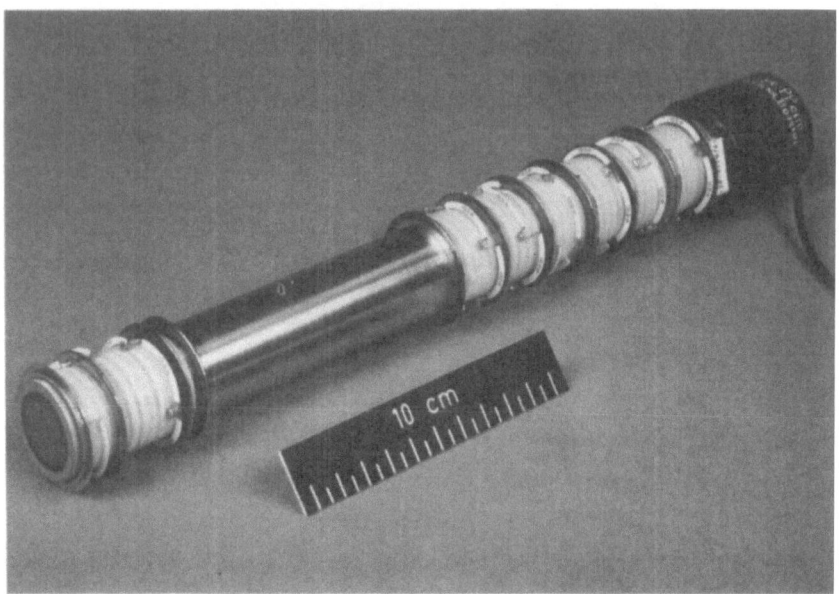

Fig. 8 A photograph of the DEP prototype.

The selection performances of the tube are illustrated in fig. 9. For timing measurements, the tube can work in a magnetic induction limited to 0.03 T. It is fed by a short light-pulse from a blue LED (fig. 9(a)). Figure 9(b) displays the output signal obtained by running all the tube grids at +7.5 V DC, the cathode being at ground, and the phosphor at +6 kV. In this mode where all photo-electrons drift through the tube without being reflected in the reflection section, a delay of only 170 ns is obtained and no image selection is possible.

If the tube works in the image elimination mode, grids G1, G2, G3 and G5 are at +7.5 V and grid G4 at –1 V so that every image is now reflected and delayed by 400 ns. The selection of an image is realized by pulsing G1 with a short negative pulse (triangular shape –200 V in amplitude, 3 ns fall time, 5 ns rise time). The resulting phosphor signal is displayed in fig. 9(c). Note that the timing values were achieved with the 1:2 scale model.

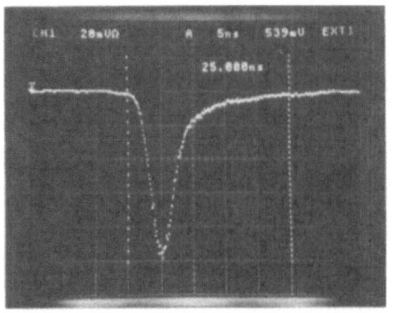

Fig. 9(a) Averaged blue LED light-pulse feeding the tube photocathode (time base: 5 ns/div., cursor separation: 25 ns).

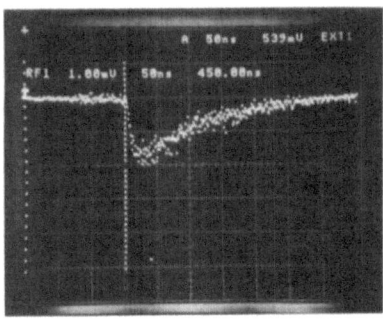

Fig. 9(b) Averaged phosphor-light signal obtained when the tube grids are at the same potential of +7.5 V. The rise time of the signal is given by the decay time constant τ of the phosphor screen (we use a P47 screen so that $\tau \sim 60$ ns (time base: 50 ns/div., right cursor position: 168 ns)).

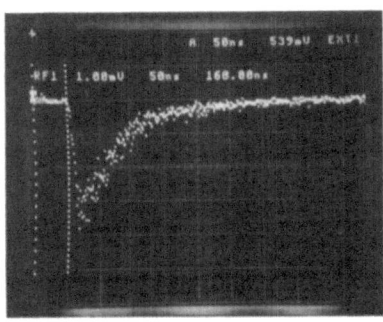

Fig. 9(c) Averaged phosphor-light signal obtained when the tube is working in image elimination mode while a negative pulse is applied on grid G1 in order to select an image (time base: 50 ns/div., right cursor position: 450 ns).

READOUT SCENARIOS

The choice of the delay tube solution for an analog pipeline has the following advantages:

– It is an imaging device by itself, and can perform parallel processing of up to 1 million microfibres. As a consequence, it yields a low cost per fibre.

– It preserves both excellent time resolution (within the bunch crossing period) and space resolution (within the microfibre diameter).

– It has a low power consumption (of the order of 1 W per tube) and thus minimizes heat production.

– It represents only a few percent of a radiation length.

– It is expected to be radiation hard.

Assuming a rejection capability of 10^3 for the first-level trigger, the mean image rate at the delay tube output is in the range of 100 kHz. The fibre occupancy has been estimated to be of the order of 1% [2] for a luminosity of $10^{34} \mathrm{cm}^{-2} \mathrm{s}^{-1}$. Each "hit" (or detected photon from the fibre) provides about 100 photons at the phosphor output (assuming a screen voltage of 15 kV, a 3 kV dead voltage caused by the aluminium layer in front of the phosphor, a luminous efficiency (in watt per watt) of 3% and 3 eV energy of the photons emitted from the phosphor). Different scenarios can be envisaged for the readout chain after the delay tube (fig. 10).

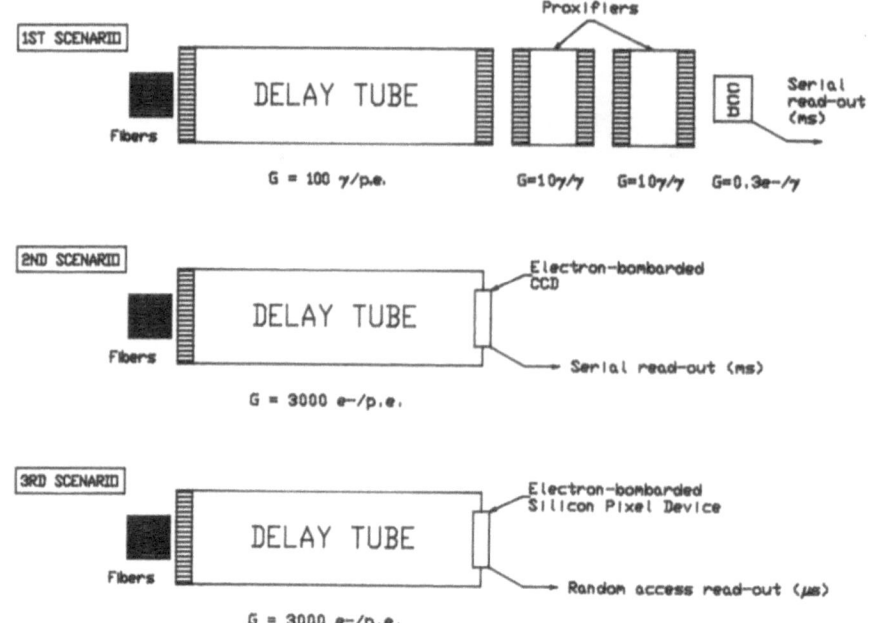

Fig. 10 Readout scenarios for the fibre tracker. The symbol G stands for the gain of the different devices.

First scenario

The light signal from the P47 phosphor of the delay tube must be amplified in order to be detected by standard CCD devices. For this amplification, the magnetic field of the tracker imposes the use of proximity-focused image intensifiers since microchannel plates do not work under these conditions. This asks for 2 proxifiers in cascade to provide the required amount of photons at the CCD input. Assuming a proxifier photon gain of 10 and a CCD quantum efficiency of 30%, the corresponding pixel signal is $3 \cdot 10^3$ electrons, which is just enough to achieve a signal-to-noise ratio of 10.

For this solution, we could use commercially available image intensifiers and CCDs. However, cascades of image intensifiers increase considerably the amount of material inside the tracker volume. Moreover, they degrade both the space (by their optical characteristics) and time (by their phosphor screen decay characteristics) resolutions of the opto-electronic chain. Finally, the CCDs serial scanning readout and their radiation sensitivity limit their use for high-luminosity hadron colliders.

Second scenario

An improvement can be realized by replacing the delay tube phosphor screen with CCDs mounted directly inside the tube. For this purpose, we need a thinned back-side CCD [EBCCD] directly bombarded with the photo-electrons of the selected images. The mean energy to create an electron-hole pair in this mode is 3.6 eV. With a photo-electron energy of 15 keV and a dead voltage (due to surface states of the CCD) of 3 kV, we expect a pixel signal of $\sim 3 \cdot 10^3$ electrons, corresponding to the previous scenario.

The main advantage of this scenario is the suppression of the proximity focused tubes and the drawbacks related to them. According to our investigations, the technique of thinned CCDs and their implementation in vacuum tubes can be mastered.

Third scenario

The problem of the CCD slow readout can be overcome by replacing the EBCCD device with a silicon pixel device working in electron-bombarded mode. One can expect the same pixel signal (3 10^3 electrons) as in the EBCCD, since the basic gain process remains unchanged. In this scenario, pixel devices can be randomly accessed and consequently, their readout speed can be increased by at least 2 or 3 orders of magnitude with respect to CCDs.

The application of this technology becomes accessible due to the development of silicon pixel devices for direct particle detection. In this context it is worthwhile noting, that the pixel surface required by our fibre tracker in this scenario is much smaller than the one needed for direct particle detection as indicated in columns 4, 7 and 8 of table 1.

CONCLUSIONS

For the readout of a scintillating fibre tracker, a novel opto-electronic delay line has been designed, providing image delay, selection and amplification. It consists of an electro-magnetically focused image intensifier working with low-energy photo-electrons. The space resolution of a first prototype has been measured, and is of the order of 25 μm for a 400 mm long tube in a 0.7 T magnetic induction. We also tested a scale model prototype. Preliminary results show encouraging performances in terms of signal delay, selection and amplification capabilities.

We are currently investigating in detail the time resolution of the device by testing the selection of a wanted image out of several consecutive images separated by 15 ns. We intend also to study the noise performance of the tube. In parallel, we are looking for the replacement of the delay tube phosphor screen by an electron-bombarded silicon detector.

Acknowledgement

We want to thank Prof. A. Zichichi for his continuous interest and support for our work.

REFERENCES

[1] A. Zichichi, The LAA Project, ICFA Instrum. Bull. 3 (1987) 17.

[2] C. D'Ambrosio et al., Scintillating fibres for central tracking of charged particles (these proceedings).

[3] J.P. Fabre et al., Conceptual design for an opto-electronic delay line, Rev. Phys. Appl. 24 (1989) 1019.

[4] K. Spangengerg, Vacuum tubes, Mc Graw–Hill (1948).

[5] I.P. Csorba, Image tubes, Sams (1985).

Gaas DETECTORS FOR PHYSICS AT THE LHC

presented by K.M.Smith

S.P.Beaumont[4],R.Bertin[2],S.D'Auria[1], C.del Papa[1], M.Edwards[6],F.Fiori[1],
J.G.Lynch[5],B.Lisowski[1],M.Nuti[3],V.O'Shea[1], P.G.Pelfer[3],C.Raine[5],
I.O.Skillicorn[5],K.M.Smith[5],R.M.Turnbull[5], and A.Zichichi[2]

1 - Dipartimento di Fisica dell'Universita' and INFN Bologna, Italy
2 - CERN, Geneva, Switzerland
3 - Dipartimento di Fisica dell'Universita' and INFN Florence, Italy
4 - Dept. of Electrical and Electronic Engineering, University of Glasgow, U.K.
5 - Dept. of Physics and Astronomy, University of Glasgow, U.K.
6 - Rutherford Appleton Lab., Chilton, Didcot, U.K.

Summary

Over the last two years the Glasgow and CERN-based LAA groups have successfully
constructed GaAs detectors for minimum ionising particles with radiation hardness and
potential speed which is more than competitive with silicon detectors. The timely development of this new technology, in particular for high radiation regions of LHC detectors near
the beam pipe and in the forward region, now requires the investment of more resources
and more intensive effort, including industrial collaboration in detector fabrication. In
addition, there is an urgent need for the development of appropriate read-out electronics
with radiation tolerance and speed characteristics to match the detectors, and with the
lowest possible power consumption, needed for the large number of channels required in
a vertex detector, for example. The present report describes the results obtained using
GaAs Schottky diode detectors which we have constructed, and the initial steps which we
have taken towards the design of a GaAs preamplifier to match the detectors.

Introduction

We first discuss the choice of semi-insulating GaAs as a detector material, the manufac-
ture of our Schottky diodes [1], and details of the response of the diodes to alpha, beta
and gamma radiation and to high energy test beam pions. Results are presented of the
effect of gamma - and neutron - irradiation on GaAs diodes and of measurements of the

New Technologies for Supercolliders, Edited by L. Cifarelli and
T. Ypsilantis, Plenum Press, New York, 1991

speed of response of the diodes, using picosecond laser pulse excitation. This status report concludes with a summary of our current understanding of GaAs diode detectors fabricated in semi-insulating material, and a discussion of the programme of work still to be done to improve our understanding of the charge transport mechanism in this material. The requirements of appropriate read-out electronics are very briefly discussed.

The Advantages of GaAs

That the time is ripe for development of GaAs detectors is supported by evidence of the radiation tolerance of integrated circuits in this material and by the improved understanding of the processes of crystal growth and wafer preparation achieved recently [2].

GaAs is a direct band-gap semiconductor, with a band-gap of 1.43 eV. The larger band-gap reduces the bulk generation current by almost four orders of magnitude compared to silicon, (with a band-gap of 1.11 eV). The radiation length is 2.3 cm, (four times shorter than silicon), but this is partly compensated by the higher specific ionisation loss of 5.6 MeV/cm, which means that for a given signal, a GaAs detector can be thinner than silicon. GaAs wafers are generally prepared by the LEC (Liquid-encapsulated Czochralski) method. Higher purity may be achieved by Molecular Beam Epitaxy, (MBE), but this is limited to layers of at most a few tens of microns in thickness by the slow rate of deposition. Vapour Phase Epitaxy (VPE) may overcome the latter limitation in the future [3]. Liquid Phase Epitaxial growth, (LPE), formerly employed for preparation of pure and doped GaAs wafers, has been almost completely superseded by MBE and VPE for most applications. Typical characteristics of a semi-insulating (undoped) GaAs wafer prepared by LEC growth are given in Table 1, together with some physical properties of the basic semiconductor material.

Several studies were made of LPE GaAs diodes as particle detectors in the early 1970's [4]. The best results at that time were rather difficult to reproduce but established that good energy resolution for X- and gamma-rays was indeed possible with small diameter, n-doped diodes. While development in LPE detectors has been inhibited by the variability of the wafer material available, recent improvements in LEC commercial crystal growth techniques, including better stoichiometric control, magnetic field melt stabilisation, post-growth annealing and stress-relieving In doping [2] have resulted in more uniform wafers of higher purity and mechanical strength, and with a greatly improved surface quality. The prospects for further improvements and for larger wafer diameters also seem bright.

GaAs devices are also of great interest because of the high electron mobility in this material, (almost six times that of silicon at best), offering the prospect of high speed particle detection and signal processing. Modern digital technologies based on GaAs have smaller gate delays and power consumption per gate than even the best silicon technologies available.

Diode Manufacture and Electrical Tests

Diodes of 500, 300 and 125 microns thickness and 1 and 3mm contact diameter were manufactured in the University of Glasgow Department of Electrical and Electronic Engineering by evaporating a Ti - Au Schottky barrier onto one side of a semi-insulating GaAs chip and an ohmic contact of Ni - Ge - Au onto the other. The sequence of evaporations for both contacts, described in Table 2, constitutes a rather reliable recipe for good contacts. No attempt was made to deposit a passivating layer on the surface of the diodes, so that the processing is relatively straightforward. The thinnest diodes are relatively fragile mechanically.

A typical diode characteristic, shown in Figure 1, reveals a reverse bias leakage current which is significantly higher than that for typical silicon detectors. Measurements on diodes with a guard ring electrode surrounding the Schottky contact suggest that the leakage is not predominantly due to surface effects, but probably more to bulk generation. Capacitance measurements are compatible with total depletion of the semi-insulating sample with almost zero bias voltage.

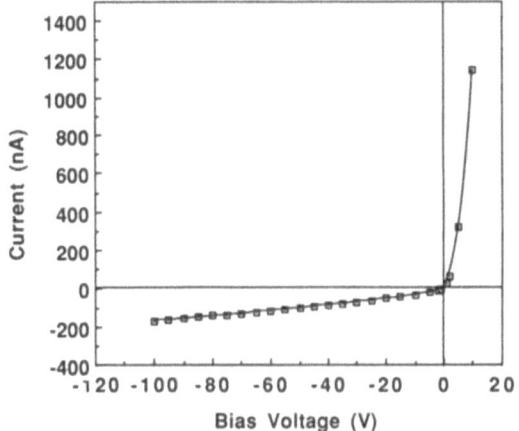

Figure 1. I-V characteristic of a typical semi-insulating substrate GaAs Schottky diode.

Tests with Radioactive Sources and in a Test Beam

Each diode was tested with alpha-, beta- and gamma ray radioactive sources. Typical spectra obtained with Am^{241}, Ru^{106} (in coincidence with pulses detected in a scintillator behind the GaAs detector), and Co^{57} are shown in Figure 2.

The charge released by each particle was known from the energy required to generate an electron-hole pair in GaAs, namely 4.2 eV. The measured charge collection efficiency increased with bias voltage as shown in Figure 3, up to a maximum of around 50% for alphas, with lower values for betas and gammas. The inefficiency is interpreted as due to very rapid trapping of the holes released by the ionising particles, the observed signal being due to the residual electrons migrating to the collecting electrode. This interpretation is supported by a comparison of the pulse height spectrum obtained from an alpha source next to the ohmic contact with that obtained with the source at the Schottky electrode. Since the alpha particle range in GaAs is only about 20 microns, the hole trapping and very short transit distance of the electrons released gave a very small signal in the former

Table 1. GaAs wafer characteristics, as supplied by the manufacturer.

Diameter (mm)	50.9
Orientation	(100)
Dopant	Nil
Wafer flatness (microns)	< 3
Wafer thickness (microns)	500 ± 25
Weight (g)	5.37
Resistivity (ohm-cm)	$7.5.10^7 - 1.2.10^8$
Mobility ($cm^2 V^{-1} s^{-1}$)	$6.8.10^3 - 6.6.10^3$
Carrier Concentration (cm^{-3})	$1.2.10^7 - 8.2.10^6$
Etch Pit Density (cm^{-2})	$2.1.10^4 - 3.0.10^4$

Table 2. Diode electrode composition. The ohmic contact metals are evaporated at a base pressure of $< 8.10^{-6}$mbar and the final annealing step at $360°$ lasts 50 - 60 seconds. The Schottky contact requires evaporation at the lowest possible pressure, typically $< 5.10^{-6}$ mbar.

Contact	Metal	Thickness
ohmic	Ni	5 nm
	Ge	25 nm
	Au	43 nm
	Ni	30 nm
	Au	50 nm
Schottky	Ti	30 nm
	Au	80 nm

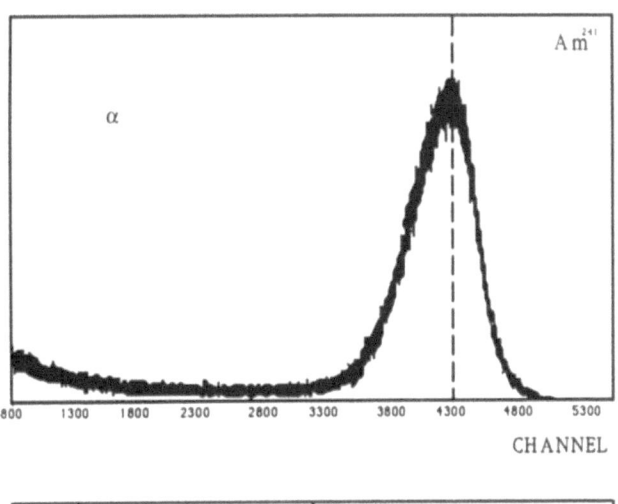

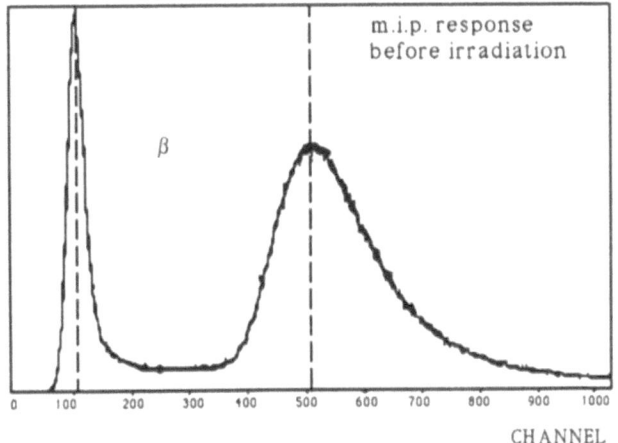

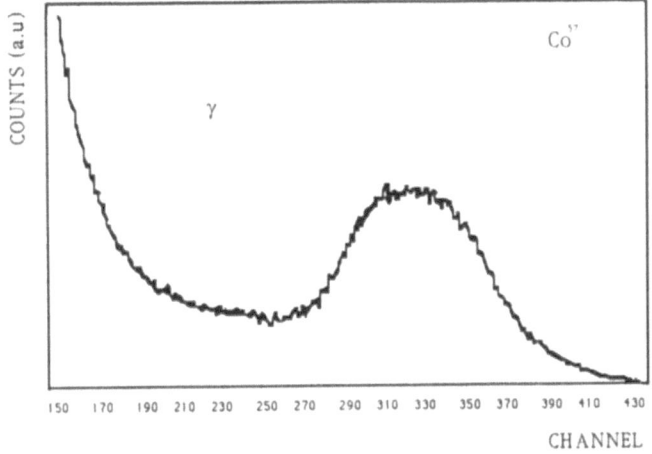

Figure 2. Pulse height spectra obtained with alpha-, beta- and gamma-ray sources in a GaAs detector.

case compared with that seen in Figure 2. Simulation of the charge trapping mechanism by a simple model gave the predicted charge collection efficiency variation also shown in Figure 3, (cf. [1], [5]).

The variation in charge collection efficiency among the diodes of the array illustrated in Figure 4 is given in Figure 5.

Figure 6 shows the pulse height spectrum obtained in a 6 GeV/c pion test beam, with a conventional scintillator telescope trigger, [1]. The efficiency for recording the passage of minimum ionising beam particles, shown in Figure 7 as a function of detector bias, indicates that the GaAs diode is essentially 100% efficient for minimum ionising particles.

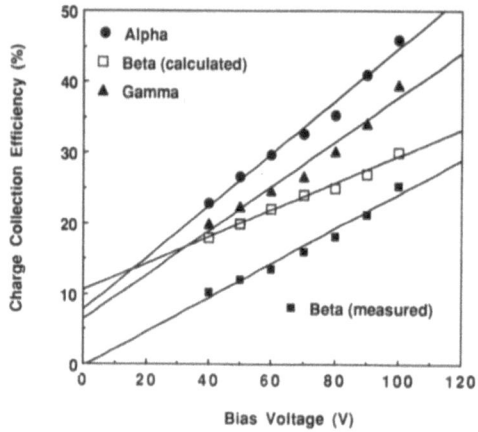

Figure 3. Charge collection efficiency variation with bias voltage, for alphas, betas and gamma rays.

Speed of Reponse

The picosecond laser facility at the Lens Laboratory of the University of Florence was used to excite a GaAs diode with a 1.5 picosecond pulse. The output signal from the diode, observed using a high speed (8 GHz) oscilloscope, is shown in Figure 8.

The very fast initial pulse is followed by a tail of around 4 nanoseconds in length, thought to be due to trapping effects in the charge tranport mechanism. A simulation of the effects of trapping, using a trap cross-section of 10^{-13} cm^2 and a de-trapping time of 245 psec gave the pulse shape, shown in Figure 9.

More detailed studies of the trapping and de-trapping process may lead to better understanding of the charge collection inefficiency discussed above.

Radiation Hardness

Several diodes, subjected to gamma irradiation up to a total dose of 17 MRad, showed only a very small change in leakage current, and the response to radioactive sources was almost completely unaffected. A set of diodes was also subjected to neutron irradiation at the R.A.L. test facility in the ISIS accelerator. Figure 10 shows the changes in leakage current, charge collection efficiency and signal to noise ratio in a number of diodes, after a total neutron fluence of 7.10^{14} n/cm^2. This exceeds the total fluence expected at a radial distance of 5 cm from the intersection point of the proposed LHC collider operating at

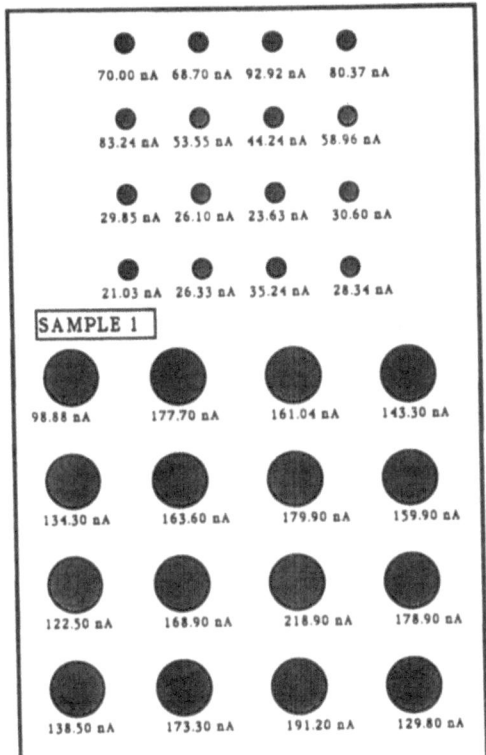

Figure 4. Diode array on a single chip, with measured leakage current at 36 V reverse bias.

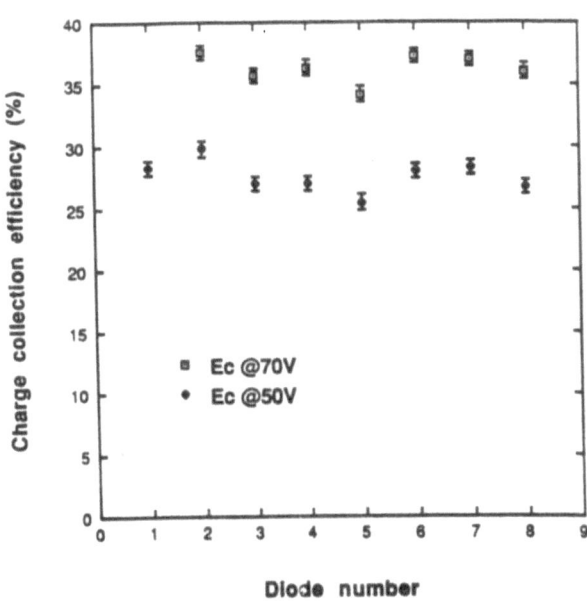

Figure 5. Charge collection efficiency variation for the diode array.

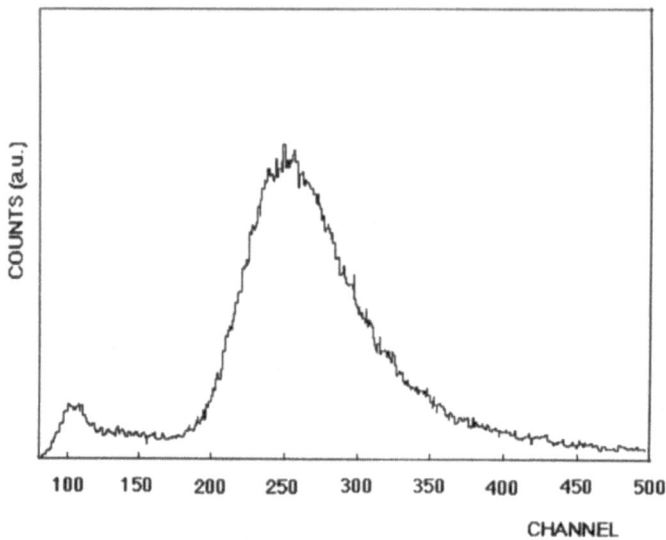

Figure 6. Pulse height spectrum obtained in a GaAs detector with 6GeV/c pions.

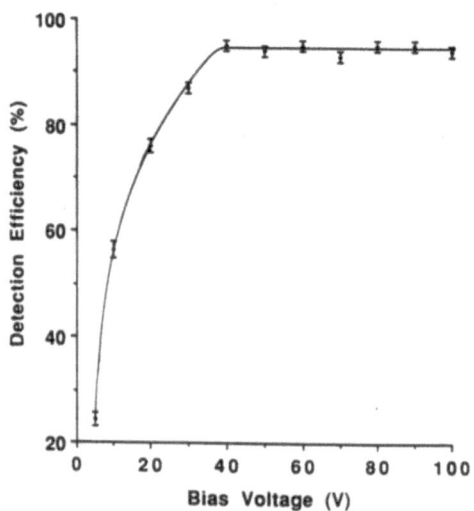

Figure 7. Detection efficiency variation with bias for 6 GeV/c pions.

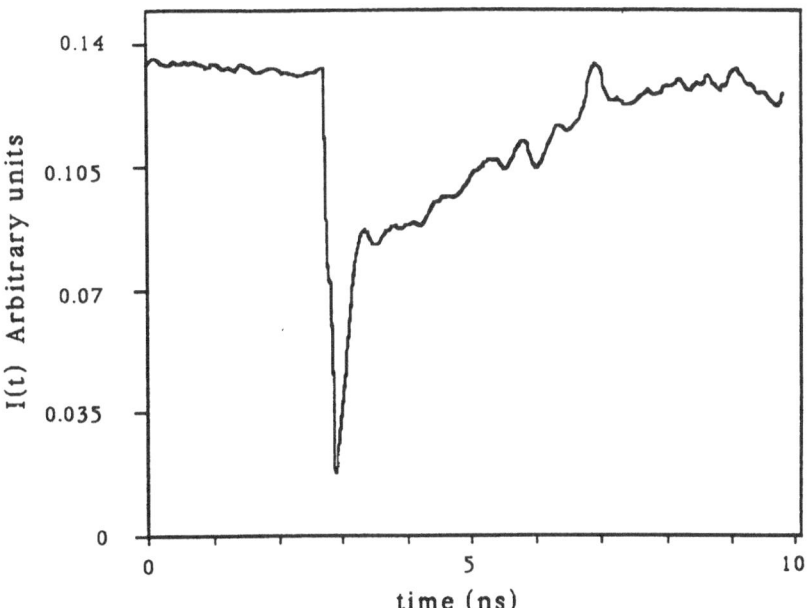

Figure 8. Oscilloscope trace of GaAs detector response to a 1.5 picosecond laser excitation pulse.

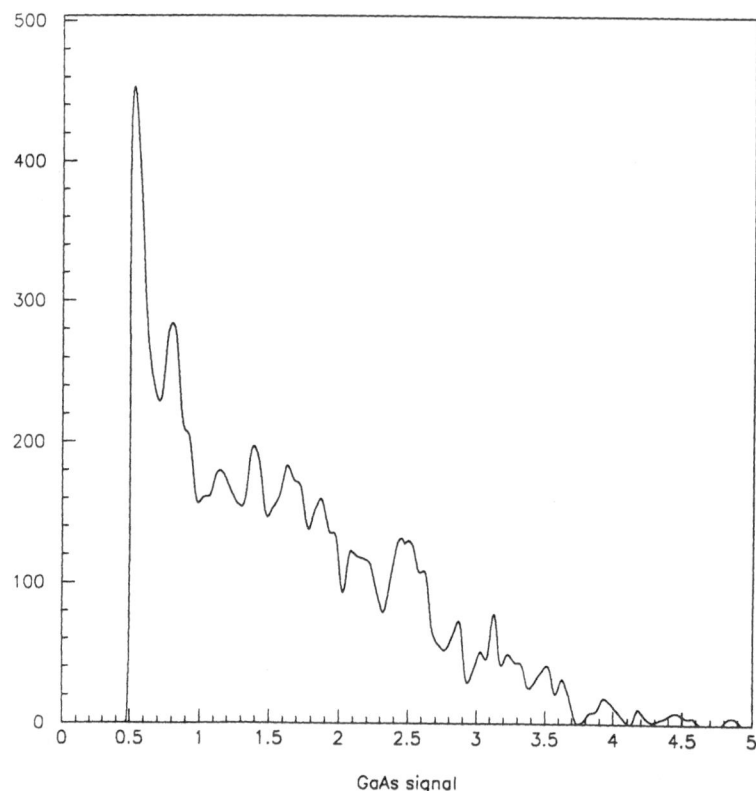

GaAs signal

Figure 9. Simulation of GaAs reponse to the picosecond excitation, using a charge transport model which incorporates a hopping mechanism between traps.

nominal luminosity for one year. (The assumed integrated luminosity for this period was taken as 10^{41} cm^{-2} s^{-1})[7].

A comparison is given in Figure 11 of the pulse height spectrum obtained with a collimated Ru-106 source, (equivalent to minimum ionising particles), before and after the neutron irradiation. The GaAs diode has clearly deteriorated, but the signals are still easily resolved, and the detector continues to be usable.

Alternatives to Semi-insulating material

We have concentrated on diodes made from semi-insulating GaAs because this is the cheapest available material, (used as a substrate in GaAs electronic device construction). Previously, LPE GaAs diodes have been used successfully for gamma ray spectroscopy. More recently, high purity V.P.E. GaAs wafers have been produced at growth rates exceeding 100 microns per hour [3]. The density of trapping centres in these LPE and VPE materials is reported to be significantly less than in semi-insulating wafers, so that the charge collection efficiency may be higher in epitaxial diodes. We hope to test samples of both these alternative materials within the next few months.

GaAs Read-out Electronics

Up to the present, very little use has been made of GaAs front-end electronics in High Energy Physics experiments. GaAs FET pre-amplifiers operating at low temperatures have recently been used in a liquid argon calorimeter read-out system, [8]. In addition, a SPICE analysis of a bipolar pre-amplifier design in GaAs predicts noise and power consumption figures which are significantly lower than corresponding values for silicon, [9]. The expected higher radiation resistance of the GaAs circuits make this a worthwhile area of study. In the longer term, the possibility exists of integrating optical read-out onto the semi-insulating wafer. This is a rapidly evolving commercial activity and could offer useful advantages in particle physics, which we shall attempt to monitor closely.

Summary of Results Obtained to Date

In summary, GaAs Schottky diodes have been shown to work satisfactorily as charged particle detectors, with essentially 100% detection efficiency for minimum ionising particles. We have established their ability to tolerate radiation loads at the level expected in more than one year of running at a radial distance of only a few cm. from the intersection point of the LHC at nominal luminosity. The diode output signal of only a few nanoseconds is very satisfactory for the new colliders. While these results are already extremely encouraging, a more comprehensive study of the charge trapping mechanisms which we have observed in semi-insulating GaAs may enable substantial improvements in the performance of diodes manufactured in this material. It is our aim to pursue this study and also to proceed to a more systematic study of the use of GaAs read-out electronics. Our plan is to design and construct, within a time scale of two years, a GaAs detector module which can be installed in a working experiment and serve as a demonstration of the viability of GaAs detectors in a real, high radiation environment.

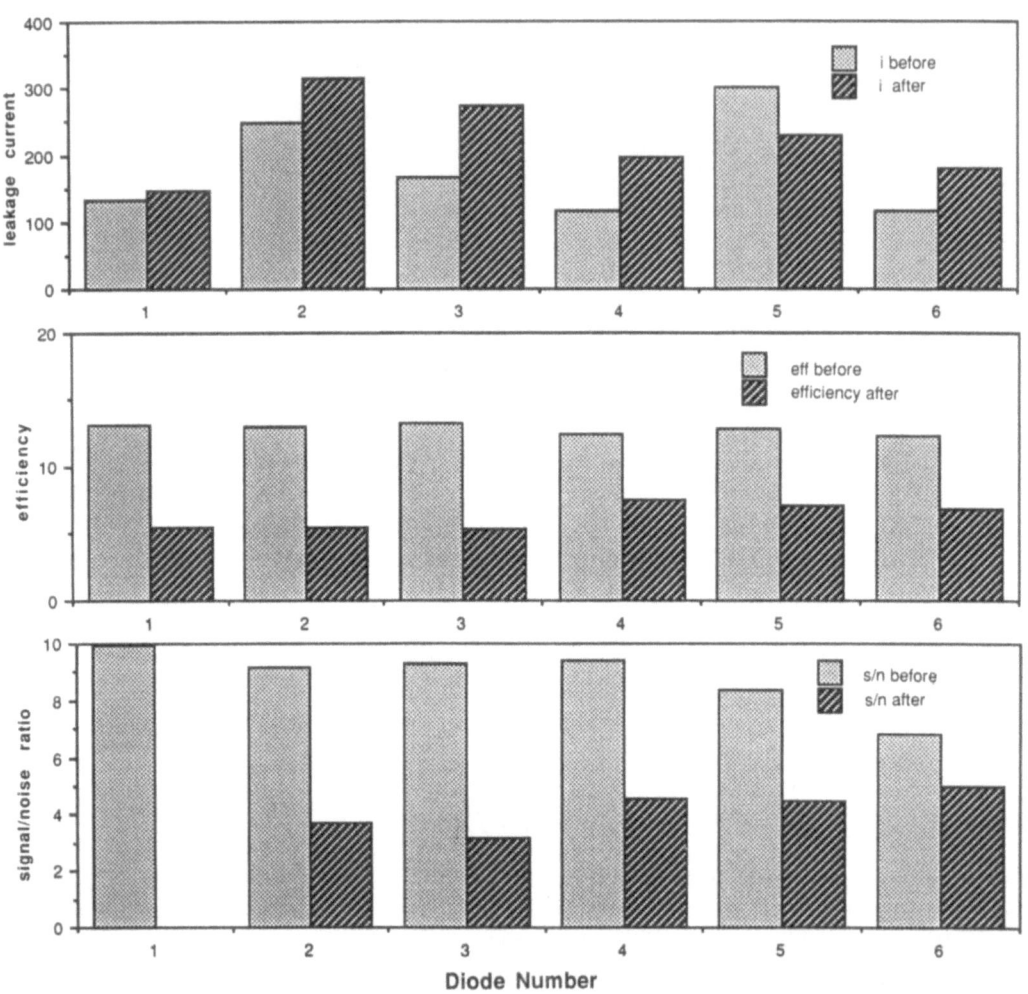

Figure 10. Comparison of reverse bias leakage current, charge collection efficiency and signal-to-noise ratio in several GaAs diodes before and after neutron irradiation by a neutron fluence of 7.10^{14}n/cm^2.

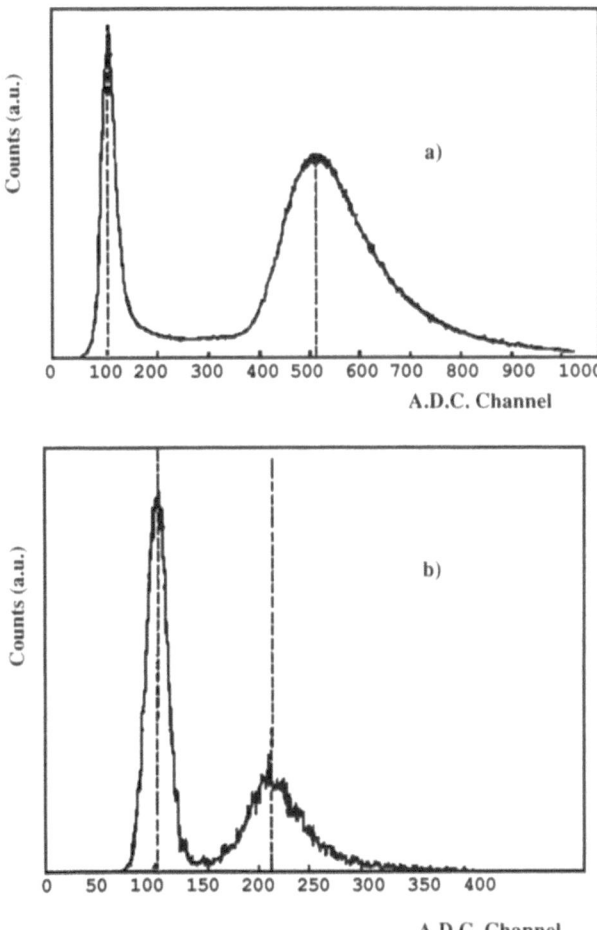

Figure 11. Pulse height spectrum from Ru-106 source, corresponding to minimum ionising particles, before (a) and after (b) irradiation of the GaAs detector.

References

[1] R.Bertin et al.,
A Preliminary Study of GaAs Solid-State Detectors for High Energy Physics Nu
Instr. and Meth. A294, (1990), 211

[2] R.N.Thomas et al.,
Status of device-qualified GaAs substrate technology for GaAs integrated circuits,
Proc. I.E.E.E. 76, (1988), 778

[3] K.Gruter et al.,
Jour. of Crystal Growth 94, (1989), 607

[4] J.E.Eberhardt, R.D.Ryan and A.J.Tavendale,
High-resolution nuclear radiation detectors from epitaxial n-GaAs,
Appl. Phys. Lett. 17, (1970), 427

[5] C. del Papa, C. Raine, *New Results on GaAs Solid-State Detectors for High Energy
Physics,*
Contributions to the 4th San Miniato Topical Seminar on Experimental Apparatus
for High Energy Physics and Astrophysics, June 1990. (To be published)

[6] M.Edwards and D.R.Perry,
The radiation hardness test facility,
R.A.L. Report RAL-90-065, August 1990

[7] G.R.Stevenson,
New Dose and Neutron Flux Calculations,
ECFA Working Group on Radiation Hardness, 5th July, 1990

[8] A.Alessandrello et al.,
*Gallium-arsenide charge-sensitive preamplifier for operation in a wide low-
temperature range,*
Nucl. Instr. and Meth. A289, (1990), 426

[9] P.H.Sharp,
Developments in VLSI,
Proceedings of the Eloisatron Workshop on New Technologies for 200TeV Physics,
Erice, September 16-20th, 1990

NEW DEVELOPMENTS IN GASEOUS TRACKING CHAMBERS

Fabio SAULI

CERN, Geneva, Switzerland

THE MULTIDRIFT MODULE

We are developing as an LAA project a vertex detector designed to operate at the very high particle rates and radiation levels expected of future high-energy hadron colliders [1,2]. The detector is made of an array of hexagonal multidrift modules, each independently providing local reconstructed track segments in space with very good angular resolution (0.5 mrad or better in the radial direction) and two-particle separation (500 µm). Each module has 132 drift cells, constituted by anodes centered in a hexagonal cell of 1.4 mm radius, defined by six cathode wires. The completed tubes are 30 mm in diameter and 80 cm long. Signals are read out from both sides, to get the longitudinal coordinate from the charge ratio.

A clean assembly room for the construction of modules has been completed. The room contains a high precision optical bench used to hold the end plates during wiring, installed under an electrostatic dust filter. The room is maintained at constant temperature and low humidity to allow proper curing of epoxies with high dielectric strength. An ultrasonic cleaning bath can accommodate up to 1 meter long modules. Fig. 1 shows the optical assembly bench.

A new mounting scheme, using two carbon fiber half shells, has proven to be successful: several final prototypes have been completed and tested. At present, one of these modules is undergoing long-term tests in our laboratory, under permanent control by a computerized data acquisition system (Fig. 2). Figs. 3 shows an examples of average gain, measured with an x-ray source on all wires, and noise rate without source at the operating voltage of 2.9 kV.

We have found a new technique to make an internal wire support. This support is essential if we want to make long tubes because of the electrostatic forces between wires. We have developed a way to form a thin insulating membrane connecting all wires anywhere along the length of the module, and keep them in place. The procedure is as follows: the wired but not yet assembled module (still on its temporary wiring supports) is immersed half way in water. A lower density liquid is poured atop and polymerizes in a thin insulating membrane. The module is then removed from the bath, cleaned and assembled as usual. An example of membrane formed in a pocket size test module is shown in Fig. 4. The

New Technologies for Supercolliders, Edited by L. Cifarelli and
T. Ypsilantis, Plenum Press, New York, 1991

211

local efficiency loss due to the membrane is limited to a zone less than 1 mm wide.

A new set of high density preamplifier cards, matched to the size of the new prototypes, has been built. A corresponding number of high density receiver modules, containing amplifiers and discriminators are also ready. This design is based on commercially available elements. We are also collaborating with several groups to pursue the design of a custom multichannel integrated amplifier, and of a fast multiple switched capacitor analogue memory for signal recording.

THE MICROSTRIP GAS CHAMBER

The microstrip gas chamber was introduced some time ago in an attempt to reproduce the field structure of multiwire chambers, at a much smaller scale [3-5]. It consists essentially in a sequence of alternating thin conductive anode and cathode strips placed on an insulating support; a drift electrode defines a region of collection of charges, and application of the appropriate potentials on anodes and cathodes create a proportional gas multiplication field (Fig.5). The multistrip chamber allows to reach large proportional gains with very good energy resolution. Detectors have typically cathode strips widths between 30 and 60 μm and anode strips between 5 and 10 μm, at a 200 μm pitch; the technique used for engraving the strips, electron lithography, insures an overall accuracy of about 0.1 μm. The active area of the devices is 80x80 mm^2. An electrode is implanted on the back plane to prevent charging up of the insulator between strips by repelling positive ions off the surface; conveniently stripped, it can be used to obtain a second coordinate of the avalanche. Drift gaps between 3 and 6 mm were used in the detectors, the shorter gaps being favoured for detection of high rates of charged particles.Fig. 6 shows the computed field lines in the multiplying region at typical values of the operating voltages. One can understand the operation of the device as follows: drift lines connecting the upper cathode to the anode strip concentrate on the anodes, even more so due to the high potential applied to the cathode strips; the high resulting electric field in the neighborhoods of the anodes results in gas amplification.

Proportional gains in excess of 10^4 can be safely reached. A typical pulse height spectrum measured on a group of cathode strips for 5.9 KeV x-rays is shown in Fig. 7; it has a fwhm of about 12%. This resolution is remarkably good when compared to typical results obtained in multiwire proportional chambers.

To check the rate dependence of the proportional gain, the chamber was exposed to an x-ray generator. At increasing values of the flux, the current, counting rate and pulse height spectrum were recorded on the cathode strips at fixed operating potentials. The result of the measurement is summarized in Fig. 8 that shows the normalized gain as a function of the detected current per unit length of strip; a typical result obtained in a multiwire chamber is also shown for comparison . As one can see, in the microstrip detector the gain is unchanged at currents an order of magnitude larger than in a MWPC; taking into account the reduction in

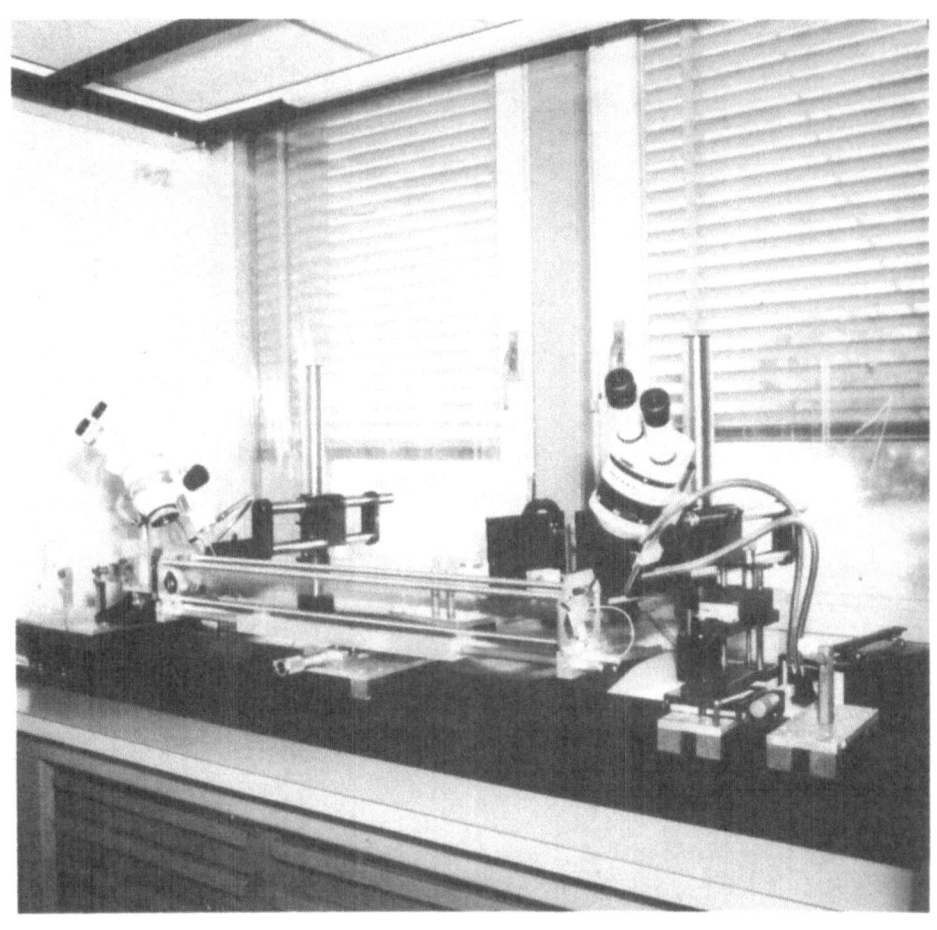

Fig. 1. Optical assembly bench for the prototype
multidrift modules.

Fig. 2. Test bench for long-term testing of the modules.

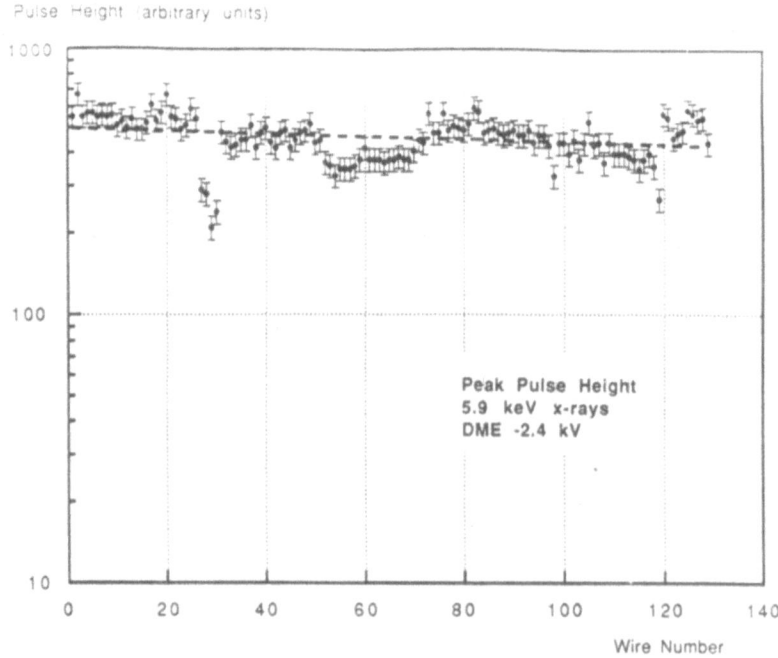

Fig. 3. Average gain as a function of wire number in a
module, measured with an x-ray source.

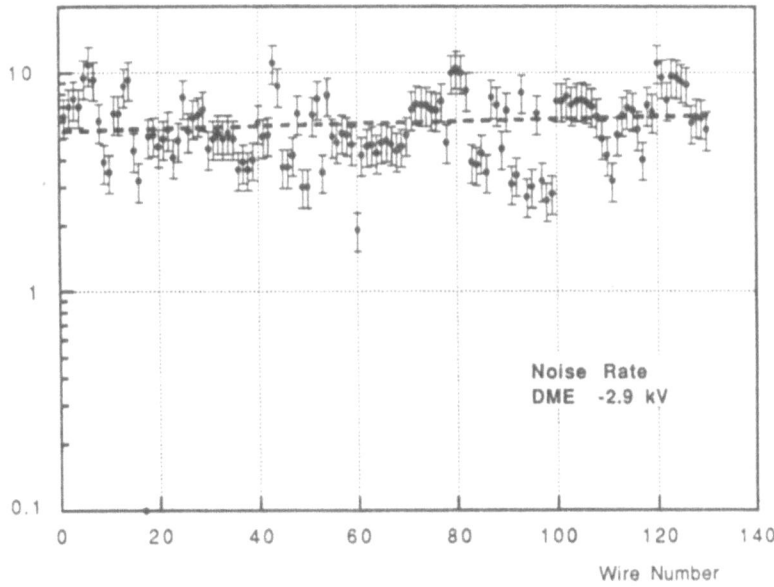

Fig. 4.   Noise rate without source as a function of wire number.

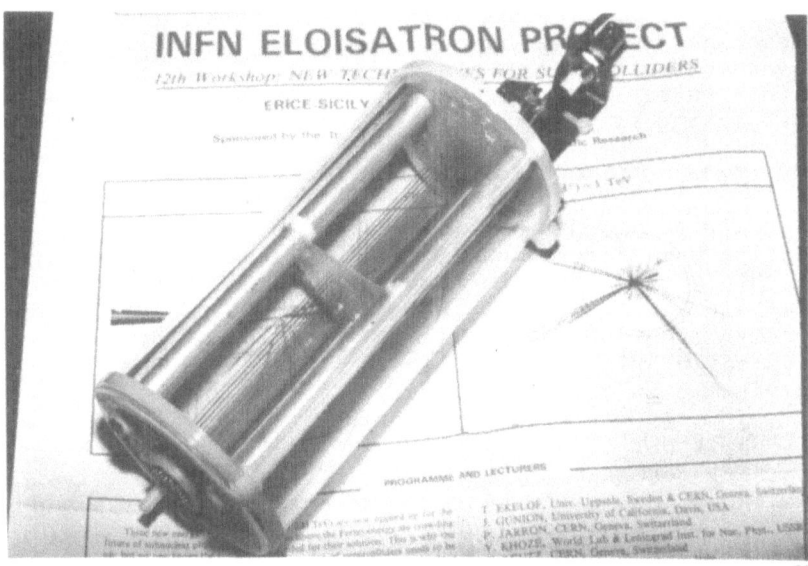

Fig. 5. Thin polymer membrane supporting the wires in a small test module.

pitch (at least by a factor of ten), one expects a rate capability in the microstrip chamber more than two orders of magnitude larger than in a MWPC, as shown by the upper scale in the figure. An actual measurement in a high intensity minimum ionizing particle beam has however still to be performed.

Localization in the direction perpendicular to the strips can be performed by recording the induced charge profile on the cathode strips and computing event per event the corresponding center of gravity. For a measurement of efficiency and localization accuracy for minimum ionizing particles, a microstrip chamber was installed in a high energy test beam at CERN, using as reference the space coordinates provided by a pair of silicon strip detectors. For each event, the induced charge profile on 10 adjacent cathode strips was recorded, thus covering a 2 mm region; the gas filling for these measurement was argon or xenon with about 10% dimethylether (DME) as quencher. Xenon was used in order to increase the energy loss and reduce the primary ionization fluctuations in the thin (3 mm thick) drift gap constituting the sensitive volume of the detector.Fig.9 shows the accuracy distribution, as measured using the solid state detector as a reference. Taking into account the estimated dispersion of the silicon strips themselves, one can infer an intrinsic localization accuracy for the gas microstrip chamber of around 30 μm rms.

The multitrack resolution depends on the rms of the induced charge profile. In our geometry, it has an average value of 125 μm. Assuming that two tracks can be resolved if the corresponding induced pulse profiles are at least two standard deviations apart, we infer a multitrack resolution of 250 μm.

The multistrip gas detector has been shown to allow fast and accurate detection of both soft x-rays and minimum ionizing particles. Its performances compare rather well to those obtained with solid state microstrip detectors; the advantages could be however a higher radiation resistance, a lower cost and a larger signal/noise ratio, this last point rather interesting in that it could lead to the use of cheaper and/or faster electronics readout.

THE COATED CATHODE CONDUCTIVE LAYER CHAMBER

The microstrip gas chamber described in the previous section has some drawbacks. Implementation of the structure on thin glass plates using integrated circuit technology limits the size and does not allow for non-planar geometries. Moreover, due to the small cross section of the anodes (typically 5x1 μm^2) an accidental discharge results inevitably in a permanent damage of the strip.

In the Coated Cathode Conductive Layer (COCA COLA) chamber [6] we have attempted to build a similar (though coarser) geometry on a plastic support, and to solve the breakdown problem by placing the anode and the cathode strips on opposite sides of a thin plastic foil (see Fig. 10). Apart from a small modification due to the dielectric constant of the support, and before charge-up process sets in, the electric field in the COCA COLA chamber is sensibly the same as in the original microstrip detector, and one would expect gaseous amplification to occur. It is conceivable that if a

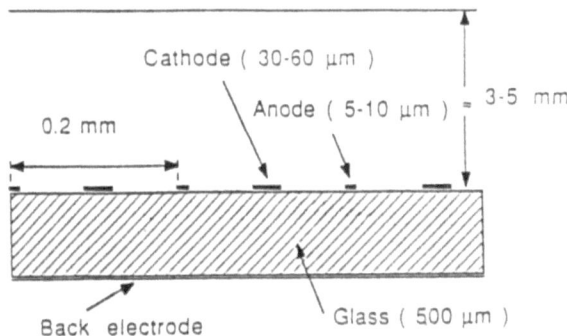

Fig. 6. Schematics of the microstrip gas chamber.

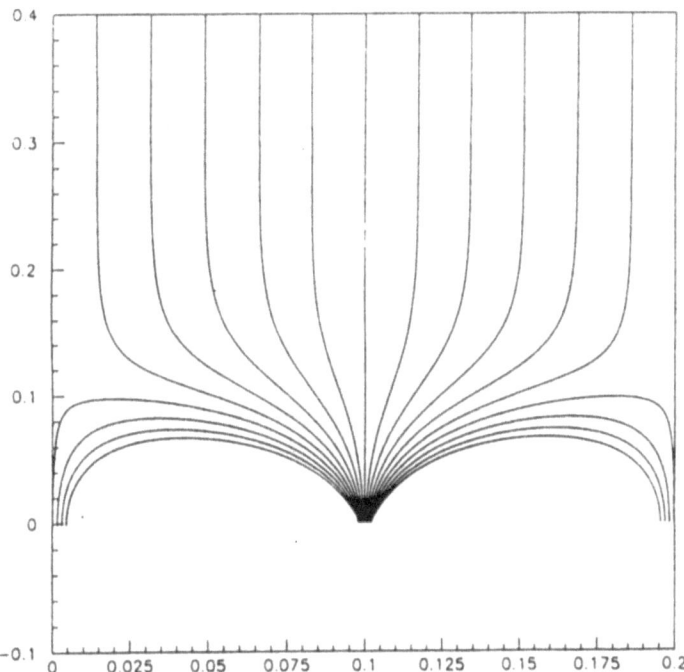

Fig. 7. Electric field in the microstrip chamber.

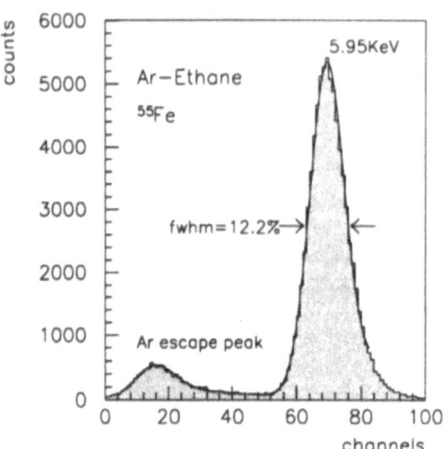

Fig. 8. Pulse height spectrum for 5.9 keV x-rays.

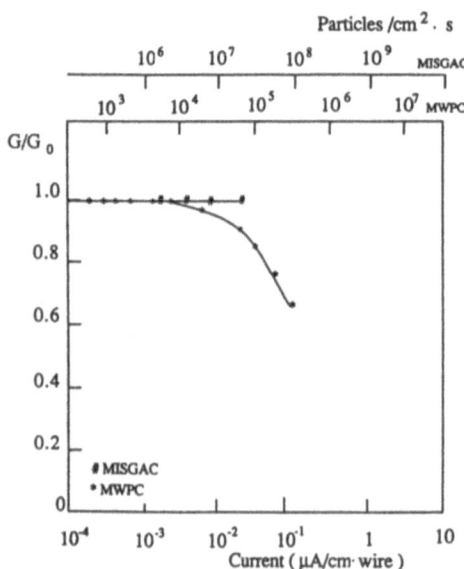

Fig. 9. Rate dependence of gain in the microstrip chamber, compared to a standard MWPC.

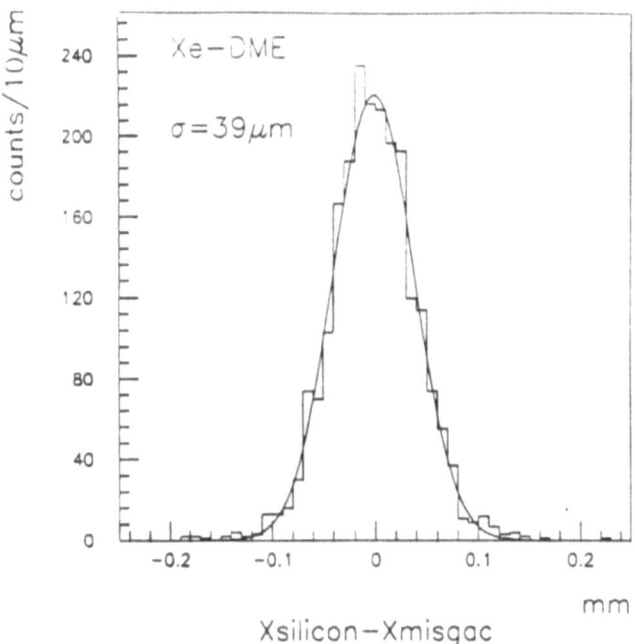

Fig. 10. Localization accuracy distribution for minimum ionizing particles measured with reference to a pair of silicon microstrip detectors.

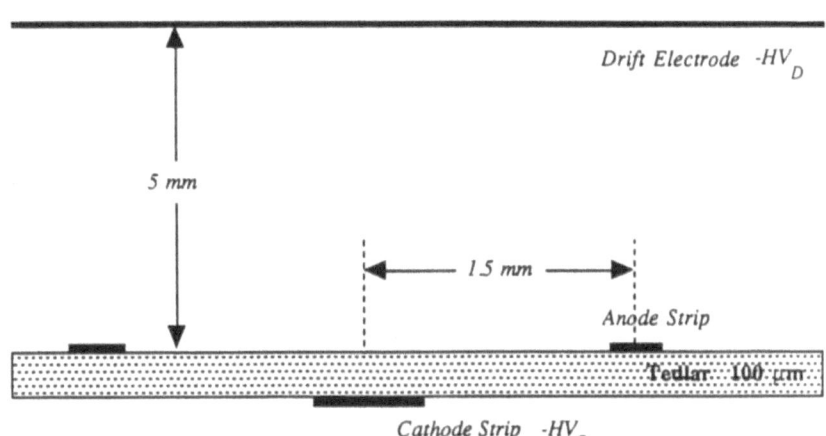

Fig. 11. Schematics of the COCA COLA Chamber. Thin anode and cathode strips are placed on the two sides of a plastic foil. Ionization electrons released in the gas are drifted towards the anode strips, where avalanche multiplication occurs.

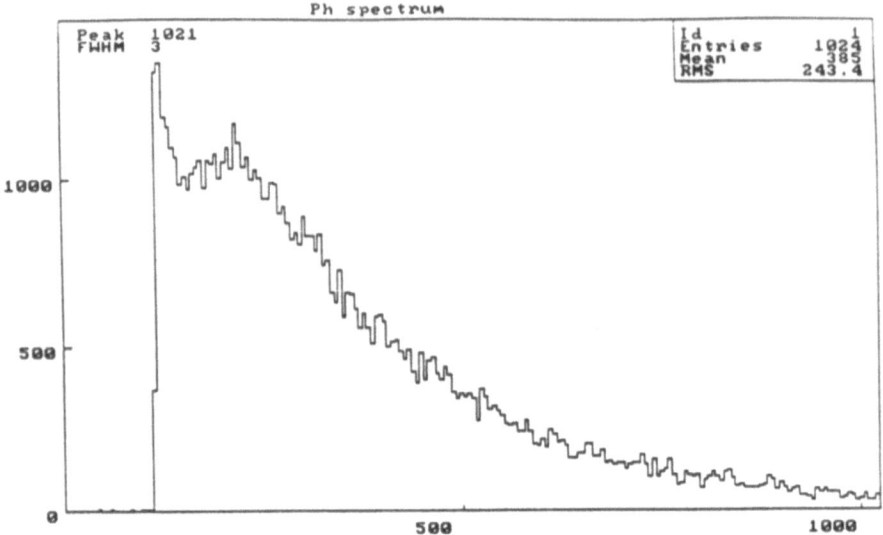

Fig. 12. Pulse height spectrum for minimum ionizing electrons recorded on one anode strip. The peak corresponds to about 0.1 pC of charge on the anode.

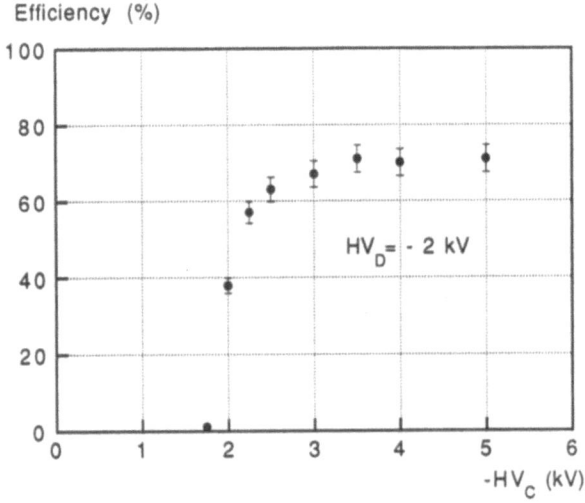

Fig. 13. Efficiency plateau on a single strip as a function of cathode voltage, measured for minimum ionizing electrons.

support is chosen with sufficiently low bulk resistivity, charges would neutralize with a time constant smaller than the ions production rate and a stable operating condition could be reached. A detector based on similar principles was described some time ago [7].

In our first attempt to realize the COCA COLA chamber we have used as support a 100 μm thick foil of white Tedlar*, having a bulk resistivity around 10^{12} ohms.cm . The nominal dielectric rigidity of the 100 μm Tedlar foil is 10 kV. Anode strips about 200 μm wide were vacuum evaporated on one side at a 3 mm pitch. A printed circuit with wider (500 μm) strips and the same pitch was mounted on the back side of the foil, to realize the geometry sketched in the figure. The active area of the detector was 10x10 cm^2, although for practical reasons only half a dozen strips were actually readout. An 8 mm thick gap, overlaying the plastic foil on the anode side and delimited by an upper electrode served to collect and drift the ionization. For most measurements, we have used an argon-methane (90-10) gas filling.

At moderate detection rates (around hundred hertz per cm^2), the operation is rather stable. Using a collimated Ru^{106} electron source in coincidence with a small scintillator behind the chamber, we have recorded the pulse height distribution on individual anode strips; an example is shown in Fig.11. The peak corresponds to a detected charge of about 0.1 pC which, taking into account the ionization loss in the gap (~70 electron-ion pairs) implies a proportional gain of around 10^4. In these conditions, for a single strip readout, we have measured the efficiency plateau as a function of cathode voltage (the anode being grounded): an example is shown in Fig. 12. We believe that the fact that we could not reach 100% is due to the poor geometry and to the use of a single scintillation counter as trigger; the fact that the plateau remains constant up to the highest voltages supports this statement.

Various phenomena that can be associated to charging up of the insulator have been observed in the detector. The most obvious one is a large drop in efficiency observed whenever the voltage to the cathodes is decreased, even by a few hundred volts. With a time constant of minutes, leaving the chamber exposed to the source, the efficiency recovers its normal value; there seems to be a slight dependence of the recovery time from the source intensity. When however the voltage is increased by the same amount, no similar effects are observed; full efficiency indeed is measured immediately even when powering the cathodes all the way up from zero potential. We explain the observation as follows: due to the very high resistivity of the foil (as mentioned before, above 10^{13} ohms.cm), ions actually accumulate on the surface until, with their distribution, they oblige the electric field to be parallel to the foil surface for most of the anode-to cathode distance. This is a condition of equilibrium, since it prevents more ions to reach the insulator. When increasing the voltage, ions produced in the avalanches easily attach to the surface to recreate an equilibrium condition. However, when the voltage is decreased, to obtain the same equilibrium one has to actually remove ions from the surface, a much

---

* Trade name of Du Pont de Nemours & Co

longer process. Note that similar arguments have been used to explain the behavior of the so-called electrodeless drift chambers [8]. While obviously not suited for high or even moderate rate applications, the COCA COLA chamber may be a valid alternative to existing low-cost devices designed to cover very large surfaces (as for example for muon detection). The very large efficiency plateaux, together with complete immunity to spark breakdown, make the detector rather suitable for use in this case.

REFERENCES

1. A.Zichichi et al, The LAA Project, CERN-LAA/88-1 (1988)

2. R.Bouclier, J.Gaudaen, I.Gouz, B.Guerard, J.C.Santiard, F.Sauli and R.Wojcik, Nucl. Instrum. Methods A283 (1989) 509

3. A.Oed, Nucl. Instrum. Methods A263 (1988) 351

4. F.Angelini, R.Bellazzini, A.Brez, M.M.Massai, G. Spandre and M.R.Torquati, Nucl. Instrum. Methods A283 (1989) 755

5. F.Angelini, R.Bellazzini, A.Brez, E.Focardi, M.M. Massai, G.Spandre, M. R.Torquati, R.Bouclier, J. Gaudaen and F.Sauli, INFN PI/AE 90/6. Pres. at 2d Int. Conf. on Advanced Technology and Particle Physics, Como (1990)

6. R.Bouclier, J.Gaudaen and F.Sauli, CERN-PPE/90-139. Pres. at 2d London conf. Position Sensitive Detectors, 1990

7. M.J.Neumann and T.A.Nunamaker, IEEE Trans. Nucl. Sci. NS-17 (1970) 43

8. J.Allison, R.J.Barlow, C.K.Bowdery, I.Duerdoth and P.G.Rowe, Nucl. Instrum. Methods 201(1982) 34

SUPERCONDUCTING STRIPS FOR MICROVERTEX DETECTORS

T.O. Niinikoski, A. Rijllart, B. Strehl and K. Winter
CERN , CH-1211 Geneva 23, Switzerland

M. Caria
INFN Cagliari, Via Ada Negri 18, I-09127 Cagliari, Italy

1. INTRODUCTION

1.1. Beyond the limits of present track devices?

One of the main physics motivations for the future large hadron colliders, the SSC in the U.S.A. and the LHC proposed at CERN, is the study of the mass generation mechanisms for particles. The Higgs mechanism requires new massive bosons, the search of which leads to the requirement of very high luminosities at the proposed multi-TeV collision energies. The design parameters for the planned LHC, for example, basing on such searches, aim at luminosities of $(2 - 4) \cdot 10^{34}$ cm^{-2}s^{-1} at the maximum colliding beam energies of 8 TeV. This, combined with the large multiplicity at high energies, will entail severe problems for any existing vertex detector. Radiation damage, spatial resolution, timing resolution and readout are among the most evident. It has been proposed that these problems could be solved by applying superconducting microstrip detectors, made of suitable radiation-hard materials, such as niobium nitride (NbN).

Other promising uses of these new detectors could be found in tracking-applications in intense secondary beams. A typical example is the deep-inelastic scattering of high-energy muons, probing the parton structure functions of free and bound nucleons. Extending these experiments to very small angles, to spin structure, and to a small fractional momentum carried by the constituent, demands a good statistical accuracy and therefore intense beams. This and the requirement of a high vertex resolution may also lead to specifications which can only be met with the proposed new superconducting detectors.

1.2. Radiation damage in the microvertex detectors of future colliders

The radiation dose in the proposed LHC machine would be about 4 - 8 MRad/year at a radius of 10 cm around the beam axis in the collision area. This dose rate would drop roughly inversely proportional to the radial distance from the beam in the vicinity of the beam tube, in the area where vertex tracking would be required.

New Technologies for Supercolliders, Edited by L. Cifarelli and
T. Ypsilantis, Plenum Press, New York, 1991

Because of the desired momentum resolution for the decay products of massive particles, and because of the large particle multiplicities at the proposed high energies, the track resolution in the vertex detector would have to be as high as possible. It would be highly desirable to be able to resolve the particles in a dense jet already in the first layer of the detector, placed at a radius of some 5 cm. Microstrip detectors are thus called for, and serious development of silicon devices has already begun, with particular concern in improving their radiation resistance.

1.3. Silicon microstrip detectors

Silicon microstrip detectors presently show serious degradation in the leakage current between electrodes at a dose rate of about 0.1 MRad. This can be somewhat corrected by changing the biasing conditions, and further improvements have been made in materials and geometry so that devices may remain potentially useful up to 1 MRad currently. Improvements beyond this could be envisioned by more substantial changes in the technology, for example by depositing the detector on a sapphire substrate, or by using amorphous silicon. The basic problem of radiation damage in high-resistivity semiconductor materials, however, limits the technology so that operation in the LHC environment may require changing the silicon microtracker devices several times during each year of operation at full luminosity.

The readout electronics of the current silicon microstrip detectors is located adjacent to the strip devices, and the damage of these circuits is limiting the silicon technology even more severely than that of the strip devices themselves. Radiation hardened electronics is becoming available, but the basic problem is again in the intrinsic damage of semiconductors, which may set a brick-wall limit at a few MRad for the operation of a practical silicon microtracker.

1.4. Superconducting microstrip detectors

It has been proposed that most of the foreseeable problems of mictrovertex detectors in the future colliders could be solved by applying superconducting microstrip detectors, made of suitable radiation-hard materials. Superconducting materials are metallic conductors where radiation-induced faults and impurities change the characteristics only at dose levels several orders of magnitude above those already exterminating the semiconductor devices. The cases in point are hard superconductors, which naturally contain large amounts of faults and impurities. These actually improve several characteristic parameters related with superconductivity; thus the critical temperature T_c and current I_c of many materials increase slightly up to 100 MRad before any significant decrease is seen at much higher doses.

The radiation damage in superconductors is reviewed in Ref. [1]. As examples we may select Nb_3Sn and Nb_3Ge which shown minor decrease of T_c and I_c at 100 MRad, and NbTi which has no damage up to 700 MRad.

The radiation damage of isolating PbIn oxide thin film has been studied by measuring the change in the normal-current tunneling resistance of Josephson junctions under electron irradiation in a 4.2 K scanning electron microscope [2]. The very thin layers of this material can take more than 50 MRad before degradation of the Josephson gate properties occurr. Isolating materials such as Al_2O_3 or SiO_2 are likely to withstand higher doses up to 500 MRad, before the high frequency low-voltage characteristics degrade beyond limits set by the strip detector application. The changes occurring in the dielectric coefficient and magnetic permeability can be partly recovered by annealing the material around the room temperature.

In this paper we shall describe the principles and methods of superconducting microstrip detectors, with view on their application in particle physics experiments. The theories of thermal transients in superconducting microstrips are reviewed, and the thermal stability criteria under steady bias current conditions are discussed in Section 2. It is shown that detectors relying on the thermal equilibrium principle may be insensitive to minimum ionizing particles. Better sensitivity and much faster transients can be expected when considering non-equilibrium phenomena in superconductors, especially in granular thin films of high resistivity.

Our first test results on NbN are reviewed and compared with earlier results on other materials in Section 3. New potential materials are listed and their characteristic properties will be systematically discussed in the light of the experimental data.

In addition to the high speed, the spatial resolution and the radiation resistance of the projected track detectors, superconductivity may also offer other interesting possibilities for future high-luminosity experiments. Most important among these are superconducting transmission and delay lines, and superconducting logic readout electronics.

Signal propagation in a superconducting transmission line is virtually free from dispersion up to terahertz frequencies. Therefore the shape of the expected very fast transients should be well conserved in the microstrip line, enabling high-quality delay lines over several metres length to be made. Signal delays of several hundred nanoseconds can thus be realized, taking into account the slow propagation which occurs in low-impedance superconducting lines.The theory of superconducting transmission lines is reviewed in Section 4.

Superconducting logic devices have impedances matching well with the proposed microstrip lines. The ultra high speed of these circuits would enable one to perform processing close to the vertex detector, reducing thus the amount of information to be transferred at each bunch crossing, and possibly allowing to perform first-level triggering for event selection. The use of superconducting Josephson-gate logic circuits for readout and data compression is discussed, and the problem of interfacing with normal electronics is pointed out in Section 5.

The paper concludes by a brief discussion on the design specifications of a superconducting microtracker in Section 6.

2. THEORY OF OPERATION

2.1. Superconducting strip counters

Superconducting strip detectors consist of narrow lines, microstrips, etched out of a thin (50 - 500 nm) superconducting film. The strip counter is a digital device, with resolution equal to the spacing of the lines. The spacing depends on the linewidths, which are limited in practice by the lithographic and etching techniques; few micron widths are accessible by standard optical techniques.

The superconducting strips can be operated at bias currents close to the limit of the intrinsic critical current densities, up to 10^7 A/cm^2. Localized energy deposited by an ionizing particle may transform a short length of the strip into the normal, resistive state, detected as a voltage pulse in the end of the line. The supeconducting state may be re-established by self-recovery if certain stability crieria are satisfied, or by reduction of the bias current by means of a resistive shunt or an active current switch.

Since Sherman [3] proposed superconducting tin strips as a new class of particle detectors, several materials have been tested [4, 5, 6, 7]. In these experiments α-particles were used, but recently Gray et al. [8, 9] could demonstrate the applicability of superconducting Nb microstrips for the detection of X-rays and fast electrons. The pulses had amplitudes between 250 μV [4] and 60 mV [9], and rise-times between 5 ns [7] and 150 ns [8]. All these authors resorted to the thermal model of Sherman to describe their results, with fair success.

In our recent study [10, 11] of a NbN microstrip, using minimum ionizing electrons, 5.5 mV self-recovering pulses were observed, with rise times below 300 ps. The self-recovering pulses contradict the thermal equilibrium model, and suggest that even faster non-equilibrium phenomena in the superconductor would give rise to voltage transients following the local heating of the electronic system in a very fast (picosecond) time-scale. Recovery from such pulses should be possible, provided certain stability criteria are satisfied.

In the following we shall discuss the stability of thin superconducting microstrips under bias currents, and the dynamics of pulse formation following a localized energy deposit by an ionizing particle.

2.2. Biasing and stability in steady-state operation

The superconducting current carried by the strip is limited by the critical current density of the material. Strips made of the best available materials, having a thickness of about 500 nm and a width of 3 μm, can carry over 100 mA currents before transition to normal state will occur. The transition begins at the weakest point of the strip and the normal zone will propagate at a characteristic speed along the line until the whole strip is normal.

If a heated normal-conducting zone is created in the strip artificially, for example by a traversing or stopping particle, the normal zone may expand or collapse depending on the value of the steady bias current. In the latter case the resulting electric signal seen in the end of the line is said to be self-recovering or self-terminating. The former event is often called a quench.

Very narrow strips of high-resistivity materials, some of which have very high critical current densities, may not withstand the heating due to the energy stored in the bias circuitry, which necessarily has some inductance, in addition to the inductive energy of the strip itself. In this case burnout of the strip may occur when the critical current is exceeded. This necessitates some care in the design of the biasing circuitry.

The critical current $I_c(T)$ of a superconducting line, having a transverse dimension large compared with the coherence length, depends on its temperature by the same law as the thermodynamic critical field, i.e.

$$I_c(T) = I_c(0)\left[1 - \left(\frac{T}{T_c}\right)^2\right] \quad , \tag{1}$$

where $I_c(0)$ is the maximum value reached theoretically at absolute zero temperature, T is the temperature of the line, and T_c is the intrinsic critical temperature of the material at zero current and zero applied field. We should warn, however, that for thin strips the above formula may not be quite exact, and the formula is in large variance with the theoretical and experimental temperature dependence when the strip has transverse dimensions small in comparison with the coherence length.

In addition, some areas of a strip carrying a high bias current may actually be slightly heated above the substrate temperature, due to dissipation in the electrical contacts, for example. In order to describe the steady-state behaviour, and in particular the stability limit of a superconducting strip under such bias conditions, we have to consider the ohmic heating of the contact areas,

$$\dot{Q} = RI^2$$

and the heat conducted from these regions to the substrate,

$$\dot{Q} = \alpha A_{th}(T_s^4 - T_a^4) . \qquad (2)$$

In these expressions I is the current through the strip, T_s is the temperature of the strip in the contact area, T_a is the temperature of the substrate, R is the electrical resistance of the contacts, A_{th} is the area over which heat is conducted to the substrate, and α is the Kapitza boundary conductance constant, which describes the phonon transmission through the interface of the strip and the substrate. These yield the following expression [10] for $I_{max}(T_a)$, the maximum supercurrent carried by the strip, as a function of the substrate temperature T_a:

$$I_{max}(T_a) = I_c(0) \frac{1 - \sqrt{t_a^4 (1 - \varepsilon) + \varepsilon}}{1 - \varepsilon} , \qquad (3)$$

where we have used the scaled substrate temperature $t_a = T_a/T_c$, and have introduced the parameter ε defined by

$$\varepsilon = \frac{RI_c^2(0)}{\alpha A_{th}T_c^4} , \qquad (4)$$

which ideally should be small if not zero, by making R vanishingly small. On the other hand, finite values of R will enable one to determine the thermal resistance between the strip and the substrate, as will be discussed later.

At zero substrate temperature the maximum current carried by the strip becomes

$$I_{max}(0) = I_c(0) \frac{1}{1 + \sqrt{\varepsilon}} , \qquad (5)$$

which enables the direct determination of ε by extrapolating to zero temperature the measurement of the maximum supercurrent with and without a resistively heated zone. As it may in practice be difficult to vary the contact resistance without warming up the cryostat, the parameters ε and $I_c(0)$ can be found by fitting the experimental data on $I_{max}(T_a)$ using the expression of Eq. (3). The upper curves of Fig. 1 show such fits to our results [10] on a NbN strip with pressed indium contacts.

Once the thermal conduction between the thin-film strip and the substrate has been determined, one may calculate the thermal healing length [12] η of the strip:

$$\eta = \sqrt{\frac{\kappa d}{\alpha'}}$$

(6)

where κ is the thermal conductivity of the strip material, d its thickness and

$$\alpha' = 4\alpha T_s^3$$

(7)

is the linearized transverse thermal conductance. Healing lengths of suitable strips vary around 1 μm, which indicates that heating in a contact area can be sinked into the substrate only under the contact spot and its nearest vicinity. We note that because thermal conduction at the superconducting transition temperature has usually an anomaly, and because the term under the square root is constant only when both α' and κ have a cubic temperature dependence, it can be expected that the thermal healing length is not really a well-defined constant in the temperature range of interest. We also note that the healing length may be of the same order of magnitude as the size of the hotspot generated by a minimum ionizing particle, which makes it difficult to handle the hotspot dynamics theoretically. The size of the hotspot will be estimated in the next subsection.

The thermal healing length can be used for calculating another important parameter, the minimum current beyond which a normal conducting length of the strip will expand [12]:

$$I(x_0) = \sqrt{\frac{\alpha' w^2 d(T_c - T_a)}{\rho_N}} \sqrt{1 + \sqrt{\frac{\kappa_S}{\kappa_N}} \coth\left(\frac{x_0}{\eta_N}\right) \coth\left(\frac{L - 2x_0}{2\eta_S}\right)} \quad ,$$

(8)

where $2x_0 > \eta$ is the length of the initial normal region, L the length of the strip, w and ρ the strip width and resistivity, and superscripts S and N refer to the superconducting and normal states. It is clear that high-resistivity materials will have lower currents at which a hot spot may recover, but a high transition temperature may enable one to compensate for this largely, because α' will grow quickly due to its T^3-dependence.

The lowest curve of Fig. 1 shows the maximum supercurrent which can be carried by the NbN strip described above, while exposing the strip to ionizing radiation from a [106]Ru source. The measured points are consistently below those measured without the source, and resulting curve is reproducible under thermal cycling. The points are well fitted with the same function as the upper curves. It thus appears that, in the case of $2x_0 < \eta$, the temperature dependence of the current limit, below which the hotspot may recover, would be similar to that of Eq. (3).

The limit of the normal zone of length $2x_0 > \eta$ will propagate at a velocity [13]

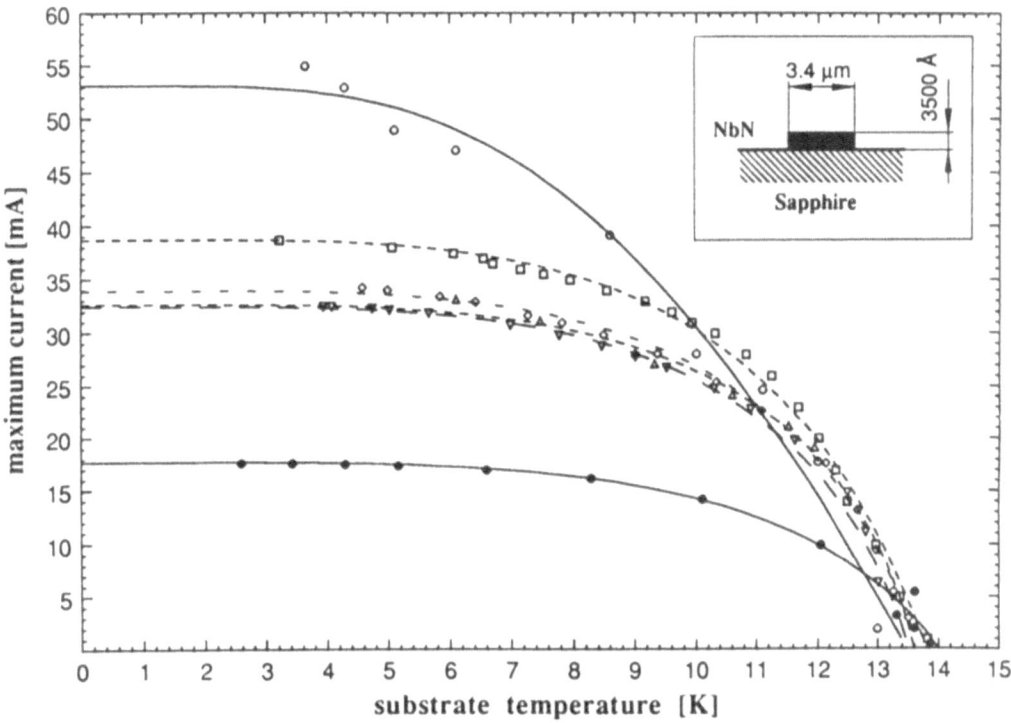

Figure 1. Maximum supercurrent carried by the NbN strip of Ref.[10] as a function of the substrate temperature. The Eq. (3) was used for fitting the data. The upper curves fit the data shown with open symbols, determined with no source in the proximity of the strip; the irreproducibility in thermal cycling is believed to be due to the variation of the pressed indium contacts used for current injection in the strip. The lowest curve fits the data plotted in the round full symbols, determined with the [106]Ru source mounted near the strip; these points were reproducible in thermal cycling. The inset shows the transverse dimensions of the strip.

$$v = \frac{\beta - 1}{C_v} \sqrt{\frac{\alpha' \kappa}{\beta d}} \quad , \tag{9}$$

where

$$\beta = \frac{T_\infty - T_c}{T_c - T_a} \tag{10}$$

and T_∞ is the asymptotic strip temperature far in the normal region, given by

$$T_\infty - T_a = \frac{\rho_N J^2 d}{\alpha'} \quad . \tag{11}$$

Here the current density J is assumed constant in the cross section of the strip.

The propagation speed of the normal zone can thus be positive or negative, depending on the current density, and the hotspot can therefore grow or collapse. For typical superconductors and desirable strip dimensions the highest speeds are around 10^3 m/s. Long strips thus quench rather slowly, and unless the bias current is reduced promptly after a quench is detected, there will be a slow rise of the non-recovering signal voltage following the rapid initial step due to the formation of the hot spot.

In the original analysis of Broom and Rhoderic [13] constant thermal resistance between the strip and the substrate was used, instead of the Kapitza resistance of Eq. (1). Gray et al. [14] have shown that this is experimentally justifiable, if all parameters are evaluated at the critical temperature of the strip.

2.3. Thermal transient models

Let us consider an ionizing particle traversing our thin-film strip, normal to the planar substrate of the strip. Neglecting any possible energy loss in the substrate or in possible passivation layers above the strip, there will an average energy deposit per unit path length dE/dx in the strip material, which will create an initial hot cylinder centred on the track of the particle. Considering the small dimension of the films, however, the energy deposit may not be entirely uniform, but we shall first assume here that this is the case after about 100 ps, during which the electronic and phonon temperatures will approach each other. Faster phenomena will be discussed in the next subsection.

According to the thermal model of Sherman [3], the radius r_0 of this 'normal current zone', created by the traversing particle, is given by

$$r_0 = \sqrt{\frac{dE/dx}{\pi(\Delta H/V)}} \tag{12}$$

where $\Delta H/V$ is the difference between the specific enthalpies in unit volume, between the superconducting and normal states, and can be deduced from the critical field according to

$$\frac{\Delta H_0}{V} = \frac{B_{ctho}^2}{2\mu_0} \tag{13}$$

The thermodynamic critical field at zero temperature B_{ctho} can be calculated from the upper critical field B_{c2} and the Ginsburg–Landau parameter $\kappa = 0.725\ \lambda(0)/l$, where $l << \xi_0$ is the mean free path of the normal electrons. For example a sputtered NbN thin film, with $T_c = 14$ K and $\rho = 130\ \mu\Omega{\cdot}\text{cm}$, has values $B_{c2} = 16.3$ T and $\kappa = 52$ according to Ref. [15]. Therefore $\Delta H_0/V = 1.2 \times 10^{11}$ MeV/cm^3, corresponding to 2.6 mJ/g. This estimate can be verified by comparison with the electronic energy difference, which is due to the superconducting gap, according to Ref. [16]

$$\frac{\Delta E}{V} = \frac{m^* \Delta_0^2}{4\pi^2 \hbar^3} \sqrt{2m^* E_F} \quad . \tag{14}$$

From Ref. [17], $2\Delta_0 = 4.08 k T_c$ for sputtered NbN, so the energy gap in the above case case ($T_c = 14$ K) is $\Delta_0 = 2.46$ meV. Assuming that the effective mass is $m^* = m_e$, the mass of the free electron, and taking the Fermi energy $E_F = 12.5$ eV (Ref. [18]), Eq. (14) yields $\Delta E/V = 3 \times 10^{10}$ MeV/cm^3. The agreement between the values given by Eqs. (13) and (14) is thus reasonable since the averaged mass at the Fermi-surface is higher than m_e. For the temperature dependence of $\Delta H/V$, we may assume the form $(1 - t^2)$.

The energy loss of electrons having an energy of 1 MeV, close to the value at which minimum ionization occurs, is about $dE/dx = 12$ MeV/cm in NbN. Using this and the above value for $\Delta H/V$, the minimum radius of the cylindrical volume which becomes normal, should be $r_0 = 61$ nm for $T_s = 5$ K. This is substantially greater than the coherence length in several type II superconducting materials, such as NbN (5 nm) and Nb (38 nm), whereas in many type I materials the hotspot will be smaller than the coherence length, notably in Sn, In and Al. In such a case the supercurrent will penetrate through the spot due to the proximity effect, with no or little Joule heating.

The minimum bias current I_{min} at which the strip would then switch to the normal state over the whole width (according to the model of Sherman) is given by

$$I_{min} = I_c(T)\frac{w - 2r_0}{w} \quad , \tag{15}$$

where w is the width of the strip. In our tests [10] we used a strip of width w = 3.4 μm; the thermal model then predicts that signals arising from minimum ionizing particles can only be observed when the strip is biased at a current between 96% and 100% of its critical value. Believing the thermal model thus leads to choosing low-T_c materials, because in these the transition enthalpy is lower and therefore r_0 is greater.

Refinements of the thermal model should take into account the fact, that at currents close to the critical value a much smaller amount of enthalpy is required to switch a unit volume in the normal state, compared with the requirement at low currents or fields. The normal spot diameter can also be enlarged by Joule heating, which will be discussed below in the terms of non-equilibrium models.

2.4. Non-equilibrium models

Non-equilibrium phenomena may play a decisive role in the initial evolution of the hot spot created by a fast traversing particle in a thin superconducting strip counter. The thermalization of the hot spot is a relatively slow process, and a rigorous treatment of the initial evolution problem would require the solution of the Boltzmann equation. This is clearly very difficult, although many simple cases can be rather well approximated by the Rothwarf-Taylor equations[19], which explain successfully the quasiparticle and phonon dynamics in tunnel junctions for example.

In addition to the quasiparticles and phonons, the electromagnetic fields need to be taken into account. This requires the simultaneous solution of Maxwell's equations. Superconductivity can be taken into account by adapting the Landau-Ginzburg equations into the time-evolution equations. We shall discuss these phenomena below mainly qualitatively and, where possible, semiquantitatively. In our discussion we shall refer to Fig. 2 which schematically illustrates the dynamics of the hotspot growth in a NbN film with $\rho = 120~\mu\Omega cm$ resistivity and $J = 10^6$ A/cm^2 current density.

Relativistic particles traverse a 300 nm thick film in 1 fs. A major portion of the ionization deposited in the material is dissipated in the conduction electrons in about 10 fs. During this time quasiparticles, with Fermi velocity of 2.1 μm/ps, may transport the heat to a distance of 20 nm, if they are not scattered. The atomic oscillations have minimum periods of about 100 fs; in the 10 fs time scale we may ignore the ionic lattice contribution to the sharing and transport of energy. The phonon system will begin to be heated by energetic electrons in about 1 ps after the passage of the particle, but for thermal excitations near gap energy the electron-phonon relaxation time is much longer, of order $\tau_{e-ph} \approx 60$ ps in NbN for example. The electron-phonon relaxation process has been discussed recently by Gray[20].

Because quasiparticles will be ultimately scattered in the material, the propagation of the initial 'normal current zone' is controllable by the microscopic structure of the material, which may reveal itself very important in the present application.

Let us assume now that in the subpicosecond timescale energy is mainly shared and transported by quasiparticles, and that the (uniform) current of density J will continue to flow in the normal conducting volume, heating thus the quasiparticles and therefore enlarging the spot radius. This is depicted in Fig. 2a, which shows the hotspot with maximum diametre of 60 nm at approximate time t = 100 fs, sufficient for the propagation of the quasiparticles within the spot but clearly insufficient for sharing a substantial part of the energy with the equilibrium phonons.

The assumption of an almost undisturbed current distribution (Fig. 2a) may be justified in the subpicosecond timescale, because eddy currents will allow changes in the self-induced flux and therefore in the current distribution only very slowly. We estimate that in NbN the time constant for magnetic flux to creep out of the hotspot is of the order

$$\frac{B}{\dot{B}} \approx \frac{B^2}{2\mu_0\rho_N J^2} \approx \frac{\mu_0 d^2}{2\rho_N} \approx 0.5 \text{ ps .} \qquad (16)$$

In the zone heated above the critical temperature, once greater in diameter than the correlation length (≈ 5 nm), the current flow will therefore be dissipative, and a voltage drop will occur in it. An electric field will also leak out of the film, if the hotspot traverses the film, as is expected in the case of fast projectiles. The electric field pulse will begin to propagate in the dielectric at a speed characteristic of the superconducting line.

In the first picoseconds the energy dissipated in the electronic system by the bias current cannot be coupled to the phonons, and one may make the simple assumption that this energy will heat and enlarge the non-equilibrium 'quasiparticle hotspot'. We estimate the growth speed of the radius in the following simple way. Equating the Joule heat generated in the hot spot of area A, with the rates of energy change due to the growth of the spot and due to the coupling of the electrons with the phonons, one obtains the equation

$$\rho_N J^2 d \cdot A = \frac{\Delta H}{V} d \cdot \dot{A} - \frac{C_e(T_c - T_b)}{\tau_{e\text{-}ph}} d \cdot A \quad . \tag{17}$$

Because of the long correlation time between the electrons and phonons in NbN, the last term is small compared with the first one on the right side, and we can get a simple expression for the growth time constant of the surface of the hot spot:

$$\tau_A = \frac{\Delta H/V}{\rho_N J^2} \approx 0.7 \text{ ps} \tag{18}$$

in the case of NbN. Assuming that the spot is circular and will not deform in the very beginning, the radius will the grow exponentially at a rate of

$$\dot{r} = \frac{r}{2\tau_A} \approx 40 \text{ nm/ps}$$

so that the initial "equilibrium" radius of 60 nm, due to a MIP, would be roughly doubled in about 1 ps. The current and flux, however, move out of the centre of the spot (Fig. 2b) somewhat faster as shown by Eq. (16), enabling the quasiparticles of the currentless centre of the spot to cool at a speed determined by the quasiparticle heat transfer to the surface of the strip (≈ 100 fs), by the phonon-quasiparticle correlation time (≈ 60 ps), by the acoustic mismatch of the phonons at the film-substrate interface, and (after phonons are heated in the whole volume) by the time for an acoustic phonon to traverse the film $\tau_{ph} = d/v_a \approx 100$ ps. The hot spot now becomes a hot ring, heated on the perimeter by the current but cooled in the center by heat transfer via the interface between the film and the substrate. The ring diameter at t = 1 ps (Fig. 2c) is of the order of the thermal healing length in NbN, and it has become slightly oval because the heated ring-shaped zone will expand faster towards the sides of the strip than along its length, due to the fact that the current density will grow on the sides whereas it will drop on the ends of the ring.

In the next picosecond the ovalized ring will grow very fast in the direction perpendicular to the original current direction, untill it will touch one of the sides of the strip (Fig. 2d). The speed may approach the Fermi velocity $v_F \approx 2$ μm/ps. The whole few-micrometre width of a narrow strip may thus be bridged by a dissipative non-equilibrium hot zone in less than 2 picoseconds, because the current density on the sides of the ovalized ring may exceed the intrinsic critical current of the material.

233

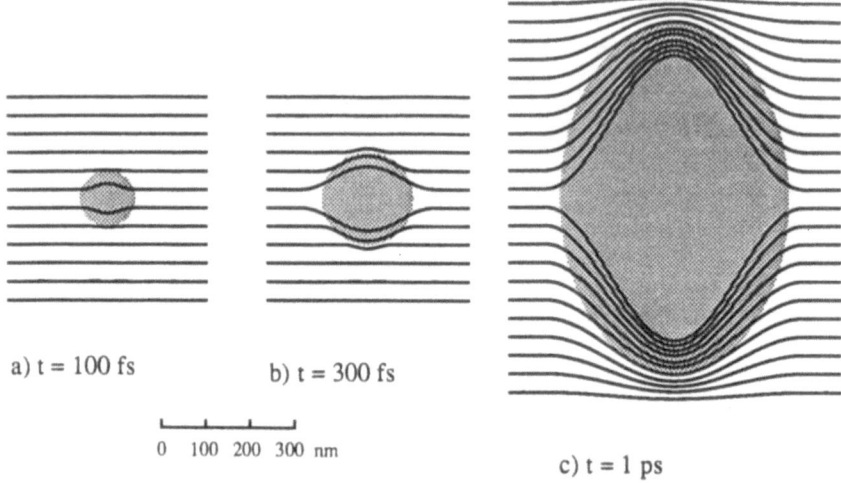

a) t = 100 fs

b) t = 300 fs

0   100 200 300 nm

c) t = 1 ps

Figure 2. Schematic illustration of the dynamics of the growth of the hotspot created by a MIP traversing a NbN strip with dimensions as shown in Fig. 1. It is assumed that the resistivity of the material is 120 $\mu\Omega$cm and the current density is $1.5 \cdot 10^6$ A/cm^2; the current density is illustrated by "current lines" which are lines parallel to the local current density vectors. Denser lines thus mean higher current density. a) The initial hotspot is circular and it is assumed to have the maximum radius given by Eq. (12). The quasiparticles have "cooled" to an energy close to the gap value, but the phonons are very far from equilibrium. The current density is almost unperturbed. b) At t = 300 fs the hotspot has roughly doubled its area due to ohmic heating. At the same time the current has moved out its centre, increasing the current density on the segments of the perimeter parallel to the current, and decreasing the current density on the segments perpendicular to the initial current. c) At t = 1 ps the circular symmetry of the spot is lost, because it has grown towards the sides of the strip at an accelerated speed. d) At t ≈ 2 ps one side of the strip has been reached, and the current has jumped to the remaining superconducting segment, where the current density quickly exceeds the intrinsic critical limit. This completes the normal current zone which now bridges the whole width of the strip.

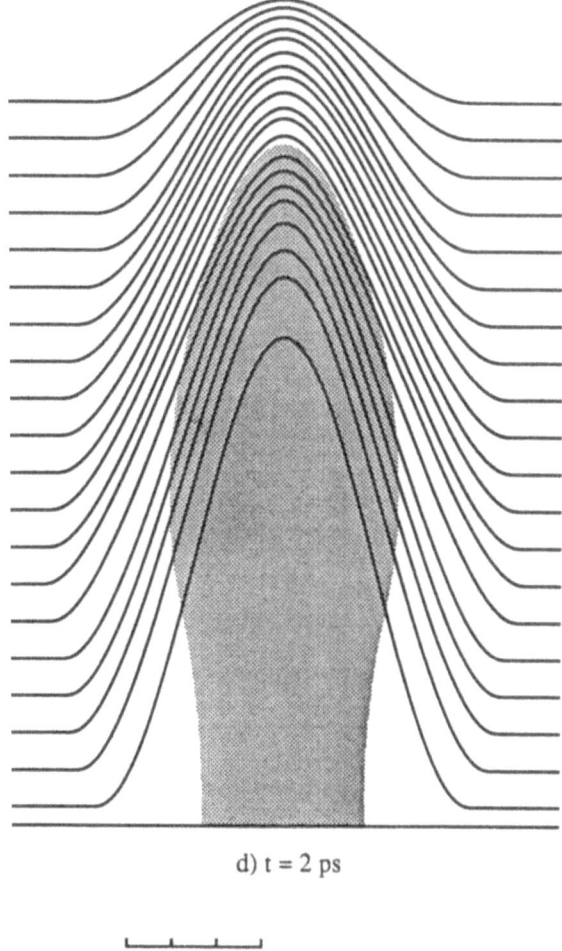

d) t = 2 ps

0   100 200 300 nm

Figure 2

It is rather logical to expect a rapid growth of the hotspot on the case of high-resistivity film carrying a high current density. It is much more difficult to explain how the hot zone may recover in such a case; intuition says that nothing might prevent the axial extension of the zone, and this is supported by the result from an equilibrium model of Eq. (8). However, our non-equilibrium model, qualitatively described above, gives two possible mechanisms providing self-recovery of the hotspot: Firstly, the current distribution across the strip is highly non-uniform just after the whole width of the strip has been bridged by the hot zone, and secondly the quasiparticles may have time to condense back into the superconducting ground state in the currentless areas, due to electron-phonon heat transfer. This may be occurring first in the proximity of the film-substrate interface, which is also strongly suggested by the results showing a strong dependence of the pulse characteristics on the substrate quality[5].

The reason for the non-uniform current distribution across the strip in the hot zone is schematically illustrated in Fig. 2d. Because usually one or the other side of the film is reached first by the hot ring, current density there may drop to zero as soon as the self-induced magnetic flux may rearrange itself. In the high-resistivity material this will be much more rapid than in the pure metals; granular films may also be better in this sense. The final heated zone will certainly have a length in the direction of the strip which is highly non-even across the strip. The current density will therefore readjust itself in a way which reflects the shape of the zone; the readjustment, however, does not happen instantaneously but within the rough time scale of Eq. (16). During this time a supercurrent may start flowing in those areas which carry no or low normal current; the dynamics of this supercurrent zone now will determine whether the original state of the strip can be self-recovered. The dynamics of a supercurrent zone is different from that of a closed zone with normal current, because the electromagnetic field behaves in a different way in these two areas, and because the relative directions of the gradients of various quasi-thermodynamic quantities (heat, magnetic flux, $\partial J/\partial t$ etc.) will be changed in these two cases. The growth of the supercurrent zone is expected to be much slower, at it may depend heavily on the geometry of the hot zone.

It may be actually required that the third dimension should be included in the quantitative estimates of the hot zone recovery, because when the film is contacted only on interface with the substrate, large vertical gradients can appear already during the growth of the non-equilibrium hot zone. This is particularly plausible when the hot zone length is of the same order of magnitude as the height of the film, as was the case in NbN [10]. The supercurrent "shortcircuit" can thus appear on the interface in an area of low normal current density; the supercurrent density in such thin layer may have a very much higher intrinsic critical limit due to the restricted geometry of the layer.

A question often asked by persons used to slower penomena, deals with the appearance of the electric signal in the strip: If a normal zone is shortcircuited or surrounded by superconducting material, how can one get an electromagnetic pulse in the strip?

The answer is yes and the explanation can be made very simple: Let us imagine a spot with normal current, giving an electric field ind the material due to Ohm's law. The field leaks out above and below the film, and actually fieldlines also reach the return conductor which partly shorts the hot zone. The resulting electric field, appearing in the isolator between the stripline and the return ground plane, is associated with displacement current and resulting magnetic field; the electromagnetic pulse will start to propagate first radially as a planar guided wave, and after reaching the sides of the strip, it will propagate along the strip in opposite directions with opposite polarities. One of the waves will reach the shorted end of the line, and it will be reflected back with opposite polarity so that it will combine with the first wave constructively. An observer (amplifier) in the other end of the line will see first the half-pulse propagating towards it, and after a time delay required for the propagation of the opposite pulse back and forth between the hot zone and the shorted end of the line, he will see a superposed step of equal height, making up the voltage drop in the resistive zone. There is thus no restriction from superconducting currents shorting the normal one, and the first edge of the signal is associated with the first dissipative hotspot rather than with the normal zone bridging the strip.

If the downstream end of the strip is open instead of short circuited (allowing, however, for the dc bias current to flow) for the high-frequency pulse, the reflected wave will not have opposite polarity. The two waves will therefore combine destructively, and the observer will see a pulse with amplitude one-half of the voltage across the normal spot, and duration equal to the time delay required for the propagation of the opposite pulse back and forth between the hot zone and the open end of the line. These timing features might be useful for extracting the information on the location of the hotspot along the strip.

3. EXPERIMENTAL RESULTS

3.1. Tests with a NbN strip

Ref. [10] describes tests made on a film with a thickness of 350 nm, deposited on a sapphire substrate by magnetron sputtering. The patterning of the actual strip geometry (see inset in Fig. 1) was made by chemical etching. The resistivity of the film was about 120 $\mu\Omega \cdot$cm, and the dimensions of the strip were 3.4 μm (width) by 18 mm (length).

The substrate was mounted on a Cu heat sink in an RF-tight housing, which itself was placed in a closed-cycle ^{3}He refrigerator precooled by a cryocooler. The upper face of the strip was in vacuum.

A ^{106}Ru source of an activity of about 400 kBq emits β radiation with characteristic energies of 3.5 MeV (79%), 3.0 MeV (12%), and 2.4 MeV (2%), accompanied by two γ-transitions of 511.4 keV (20.6%) and 621.6 keV (10.7%). The source was placed as close as possible to the strip, and was separated from it by only the thickness of the layer of varnish (0.8 mm) that was necessary to prevent contamination of the cryostat.

Handling of the substrate caused some breaks on the strip; the longest uninterrupted piece was about 2.2 mm. Contacts to this remaining strip were made by pressing small In pearls directly onto the NbN film.

In addition, a Nb sheet was placed on top of the strip to act as a superconducting ground plane. One end of the strip was shorted to ground by a small piece of copper.

The characterization of the strip was done by measuring the maximum current I_{max} that could be passed through it before switching to the normal state, as a function of temperature. The results are shown in fig. 1, where the open symbols represent different sets of measurements without the radiation source and the full circles show one set of data taken with it. When approaching lower temperatures, the I_{max} levels off owing to the heating of the strip by the bias current. This point of view is corroborated by the measurement of the voltage drop and d.c. resistance of the strip below the critical temperature, yielding contact resistance values between 3 Ω and 10 Ω. It should also be mentioned that, although the values measured for

I_{max} were reproducible in each single run, thermal cycling to room temperature generally resulted in a different curve, indicating that the contacts had changed.

The intrinsic critical current and surface boundary resistance were determined by fitting the data of Fig. 1 with the Eq. (3). A first value for $I_c(0)$ is obtained from the measurements of I_{max} by extrapolating to zero substrate temperature. As shown in Fig. 1, the flattening of the curve I_{max} versus T_a at lower temperatures is fairly well reproduced by this model. By fitting the above expression to our data, using $I_c(0)$, ε and T_c as variable parameters, we found consistent and reproducible values for $I_c(0)$ and T_c, yielding the intrinsic critical current of our strip, $I_c(0) \approx 120$ mA, and the critical temperature, $T_c \approx 13.8$ K. This gives a critical current density of 10^7 A/cm^2, which is close to the values published in Refs. [21] and [22]. Furthermore, a value for α can be deduced assuming that A_{th} is equal to the strip area covered by the In contacts; we get $\alpha = 190$ W·m^{-2} K^{-4} This can be compared with results from direct thermal experiments; the values of $\alpha = (4R_K T^3)^{-1}$ vary between 60 W·m^{-2} K^{-4} and 140 W·m^{-2} K^{-4} [23] for combinations of materials similar to ours.

We estimate the electronic and phonon heat conductivities $\kappa_e = 0.35$ W/(Km) and $\kappa_{ph} = v_a l_{ph} T^3 = 0.55$ t^3 W/(Km). These give the thermal healing lengths in the superconducting and normal states $\eta_S = 0.31$ μm and $\eta_N = 0.40$ μm, and yield $I(x_0) = 2$ mA by using Eq.(8). Basing on the thermal equilibrium model, signal recovery at bias currents above 2 mA would thus be impossible.

The bias current was fed into the strip through the same cryogenic coaxial line which was used to transmit the signals to the room-temperature electronics. The preamplifier has a gain of 20 dB and its passband extends from 10 MHz to 2 GHz. The Tektronix 7104 oscilloscope, with a 7A29 vertical amplifier and a 7B10 time base, has a bandwidth of 1 GHz.

The signals were recorded with a Tektronix CCD digitizing camera, triggered by the light from the oscilloscope. In this way, single-shot signals, with rise-times down to 300 ps, could be digitized, stored, and studied in a personal computer.

At a temperature of about 5 K and a bias current of 18 mA, which is close to the maximum current (see fig. 1), amplified negative pulses were found with an amplitude of about 60 mV and positive pulses with about 10 mV. Figure 3 shows three typical signals taken with different time scales. The overall shape of these signals was determined by the passband of the detection electronics (10 MHz–1 GHz).

The fact that the pulses persisted up to the critical temperature is a strong hint that they are definitely originating from the NbN.

After thermal cycling and cooling down to 3 K, a check was performed to see if these signals could be found also at currents lower than 18 mA. They could indeed be detected at currents down to 6 mA — even though there had been no signal when the current had been ramped-up the first time. When the current was ramped-up once more from a value lower than 6 mA, the above finding was confirmed, i.e. no pulses were detected at low currents. This hysteretic behaviour indicates that the existence of these fast signals is somehow dependent on the history, suggesting that the self-induced trapped flux could cause the effect.

The rate of the observed large signals agrees roughly with the calculated rate of the fast electrons from the source, about 200 s^{-1}.

Closer examination of the results shows in fact two different kinds of fast, self-recovering signals. The larger ones are correlated with the curve I_{max} versus T_a of the NbN, and the others are correlated with the I_{max} versus T_a of indium (see Ref. [10]).

As the strip was biased with currents between 6 mA and 18 mA only, the critical current around the initial hot spot was not exceeded, basing on Eq. (15). It is then inconceivable that the observed voltage signals could result from transition to normal state over the full width of the strip, in the sense of the thermal equilibrium model.

In addition, if the strip would switch to the normal state over the whole width in the region of the initial hot spot, one would expect to observe an expanding normal zone. Eq. (12) gives $x_0 = 2r_0 = 120$ nm in our case, suggesting that a bias current of about 2 mA would be sufficient to make the original normal region grow. One therefore could not expect to see self-terminating pulses under any circumstances, because Eq. (15) requires a minimum current over 100 mA in order to spread the normal zone over the width of the strip. As the signals do recover at bias currents up to 18 mA, which is almost one order of magnitude more than the above result based on Eq. (8), we can rule out the simple model of Sherman for the description of the results.

The magnitude of the larger signals, however, is not in such serious contradiction with the thermal model. The signal amplitude of 5.5 mV corresponds to a length of 300 nm of the normal zone; this is only two to three times larger than the $2r_0 = 120$ nm given by Eq. (12).

Assuming that the voltage transient $\Delta V(t)$ has linear growth, we may estimate the contribution of the electrical power dissipation in the final non-equilibrium hotspot at $t \approx 100$ ps. Taking $\Delta V(300 \text{ ps}) = 5$ mV from Ref. [10], we find

$$\Delta Q_{Joule} = \frac{I \Delta V d}{2v_a} \approx 30 \text{ keV} \tag{19}$$

in the case of the 350 nm thick NbN strip. This is almost two orders of magnitude larger than the amount of energy lost by the minimum ionizing electron, about 400 eV, and although the result must be taken cautiously because serious oversimplifications were made, it may help explain how non-equilibrium phenomena may enable one to see fast self-recovering voltage transients in high-resistivity strips of much larger width than what the thermal equilibrium model would require. The energy resulting from Eq. (19) would actually be able to drive a 300 nm length of the strip in the normal state, suggesting that the nonequilibrium transient, with duration time scale determined by the acoustic phonon velocity and strip thickness, would result in the pulse height observed in the tests of Ref. [10].

3.2. Tests with other materials

Early tests using alpha-particles on indium, tin and aluminium were inconclusive regarding the detection of minimum ionizing particles, because extrapolation to energy deposits smaller by a factor of about 200 was uncertain. Recent tests by the Argonne group[9, 24] of niobium and granular-aluminium strips, using 1 MeV electrons and 6 keV X-rays, show evidence for the detection of minimum ionizing particles. However, there seems to be no plateau of count rate, which is an indication of a low detection efficiency. The signals were non-recovering and could be explained by the thermal equilibrium model, which also could provide a mechanism for the apparently poor efficiency, based on the extremely small strip width and the resulting distribution in the critical current values along the line.

The tests were conducted using rather slow electronics, and it was already speculated by the authors that possible fast non-equilibrium signals could be missed, because of their speed and expected small amplitude in the chosen materials.

The ideal materials for fast self-terminating non-equilibrium signals have high resistivity and high critical current density. Both requirements lead to high transition temperatures, and to materials which are regarded as dirty superconductors. Granularity of the material may help in adjusting the quasiparticle transport properties. Granular high-T_c super-conductors, operated around 4 K, could therefore offer interesting possibilities in microstrip detectors.

4. SUPERCONDUCTING TRANSMISSION LINES

Because an important potential advantage of superconducting strip counters is the combination of fast pulses with slow propagation down the stripline, making serial readout possible, we shall briefly review below the theory of superconducting transmission lines.

The impedance of a normal strip line is given by

$$Z_L = Z_0 \frac{h}{\sqrt{\varepsilon_r^{eff}} \ w^{eff}} \quad , \tag{20}$$

where $Z_0 = 120\pi \ \Omega$, and h is the thickness of the isolating film on the ground plane and ε_r^{eff} its effective dielectric coefficient at the frequency of operation. The effective width w^{eff} is due to the edge leak fields and it is calculated from an empirical formula

$$\frac{w^{eff}}{h} = \frac{w}{h} + 2.46 - 0.49 \frac{h}{w} + \left(1 - \frac{h}{w}\right)^6 \quad , \tag{21}$$

where w is the geometric width of the strip. In a typical case case $w \sim 2 \ \mu m$ and $h \sim 20$ nm; with $\varepsilon_r^{eff} \sim 80$ we find $Z_L \sim 0.3 \ \Omega$. Such a low impedance will need to be matched to normal transmission lines (50 Ω coaxial cable) when using normal electronics for detector testing, and it would be tempting to use a stripline impedance transformer to perform the matching.

In practice the width of the strip is always much larger than the dielectric height in our case, and therefore the fringing field correction is at most a few percent.

In superconducting materials the penetration of electric and magnetic fields in the conductor behaves in a different manner, and the above formulas cannot be applied directly in the case of isolator thickness $h \sim 20$ nm. Kautz[25] has analyzed such lines in terms of the theory of Mattis and Bardeen[26] for the complex conductivity of a BCS superconductor. At frequencies well below the gap frequency (~ 1 THz) the impedance is constant and can be approximated by

$$Z_{SC} = Z_0 \frac{h}{w^{eff}} \sqrt{\frac{1 + \frac{\lambda}{h} \coth \frac{h}{\lambda}}{\varepsilon_r}} \quad , \tag{22}$$

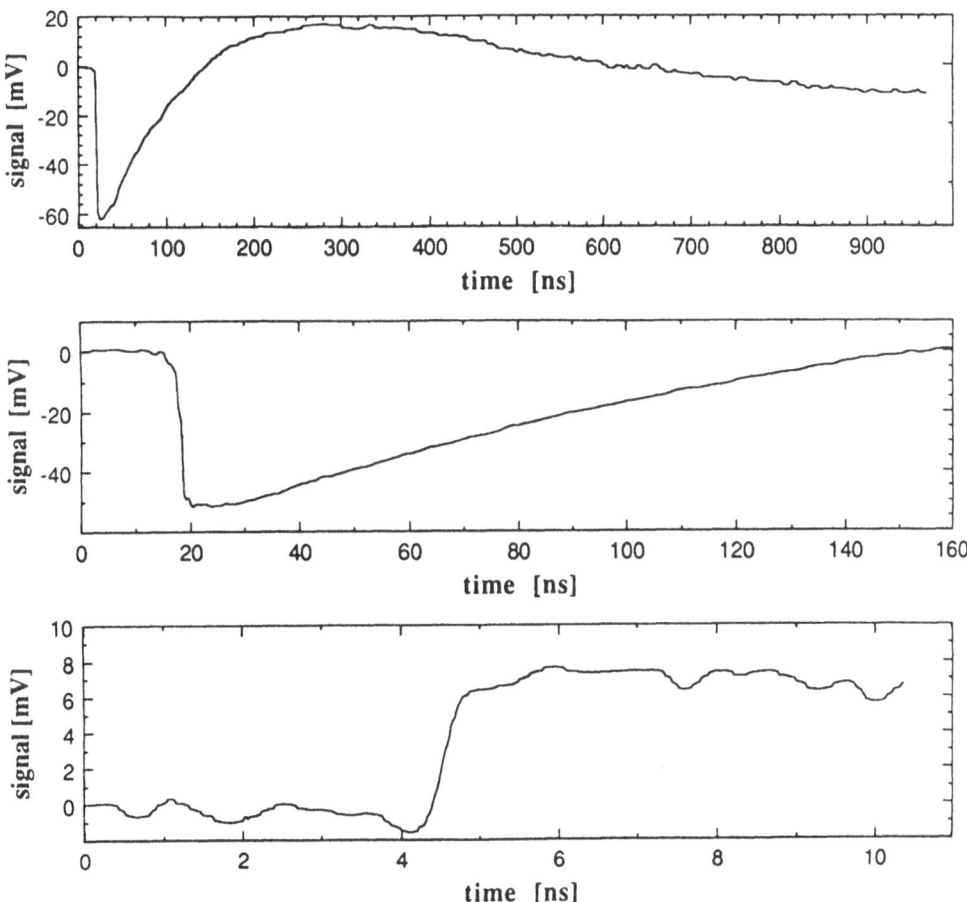

Figure 3. Three typical signals due to MIP's traversing the strip shown in Fig. 1 (from Ref.[10]). The first one shows the decay time due to the preamplifier high-pass cutoff at about 10 MHz. The third one shows the oscilloscope rise time of about 300 ps, due to the 1 GHz passband of its vertical preamplifier. The two first signals may indicate that a slow recovery of the hotspot is beginning at about $\Delta t = 20$ ns from the rising edge.

where $\lambda \sim 50 - 500$ nm is the magnetic field penetration depth in the superconducting material.

The speed of signal propagation (phase velocity) in a normal strip line is approximately

$$v_N = \frac{c}{\sqrt{\varepsilon_r^{eff}}} \tag{23}$$

If the conductors are in the superconducting state, the signal propagation is slower than this because the magnetic field can penetrate about $\lambda \sim 50 - 500$ nm into the metal, while electric fields are almost completely excluded[27] This results in the signal propagation speed[28] which is constant at frequencies below the gap frequency and is given approximately by

$$v_{SC} = \frac{c}{\sqrt{\varepsilon_r \left(1 + \frac{2\lambda}{h}\right)}} \tag{24}$$

For thin isolators ($h \sim 1 - 2$ nm in Josephson junctions) this can be less than one percent of the speed of light, and can be of order few percent of c in the case of 20 nm TiO_2 dielectric.

The theoretical[28] attenuation in superconducting microstrip lines is very small, less than 10^{-2} dB/m at frequencies below the value corresponding to the energy gap, provided that the operating temperature is below 0.2 T_c. This and the very low dispersion has been beautifully demonstrated using sub-picosecond pulses and optical sampling techniques[29].

The impedance matching of the superconducting transmission line may be necessary not only if normal 50 ohm transmission lines and electronics will be used, but also for matching the strips with specific superconducting logic circuits. Thin film impedance transformers could be fabricated on the same substrate as the detector, in order to conserve the widest possible bandwidth in the circuitry.

In addition to the matching circuits required for adopting the impedance of the signal strips with the transmission lines and readout, also other stripline components could be built on the detector substrate. The potentially useful circuits include at least directional couplers, splitters, combiners, circulators, isolators, loads, shorts etc. The motivation for such functions on the substrate is mainly in the possibility of analog handling of the signals before entering into the superconducting logic, with a view on possibly simplifying the readout logic.

5. SUPERCONDUCTING LOGIC CIRCUITS

Superconducting logic is a well studied emerging technology. Its advantages are not only ultra-high speed and remarkable compactness, but also extremely low power consumption. Operating in the microwatt range, the superconducting logic would provide almost negligible heat load in the interior of the detector, whereas this may be a major problem with conventional semiconductor logic.

The superconducting tunnel junction (STJ) with zero-voltage Josephson current works as a perfect gate for logic circuits. The two states, Josephson and Resistive, transmit or reflect clock pulses which may be scaled, shifted, added etc. in the subsequent circuitry. The transition between the two states of a gate can be controlled by a capacitively coupled input or by direct current injection; the signal from a strip detector may be sufficient to set or reset a gate with suitable characteristics, as the pulse energy required is many orders of magnitude below those required for fast semiconductor gates.

The practical realization of STJ logic functions was developed more than 10 years ago by IBM[30]. The current injection logic (CIL)[31] uses several STJ's for each logic function, and is capable of reasonable fan-in and fan-out so that large logic arrays can be designed. The single logic levels are not clocked but latched. The total IBM gate[31] delays were 27 to 45 ps for OR and AND functions, with power dissipations 1.5 to 6.0 μW. Data are read at initial latch and propagate asynchronously through the logic, and are again synchronized at output latches. This suits very well for operation in the readout logic of a collider detector, where the latch frequency can be the accelerating frequency or its harmonic. The vertex detector could thus be read out between bunches separated by no more than one nanosecond.

IBM have stopped the superconducting logic development, while Japanese companies have continued work in this field. Fujitsu has recently announced a 4-bit microprocessor integrating 5000 STJ's on a single chip. NEC has built a 1 k DRAM with about 1 million STJ's on board. The companies are expected to announce a 16-bit STJ microprocessor in the next few years.

The STJ gates cannot be operated at high fields, which is a handicap for vertex detectors. It could, however, be envisaged that the readout logic could be sufficiently shielded from the magnetic field by miniature superconducting shields and/or compensation coils. Furthermore, high-resolution microvertex detector could operate at such moderate fields that shielding could be relatively unsophisticated.

The construction of a scaler based on a STJ gate is thus rather straightforward. By counting clock pulses at 10 GHz, for example, controlled by a STJ gate, a time-to-digital conversion can be made with a resolution compatible with the detector rise time. If a detector stripline is terminated with a short circuit, the output pulse is clipped so that its length corresponds to the double propagation time from the track hit to the shorted end. With propagation speed of 0.06·c this leads to 1 mm hit resolution along the length of the strip. This information could be used for resolving the ambiguity of multiple hits in a detector with two crossed layers of strips, greatly simplifying the readout of the detector.

The STJ memory could store the hit information until trigger processors will ask for the readout outside the detector. Some of the information could be pre-processed in the STJ logic for use in the trigger decision. In this way part of the processing required for a particular trigger can be speeded up, firstly by enabling parallel processing, and secondly by the clock speed about two orders of magnitude above silicon technology.

6. DISCUSSION ON DETECTOR DESIGN

6.1. Position resolution

A speculative superconducting strip tracker could reach a 2 μm transverse resolution by using interleaved overlaid layers of strips, separated by a ground plane. The position resolution along the strip could be a few mm, helping to resolve multiple hits without many crossed detector planes. This may be an important feature in the case of very high multiplicities and dense jets.

6.2. Momentum resolution

One of the tempting ideas in future large collider detectors is to try to miniaturize the central tracking devices so that the total detector volume and hence mass could be kept reasonable. This pushes to perform high resolution tracking as close to the vertex as possible. Supposing that the radiation damage problem can be resolved by the proposed superconducting strip devices, we may speculate on conceptual design ideas in order to set reasonable goals for the specifications for these devices.

Let us consider the compact low-field tracker proposed by T. Ypsilantis[32] which has a coil diameter of 0.5 m and field B = 0.5 T. The momentum resolution for tracks perpendicular to the field, with N points measured with precision of δx along the track length of L' is[33]

$$\frac{\delta p}{p} = \frac{\delta x}{L'^2/(1\ m)} \sqrt{\frac{720}{N+5}} \frac{\left(\frac{p}{GeV/c}\right)}{0.3Z \left(\frac{B}{1\ T}\right)} \tag{25}$$

where Z is the charge of the particle. Assuming N = 5, L' = 0.2 m and δx = 2 μm we get

$$\frac{\delta p}{p} = 28\% \left(\frac{p}{100\ GeV/c}\right) \tag{26}$$

which would enable one to distinguish clearly the tracks of rare high-momentum leptons from soft hadronic background.

6.3. Detector thickness and coil design

The superconducting strip thickness is negligible and the substrates could be made of high-conductivity silicon of 100 μm thickness. The total thickness of the vertex detector could thus be about 0.5 mm of Si, together with 0.5 mm of Cu for supporting and cooling the substrates. The 0.5 T coil could be superconducting, which would avoid vacuum enclosure walls within the track device. Such a field can be created with a coil of NbTi/Cu wire with total coil thickness of 0.5 mm, with an Al support of about 1 mm thickness; these numbers are based on a smaller coil[34] operated at 1 K temperature and cooled by a film of superfluid helium only. The detector and the coil would therefore each add about 4% of the radiation length to the thickness of the beam tube.

7. ACKNOWLEDGEMENTS

The authors would like to thank F. Dydak for support and encouragement, and T. Ypsilantis for interesting discussions and helpful advise on high-luminosity vertex tracking detectors.

REFERENCES

1. F. Ruller-Albenque, Effets d'Irradiation dans les Superconducteurs, Laboratoire des Solides Irradiés Preprint (1987).
2. A.K. Drukier, T.A. Girard, R. Gross, R.P. Huebener, U. Klass, A.D. Silva, M.L. Gros and B. Turrell, "Tests of Radiation Hardness of Superconducting Particle Detectors", in Low Temperature Detectors for Neutrinos and Dark Matter II, Annecy-le Vieux 1988; L. Gonzalez-Mestres and D. Perret-Gallix, ed., Editions Frontieres, Gif-sur-Yvette (1988) p. 357.
3. N.K. Sherman, Phys. Rev. Lett. 8:438 (1962).
4. D.E. Spiel, R.W. Boom and E.C. Crittenden, Appl. Phys. Lett. 7:292 (1965).
5. E.C. Crittenden and D.E. Spiel, Superconducting Thin-Film Detector of Nuclear Particles, J. Appl. Phys. 42:3182 (1971).
6. K.W. Shepard, W.Y. Lai and J.E. Mercereau, J. Appl. Phys. 46:4664 (1975).
7. N. Ishihara, R. Arai and T. Kohriki, Proposed High-Resolution Vertex Detector Using Superconductive Thin Strips, Jap. J. Appl. Phys. 23:735 (1984).
8. R.G. Wagner and K.E. Gray, "Superconducting Thin Film Vertex Detector for SSC Applications", in Proc. Int. Workshop on Superconductive Particle Detectors, Turin 1987; World Scientific, Singapore (1988) p. 204.
9. A. Gabutti, K.E. Gray and R.G. Wagner, Granular-aluminum superconducting detector for 6 keV X-rays and 2.2 MeV beta sources, Nucl. Instrum. Methods A289:274 (1989).
10. B. Strehl, T. Niinikoski, A. Rijllart, K. Winter and M. Caria, Observation of Subnanosecond Transients in a Superconducting Microstrip Exposed to Minimum Ionizing Radiation, Phys. Lett. B242:285 (1990).
11. T.O. Niinikoski, A. Rijllart, B. Strehl, K. Winter and M. Caria, "Superconducting Strips for Microvertex Detectors", in Proc. 25th Int. Conf. on High Energy Physics, Singapore 1990; K. K. Phua and Y. Yamaguchi, ed., World Scientific, Singapore (1991) .
12. W.J. Skocpol, M.R. Beasley and M. Tinkham, J. Appl. Phys. 45:4054 (1974).
13. R.F. Broom and E.H. Rhoderick, Br. J. Appl. Phys. 11:292 (1960).
14. K.E. Gray, R.T. Kampwirth, J.F. Zasadzinski and S.P. Ducharme, Thermal Propagation and Stability in Superconducting Films, J. Phys. F: Met. Phys. 13:405 (1983).
15. J. Baixeras, P. Andro and M. Cazabat, Transport Properties of Superconducting and Non-Superconducting Sputtered Thin Films of NbN, IEEE Trans. Mag. 11:1464 (1975).
16. E.M. Lifschitz and L.P. Pitaevskii, "Statistical physics", Pergamon Press, Oxford (1980).
17. K. Komenou, T. Yamashita and Y. Onodera, Energy Gap Measurement of Niobium Nitride, Phys. Lett. A28:335 (1968).
18. P. Marksteiner, P. Weinberger, A. Nickel, R. Zeller and P.H. Dederichs, Electronic Structure of Substoichiometric Carbides and Nitrides of Zirconium and Niobium, Phys. Rev. B33:6709 (1986).
19. A. Rothwarf and B.N. Taylor, Phys. Rev. Lett. 19:27 (1967).
20. K.E. Gray, "Nonequilibrium Superconductivity for Particle Detectors", in Proc. Int. Workshop on Superconductive Particle Detectors, Turin 1987; World Scientific, Singapore (1988) p. 1.
21. J.R. Gavaler, A.T. Santhanam, A.I. Braginski, M. Ashkin and M.A. Janocko, Dimensional Effects on Current and Field Properties in NbN Films, IEEE Trans. Mag. 17: 573 (1981).
22. R.T. Kampwirth and K.E. Gray, NbN Materials Development for Practical Superconducting Devices, IEEE Trans Mag. 17: 565 (1981).
23. O.V. Lounasmaa, "Experimental principles and methods below 1 K", Academic Press, New York (1974).

24. A. Gabutti, R.G. Wagner, K.E. Gray, R.T. Kampwirth and R.H. Ono, Superconducting Detector for Minimum Ionizing Particles, Nucl. Instrum. Methods A278:425 (1989).

25. R.L. Kautz, Picosecond Pulses in Superconducting Striplines, J. Appl. Phys. 49:308 (1978).

26. D.C. Mattis and J. Bardeen, Theory of the Anomalous Skin Effect in Normal and Superconducting Metals, Phys. Rev. 111:412 (1958).

27. M. Tinkham, "Introduction to Superconductivity", McGraw-Hill, New York (1975).

28. R.L. Kautz, Picosecond Pulses on Superconducting Striplines, J. Appl. Phys. 49:308 (1978).

29. W.J. Gallagher, C.-C. Chi, I.N. Duling III, D. Grischkowsky, N.J. Halas, M.B. Ketchen and A.W. Kleinsasser, Subpicosecond Optoelectronic Study of Resistive and Superconsuctive Transmission Lines, Appl. Phys. Lett. 50:350 (1987).

30. J. Matisoo, Overview of Josephson Technology Logic and Memory, IBM J. Res. Develop. 24:113 (1980).

31. T.R. Gheewala, Design of 2.5-Micrometer Josephson Curent Injection Logic (CIL), IBM J. Res. Develop. 24:130 (1980).

32. T. Ypsilantis, private communication (1990).

33. Particle Data Group, Phys. Lett. B204:62 (1988).

34. T.O. Niinikoski, "Recent developments in polarized targets at CERN", in High Energy Physics with Polarized Beams and Polarized Targets, Argonne, Ill. 1978; G. H. Thomas, ed., AIP Conference Proceedings No. 51, American Institute of Physics, New York (1979) p. 62.

STATUS OF THE LEADING PARTICLE DETECTION

G. Anzivino[1], R. Ayad[2], M. Chiarini[2](speaker), P. Ford[3],
M. Hourican[3], H. Larsen[3], B. Lisowski[4], T. Massam[4], C. Nemoz[3],
J. Shipper[3] and Y. Zamora[2]

1) INFN, Laboratori Nazionali di Frascati, Italy
2) World Lab, HED Project, Geneva, Switzerland
3) CERN, LAA Project, Geneva, Switzerland
4) INFN, Bologna, Italy

1.Introduction

As it has been proved in pp collision at the CERN-ISR several years ago[1], the subtraction of the energy carried away by the "leading" particles (i.e., the particles which take one or more of the incident hadron valence quarks) from the total interaction energy is of paramount importance to determine the real energy involved in multi-particle production and, therefore, to compare multi-particle production in different interactions (purely hadronic, DIS, e^+e^-).

Moreover, in future Multi-TeV hadron colliders a signature for new physics will be the presence of a large "missing energy". The detection of the leading particles will allow to measure it more precisely.

Leading proton detection alone covers several fields of investigation which are characterized mainly by the degree of coherence of the proton (see Fig.1). In the highly coherent region ($x_F = 2p_L/\sqrt{s} > 0.8$) it covers elastic scattering and diffractive processes. In the intermediate range ($0.3 < x_F < 0.8$) it covers the *"Leading Physics"*, where, in the regime of multi-particle production, a leading proton carries away the greatest share of the longitudinal momentum. In this region the average momentum carried by the leading proton is a smooth function of the numbers of quarks which are common to the incident and to the final state particle.

New Technologies for Supercolliders, Edited by L. Cifarelli and
T. Ypsilantis, Plenum Press, New York, 1991

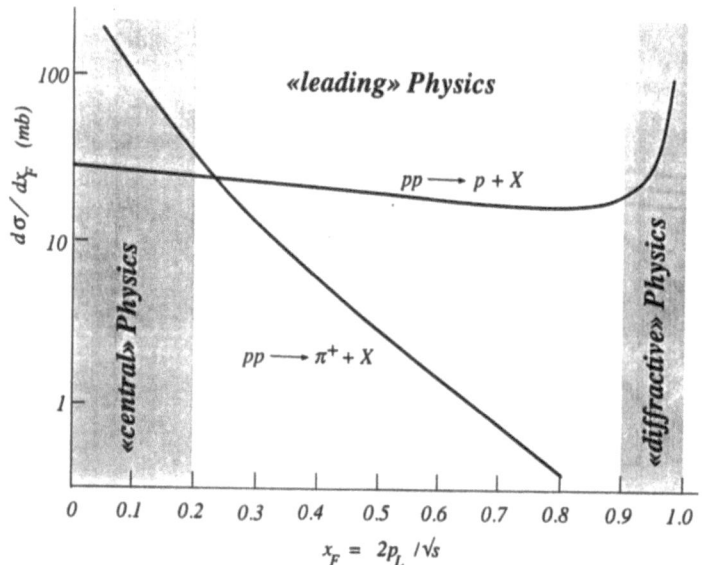

Fig.1. Comparison between longitudinal momentum distributions of the proton and the
pion produced in pp interactions.

One of the components of the LAA Project, the Leading Particle Detection group, has started two years ago to study the problem of measuring particle momenta in the very forward region of a multi-TeV hadron collider.

The need, caused by the very high background and data rates, to record all data from each and every beam crossing without ambiguity or superposition of the data from adjacent beam crossings is one of the critical points. This becomes increasingly difficult as the beam crossing interval decreases well below 100 ns. The other crucial point is the control of the effects of radiation damage to detectors and electronics. Up to now, the task of the Leading Particle Detection group has been the demonstration of the solution to these problems in a working spectrometer : HERA was chosen as the first available testing ground.

2. The prototype: Research and Development

An accelerator is a sequence of bending and focussing elements. The constraints on their positions and the spaces between them must fit in with the overall machine design, even when they are being used as a spectrometer. To make a really efficient spectrometer would require the design of special beam transport elements with wide apertures and of special vacuum tanks.

To try in a real prototype our ideas, it was not possible to start redesigning the nearly finished accelerator (HERA), so the "exercise" was limited to identify the candidate spectrometer regions, to modify the vacuum pipe and to adjust the magnet positions so far as possible to improve the overall acceptance of the detector. The lesson, which we have learned so far, on interfacing the Leading Particle Detector to HERA [2], has been specific to an e-p machine. It can, nevertheless, be applied also to other accelerators.

The achievements and the records obtained in these years of R&D work on our prototype detector – the Leading Proton Detector of HERA – are the answers to five fundamental requirements:

1) optimization of the acceptance in difficult accelerator geometry,
2) radiation hardness of all electronics and construction materials,
3) miniaturization and integration of components,
4) high precision in position measurements,
5) capability to acquire high event rates.

Fig.2 shows a plan view of the spectrometer which is being implemented at HERA: S1-S6 indicate the position of the so called "stations" of our apparatus. The problem of availability of space at HERA compelled us to use, in the first three stations, a detector system which covers only half of the angular acceptance, as can be seen in Fig.3.

In each station we can identify three main components: six planes of silicon microstrip detectors , the mechanical system to support and to move those detectors, and the electronics to read the microstrips.

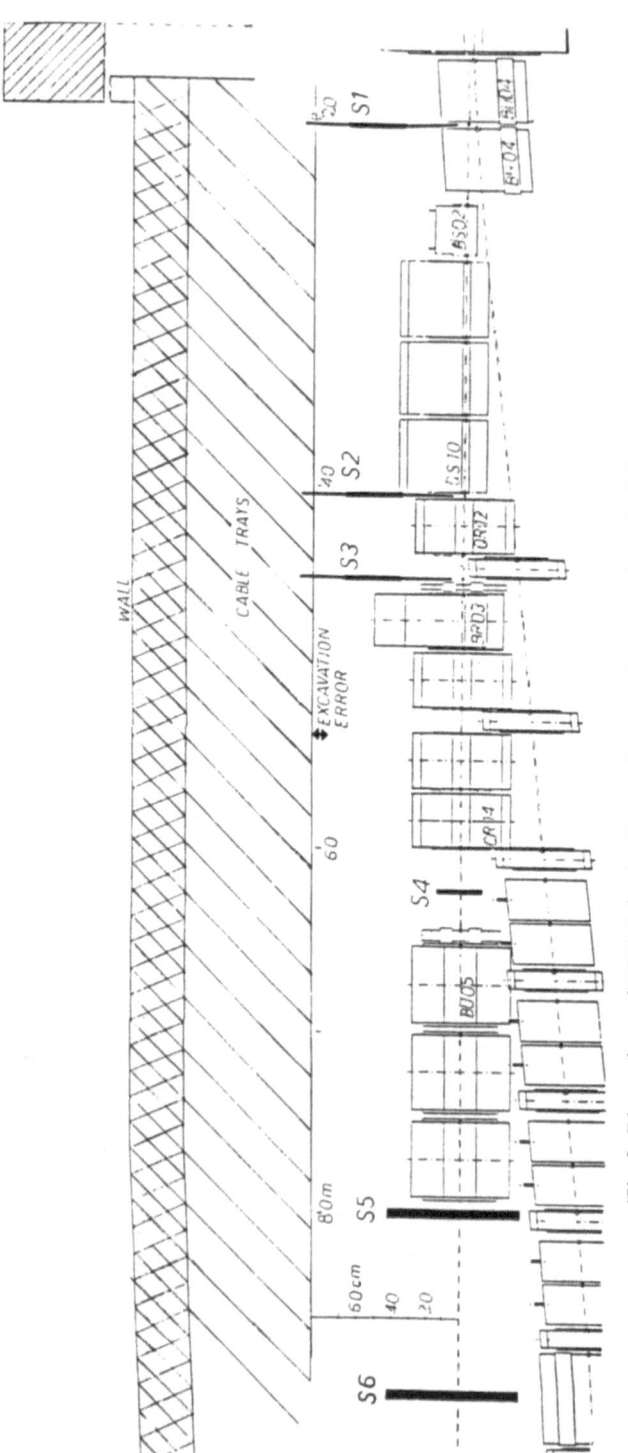

Fig.2. Plan view of HERA in the forward proton beam region. S1-S6 are the Leading
Proton Detector stations.

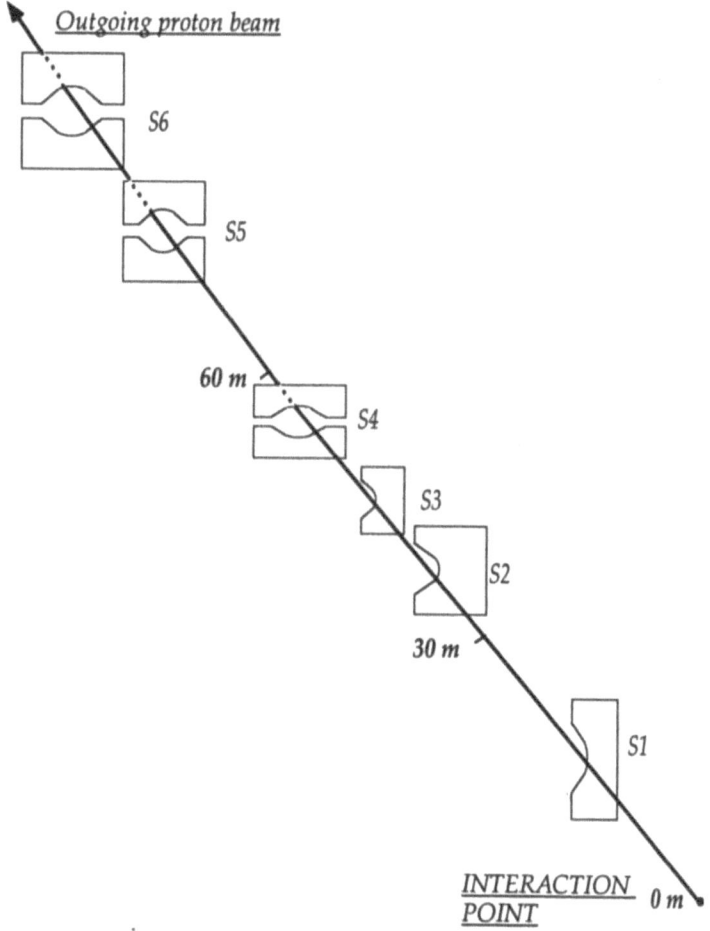

Fig.3 General layout of the spectrometer along the beam line at HERA (detectors are not to scale). Dimensions and shapes are different in each station.

Fig.4 shows the sequence of the six planes of silicon microstrip detectors: there are three strip inclinations (0°, +45°,-45°) and the pitch is such as to obtain a 25μm resolution. The dimensions of the detectors have been chosen to fit the presently available 4" wafer technology.

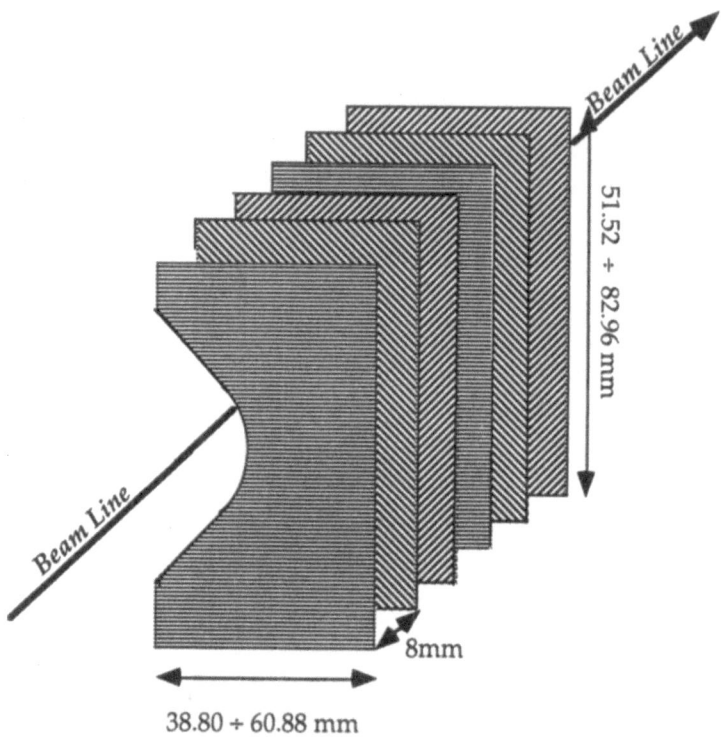

Fig.4 The six planes of silicon microstrip detectors, constituting a "station".

The detectors have a rectangular shape with an elliptical cut-out which matches the nominal 10σ beam profile: this means that each detector have to be cut with a particular shape adapted to the beam profile. Moreover, the cut has to be "clean": this means that cracks on silicon had to be below 50μm.

At the beginning of our work, the available methods for cutting silicon detectors were, essentially, two: diamond saw and LASER cutter. After some tests we decided to use a diamond saw: in Fig.5 there are two macro-photoes of the silicon which was cut by a saw with a 220 μm diameter wire and a 240m/min cutting speed.

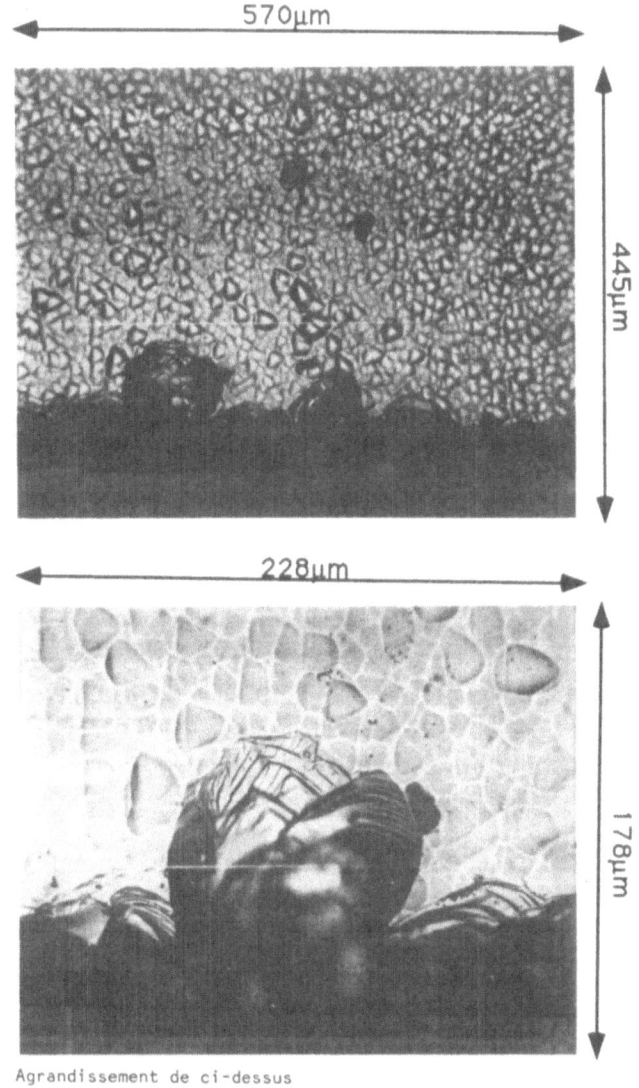

570μm

445μm

228μm

178μm

Agrandissement de ci-dessus

Fig.5.Cutting test results. In this case a diamond saw has been used: the wire diameter was 220 μm and the cutting speed was 240m/min.

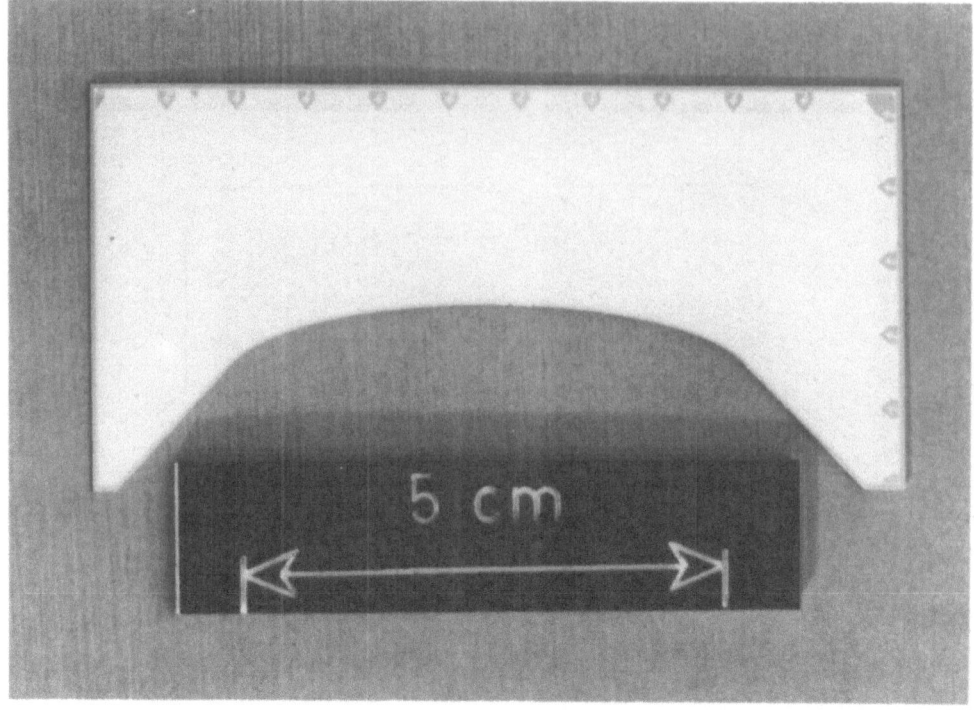

Fig.6 One of the silicon microstrip detectors produced by Micron, and cut with a diamond saw.

At the same time, we contacted some firms to explore their availability in helping us in this totally new field. Only two of them – Micron, England, and Intertechnique, France – were ready to undertake this work. After one year of tests, Intertechnique has given us several silicon microstrip detectors which have a very good standard, and which were cut using a LASER.

Micron provided us with several other detectors cut with a diamond saw: one of these is shown in Fig.6.

A problem which becomes important even at the relatively low energy of HERA, is the radiation resistance of the detectors. Waiting for new materials (GaAs), we are bound to use silicon. To understand better the behavior of the silicon microstrip detectors under irradiation, several tests, using different radiation sources, ^{60}Co γ-source and neutrons, are being carried on. The sample irradiated with γ's seems to survive to an applied dose of 17.97 Mrad[3]. Figure 7 shows the pulse height spectra for a group of ten strips, before and after irradiation.

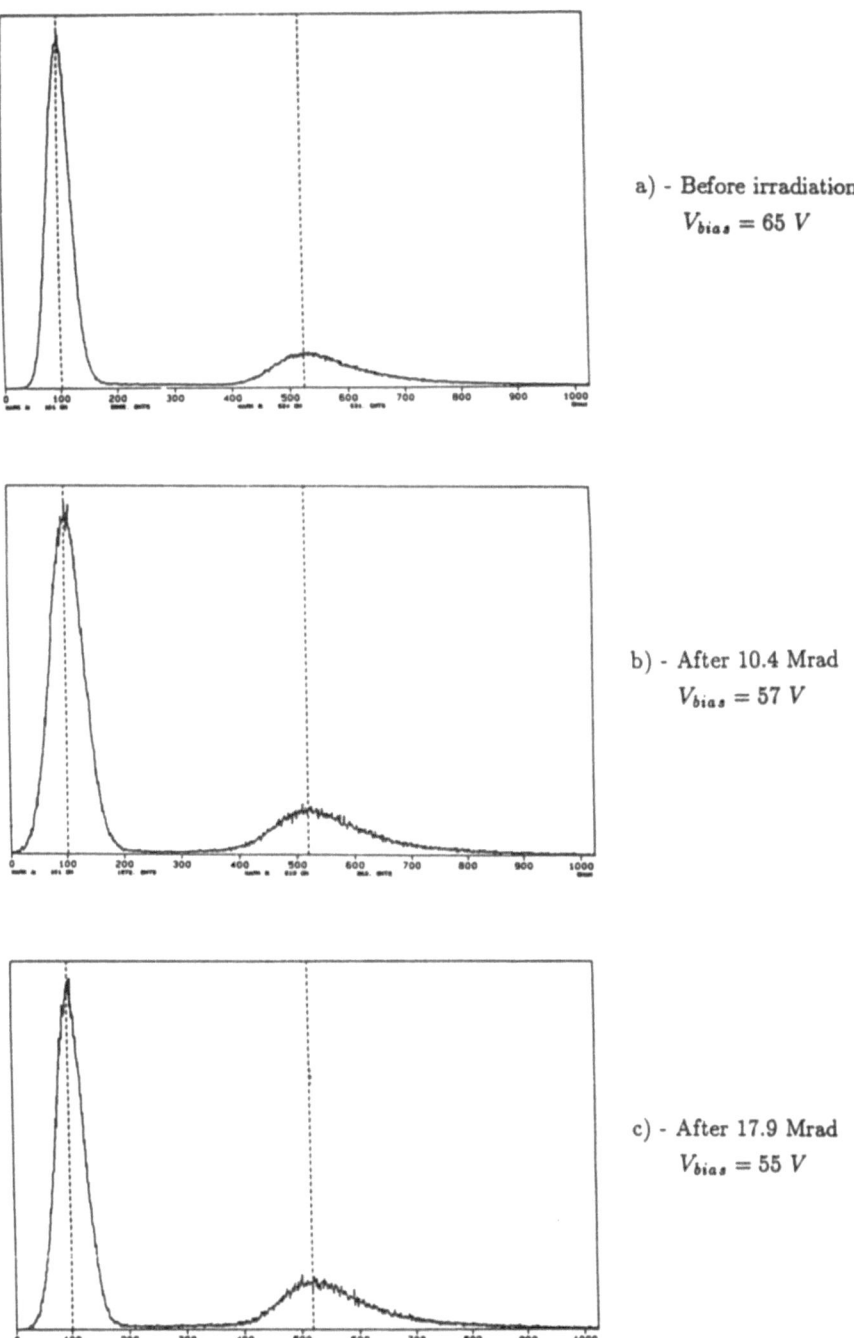

a) - Before irradiation
$V_{bias} = 65\ V$

b) - After 10.4 Mrad
$V_{bias} = 57\ V$

c) - After 17.9 Mrad
$V_{bias} = 55\ V$

Fig.7 Pulse height spectra for ten strips. The sample was irradiated with a ^{60}Co γ-source.

Our R&D work in mechanical design and construction was intended to solve the problems of acceptance optimization and radiation hardness.

In the prototype, the detectors will be mounted in retractable pots: in this way the detectors will be close to the beam, but kept at atmospheric pressure. Pots have been designed to have a "simple" drive system compatible with the need for precise positioning (20μm) of the detector and for large displacements for radiation protection. Two movements are possible: one, along the axis, is used to displace the detectors to a safe radial distance or to bring them very close to the beam with a careful control over the last 5 cm; the other is orthogonal to the beam so as to allow a precise positioning in case of small errors (± 2mm) in the beam position within the machine. In Figs.8a) and 8b) the general designs of mechanical system are shown. Figure 8a) shows how the pot is inserted into the beam-pipe; figure 8b) shows the cut view of the mechanical assembly for station S4 of the prototype. Fig.9 is the photo of the system built for station S4..

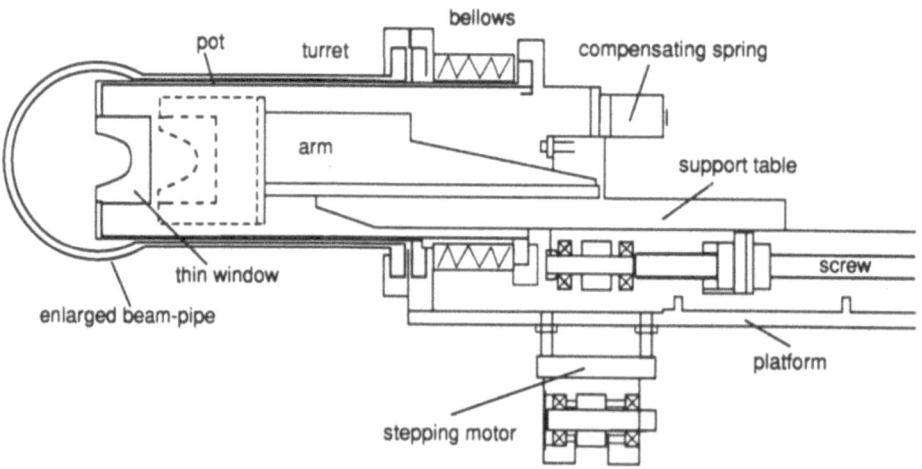

Fig.8a) The beam-pipe and the pot.

In the prototype, S4 is a complete station (see Fig.2): one set of detectors comes down from the top and the other one comes up from the bottom, both towards the beam. In fig.9 one can single out easily the "dummy pipe" used during testing at CERN, the constant tension springs, the two drive systems, the bellows and the flanges shown in the design.

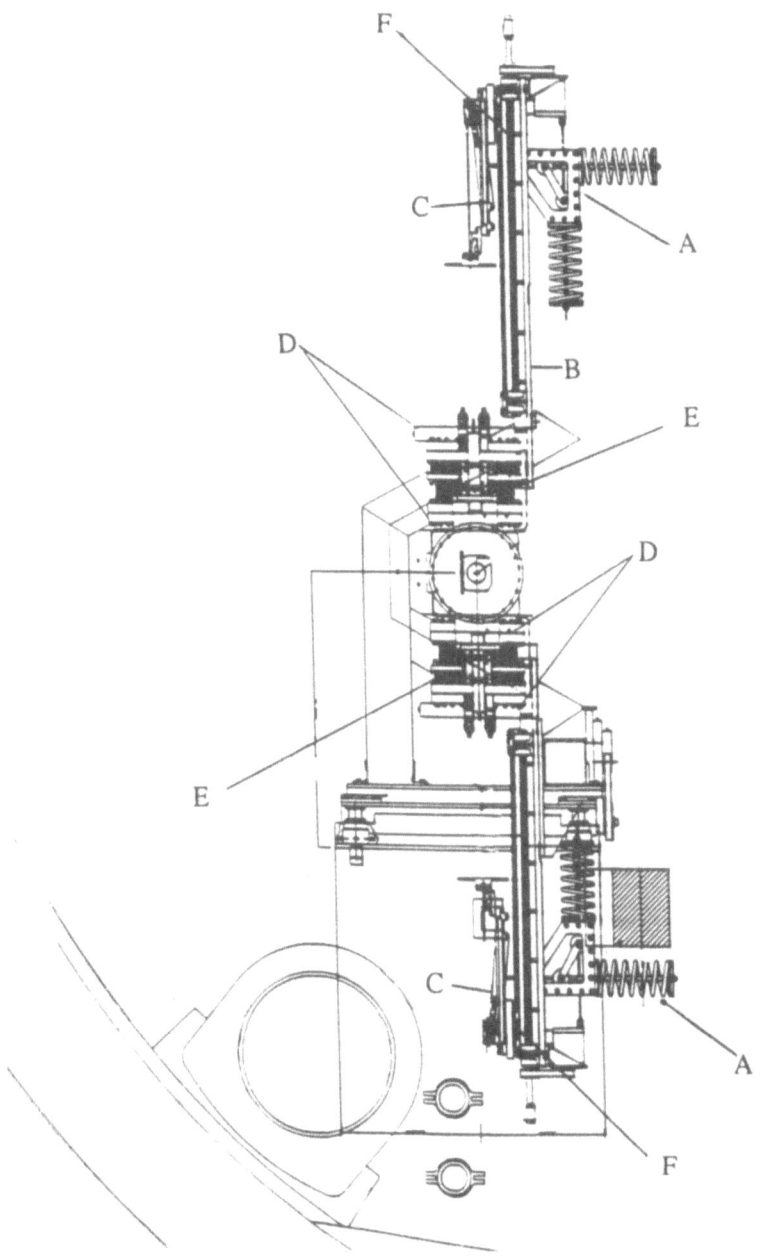

Fig.8b) Cut view of the mechanical assembly for S4. In this picture the pot, shown in previous figure is not drawn. A) constant tension spring, B) platform, C) arm, D)flanges, E)bellows, F) drive system.

Fig. 9. Station S4: mechanical system.

The constant tension springs are our answer to radiation problems. In fact, in order to balance the vacuum existing inside the beam-pipe, O-rings and lubricants cannot be used, as radiation will modify their characteristics.

Therefore, we were compelled to develop a purely mechanical system which does not degrade under radiation.

Fig.10. Constant Tension Springs.

2.3 Electronics

Another very important component of the apparatus is the electronics developed to read the microstrips. Here, the R&D work followed two lines: the development of the front-end chips and the development of a multi-layer printed circuit board to hold the chips and the detectors.

In the prototype detector we have around 5000 strips per station: we can not take all signals out of the pot and this implies the need for multiplexing of the signals. We will use two VLSI chips mounted, like the detector, on a multi-layer printed circuit board (see Fig.11).

The total charge produced by a particle hitting the detector is about 4fC with a collecting time of 20nsec, so the current is 20nA. Moreover, the chips will be exposed to an high level of radiation. These two constrains determine the chip performances: low noise, low power dissipation, high speed, and high radiation resistance.

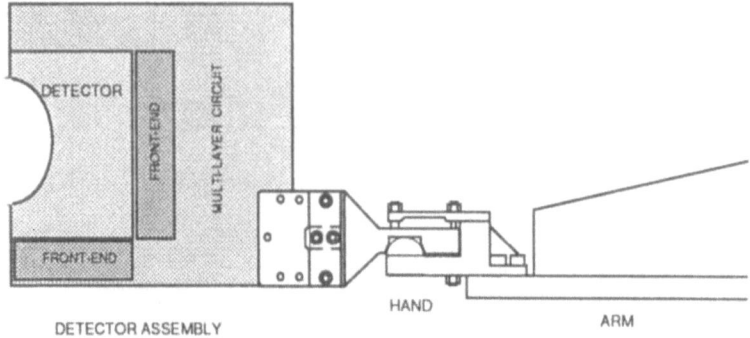

Fig.11 The mounting scheme: the detector and the front-end chips are glued on the multi-layer printed circuit board.

Figure 12 shows the general architecture of the front-end chips. Different technologies are used for the analog and digital parts. Therefore, two independent chips, each 64-channels wide, have been built: the Analog Amlplifier and Comparator Chip (AACC) and the Digital Time Slice Chip (DTSC). The AACC has been built using dielectric isolated silicon bipolar technology: this technology is known to be radiation hard (we can resist at 3 MRad) and allows us to achieve sufficient speed at a reasonable noise level. The second chip (DTSC) has been designed in CMOS 2μm technology. This technology is not intrinsically radiation resistant, so the chip will be hardened with a special industrial process at UTMC, USA[4].
The front-end chips need some input signals (threshold voltage to comparator, reference voltages, first level trigger, etc.) and, at the same time, the output signals have to be sent out to the read-out: system. All these needs have been satisfied by the multi-layer printed circuit board shown in Fig.13. This multi-layer circuits will be also used as a support for the silicon microstrip detector. A cut view of the multi-layer circuit is shown in Fig.14: it is a "sandwich" of epoxy, Copper and Copper-INVAR-Copper. These materials have been chosen to guarantee a good flatness and to have a thermal expansion comparable with the Silicon's one. The limited space and the high electronics density makes the cooling of the multi-layer and of the attached detector a very delicate task. We are studying the possibility to insert another "special" layer in the printed circuit board. Water at 20°C will flow inside a pipe running in this layer under the chips, where the temperature is expected to be the highest.

3. Conclusions

Table 1 summarizes the present achievements of LAA in the R&D on Leading Particle Detection.

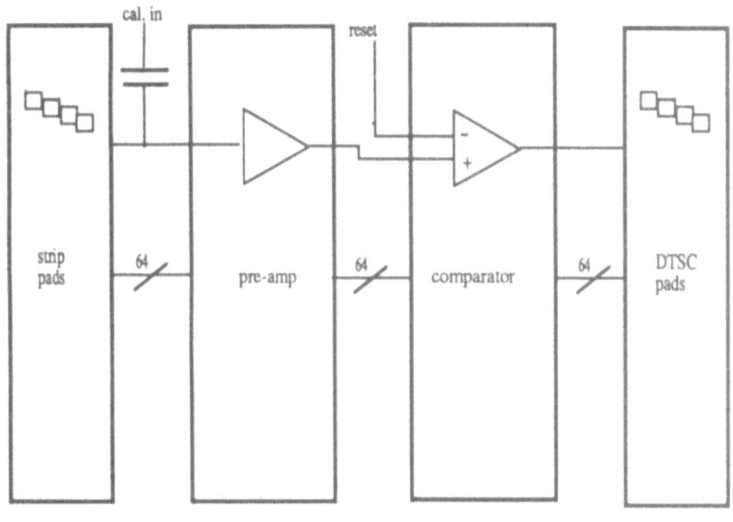

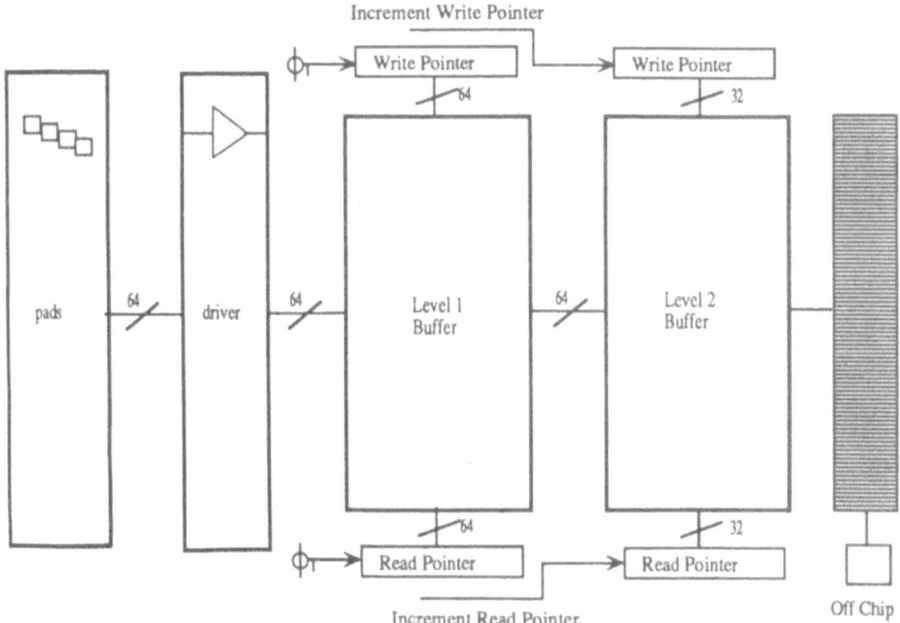

Fig.12 Architecture of the front-end chips.

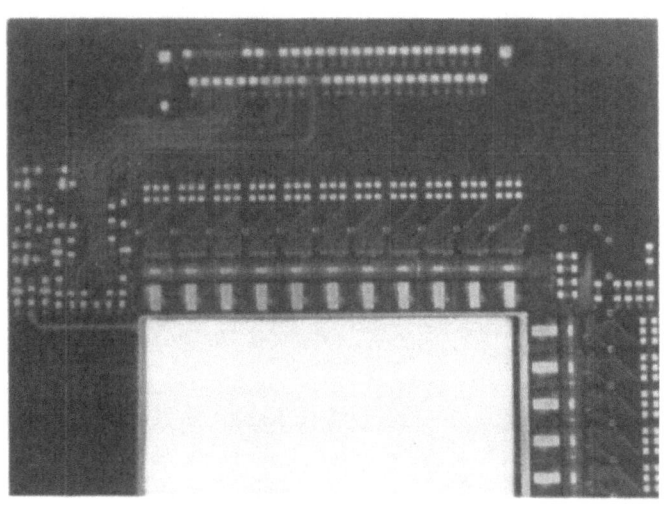

Fig.13 The multi-layer PCB.

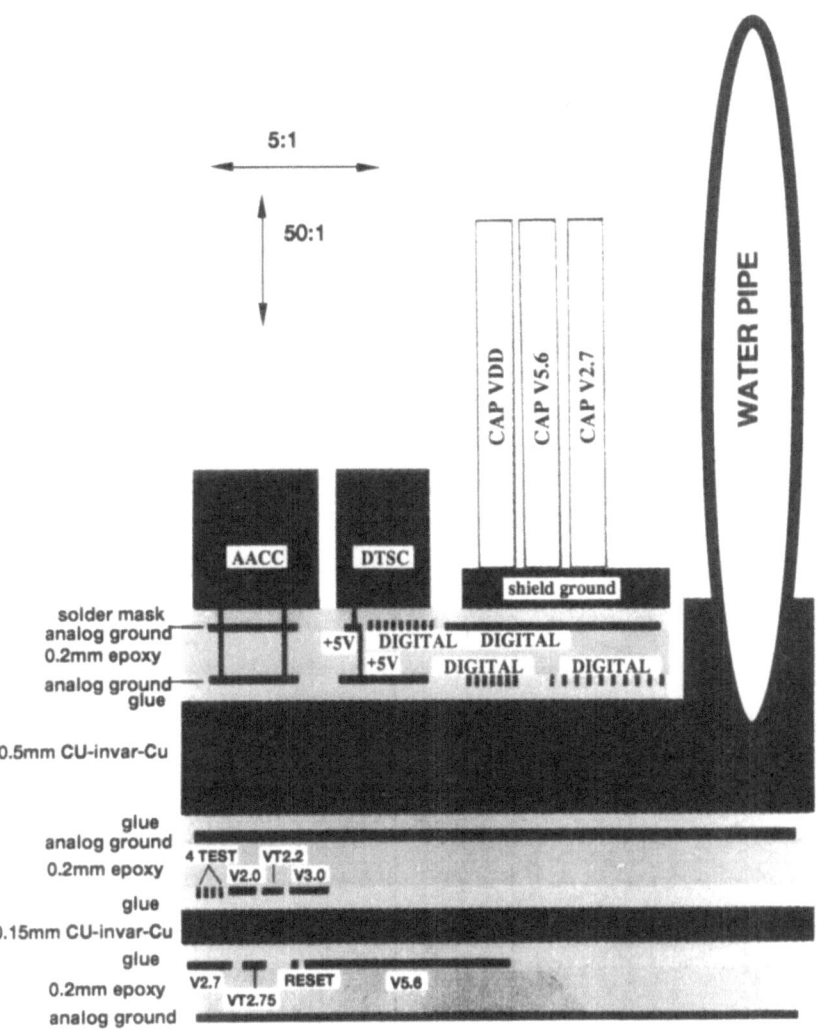

Fig.14 Cut view of the multi-layer PCB (first prototype).

Table 1. Records and achievements

BEFORE	LAA ACTION	AFTER
Not possible to buy large, specially shaped detectors.	We made our own cutting tests.	Manufacturers undertakes the work using different techniques.
No front-end electronics design existed for high-rate, high-frequency machine.	Make commercial enquires in Europe without success. Team up with Santa Cruz.	First fast amplifier design in rad-hard SOI made for strip detectors. First prototypes of CMOS digital strip detector circuits for 10 MHz pipelined, buffered and multiplexed data flow. First rad-hard version of the pipeline/buffer system.
Traditional mounting system left detector at the mercy of machine development and filling.	Design-construction program for mechanics allowing distant removal of detectors.	Prototype mechanics system built with detector retraction to 400 mm at a speed of 15mm/sec.

Table 1. Records and achievements (continue)

BEFORE	LAA ACTION	AFTER
Force compensation of pots was hydraulic.	Worry about rad-hard mechanics for new generation accelerators.	Prototype mechanics with rad-hard features. All metal guides and bearings. Replace hydraulics with all mechanical constant-tension springs.
Traditional near-elastic proton spectroscopy.	Study a multi-station spectrometer to cover a wide range of x_L.	Prototype spectrometer for HERA. Full x_L range extends down to 0.3.
Traditional PCB	High concentration of front-end electronics needs very special multi-layer circuit.	Production of possibly the most advanced front-end PCB. High thermal conductivity metal-layer substrata, matched to Si for thermal expansion serving as a self-supporting mount and also as a mount for the detector.

References

1) M. Basile et al., "Universality features in (pp), (e+e-) and DIS processes", Nuovo Cimento, **79**, 1, (1984).

2) "The ZEUS Detector", Status Report 1989, The ZEUS Collaboration, DESY/PRC 89-01, March 1989.

3) G. Anzivino et al, "Test of radiation hardness of a silicon microstrip detector under gamma irradiation", (in preparation).

4) H.F.W.Sadrozinski et al., "Test of radiation hardness of CMOS transistors under neutron irradiation". SCIPP 89/06 March 1989.

QCD PHYSICS OF JETS IN THE TeV RANGE

V.A. Khoze

TH Division, CERN, Geneva, Switzerland; INFN Eloisatron Project, Italy
and Leningrad Institute for Nuclear Physics, Gatchina, USSR

ABSTRACT

The main aim of this talk is to demonstrate the present status of the analytical perturbative QCD approach to jet physics in hard processes. We concentrate on the colour coherence phenomena in the jet-like final states and the possibility to use them as valuable additional tool, helping to extract and to study manifestations of new physics.

1. INTRODUCTION

This talk summarizes some recent advances in applications of the perturbative approach (PA) to the description of jetty final states in hard processes (HP). It is complementary to the talk given by G. Marchesini[1] and presents the quantitative predictions for jet characteristics in hadronic collisions in terms of analytical calculations (see Refs. 2-4). The main idea is to reorganize the perturbative expansion in such a way that its zero-order approximation is systematic and involves an arbitrary number of produced particles. This can be achieved through the iteration of basic A→B+C branching. This approximation is the Modified Leading Logarithmic Approximation (MLLA), which takes case of both double logarithmic and single logarithmic effects in a systematic way[5,6].

The interest in the detailed studies of jets in HP is twofold: on the one hand, they are important for testing both perturbative and nonperturbative QCD, for design of experiments and the analysis of their data. On the other hand, the characteristic features of jet-like states could provide a valuable additional tool, helping to extract and to study manifestations of new physics.

In the last decade hadron jet physics has been intensively studied at both e^+e^- and hadronic accelerators (see Refs. 7-9). It will be one of the central problems of investigation for the $pp(\bar{p})$ e^+e^-, and ep colliders of the future.

The existing data show that hadronic event characteristics calculated at the MLLA parton level agree very well with the measured ones. This supports the hypothesis of the Local Parton-Hadron Duality (LPHD)[6,10] (see also Refs. 11) and provide great encouragement for future experiments aimed at the multijet production at very high energies.

According to the LPHD concept, the conversion of partons into hadrons occurs at a low virtuality scale, independent of the scale of the primary HP, and involves only low momentum transfers, leading to a close similarity between parton and hadron distributions.

The PA (MLLA + LPHD) attempts to describe the global features of the hadronic systems, such as the mean multiplicities and multiplicity distributions, angular pattern of energy and multiplicity flow, inclusive energy spectra and correlations of particles, etc., without making any reference at all to a fragmentation scheme. The nonperturbative effects are reduced to normalizing constants relating hadronic characteristics to partonic ones (see Ref. 12).

The perturbative predictions for these characteristics of the HP are in good agreement with the results of Monte Carlo Simulation programs (see Ref. 1 and references therein) based on coherent QCD parton shower picture.

We shall focus below on the manifestations of the colour coherence phenomena. The rediscovery of coherence in the context of QCD[13] has led to a dramatic revision (see, e.g., Ref. 12) of theoretical expectations for soft particle distributions.

Thus, the coherent effects in the intrajet partonic cascades, (see Fig. 1a), resulting, on average, in the angular ordering (AO) of sequential branching, gave rise to the hump-backed shape of particle spectra-one of the most striking PA predictions[14] Not the softest particles but those with the intermediate energies ($E_h \sim E^{0.35}$) proliferate most abundantly in the QCD cascades.

Due to the interjet coherence (see Fig. 1b), which is responsible for the drag effects in the multijet events (see Refs. 12,15-18), a very important physical phenomenon can be experimentally verified, namely, the fact that the particle production is in accordance with the QCD radiophysics of hadron flows.

It is entirely unremarkable that the quantum mechanical interference effects should be observed in QCD. However, it is of importance that the experiment demonstrates that such interference effects survive the hadronization phase.

The MLLA formulae for partonic distributions essentially simplify in the limiting case, when the formal boundary Q_0 between the perturbative and nonperturbative phases of jet evolution is chosen equal to Λ (the so-called limiting distributions, see Refs. 6,12). In some sense such a choice is a specific attempt to model confinement.

Whether or not the jet energies are sufficiently large for PA to be applied is , of course, a question to experiment. So far the experimental data in e+e- collisions (see Refs. 7,9,19) have demonstrated that the PA works unexpectedly well. At least, starting from the Z^0 energy, the finite energy corrections do not change sizeably the perturbative scenario.

For hadronic collisions the underlying HP physics is more complicated, and the energy range for applicability of the PA is now less clear (see Refs. 1).

However, the nature of jets, based on the dominant role of the QCD bremsstrahlung processes, is the same for both reactions. Therefore, at least, at the energies of the future hadronic supercolliders the main physical phenomena in jet physics and characteristics of final states should also be under control of the PA. Presumably at the LHC-SSC energies the colour coherence effects should be well distinguishable from the minimum bias background.

Finally, let us enumerate the virtues of high $p_\perp$ hadronic reactions for studies of QCD jet dynamics.

1. A diversity of hard interactions at small distances; by varying the experimental conditions (triggers) one may extract the dominant subprocess and turn from one subprocess to another.

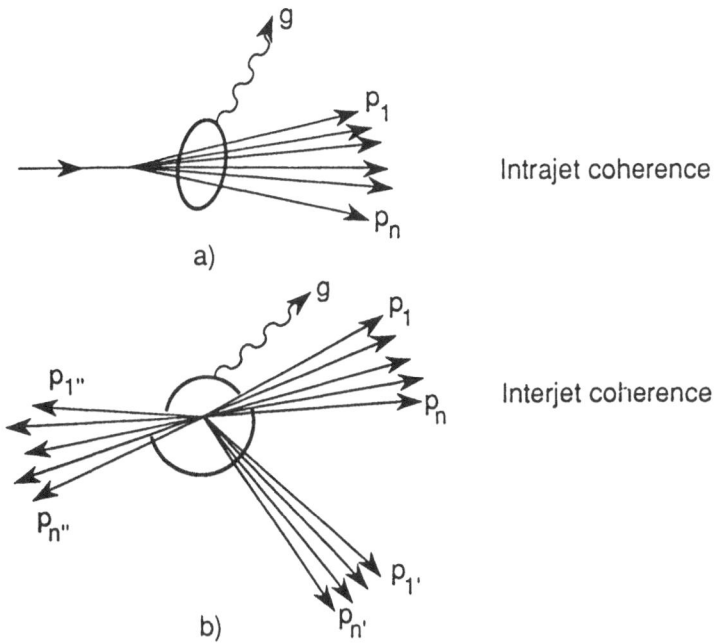

Fig. 1. Two types of coherence effects: a) intrajet coherence; b) interjet coherence.

2. Dependence of length and height of the hadron spectra on the different parameters: the length is determined by the total energy of the collision, and the height and the plateau structure - by the process hardness (trigger $E_\perp$). Thus, information becomes available that is inaccessible in e+e- annihilation.

3. There is the purely practical argument that just in high $p_\perp$ hadronic collisions the largest possible energies (hardness) will be reached. These reactions are also the best source of high energy gluon jets. Detailed studies of such processes are necessary for designing the future experiments and the analysis of their data, for finding new heavy objects. In particular, interjet coherence effects could provide a valuable additional tool, helping to extract and to study new phenomena.

The variety and complexity of colour antennae typical for high $p_\perp$ processes, complicate the picture of final hadronic distributions. However, at very high energies, when particle multiplicities become large enough, the possibility arises of using a detailed analysis of hadron flows on an event by event basis to extract information about colour transfer at small distances.

2. QCD PROPERTIES OF AN "INDIVIDUAL JET"

In spite of the importance of interjet coherence phenomena the notion of an individual partonic jet makes sense if one does not deal with the azimuthal effects, but considers only multiplicities, energy spectra and correlations, etc. In this case, all the influence of the jet ensemble on a given jet may be encoded in a single parameter θ_0, the jet "opening angle." This, in essence being the angle between the considered jet and the nearest other one.

Jet characteristics prove to depend not on the jet energy E but on the hardness Q of the process producing the jet, $Q=E\,\theta_0$ at $\theta_0 \ll 1$.

One can study the properties of an individual quark jet when measuring the different inclusive distributions in the process e+e- $\rightarrow$ hadrons. For example, the inclusive particle spectrum for e+e- annihilation is the sum of two q-jet distributions

$$\frac{1}{\sigma}\frac{d\sigma}{d(\ln\frac{1}{x})}=2\vec{D}_q^h(l,Y) \tag{1}$$

where

$$l=\ln\frac{1}{x},\ Y=\ln\frac{E}{\Lambda},$$

$x=\frac{E_h}{E}, 2E=W$ is the total c.m.s. energy

Formally, the decay into two gluons of the C - even ultra-heavy quarkonium states, $\chi_Q = Q\,\bar{Q}$, might define the individual gluon jets.

Let us discuss the collimation of a QCD jet around the parent parton j momentum ($j =$ q, g)[12]. This phenomenon is of special interest when considering the space-energy portrait of events as a natural partonometer for registering HP dynamics.

The angular cone $\Theta_{\in j}$, in which an energy fraction $\in j$ is deposited, is given by:

$$\frac{\Theta_{\in j}}{\Theta_0}=\left(\frac{E\Theta_0}{\Lambda}\right)^{-\gamma_j(\in j)} \tag{2}$$

where , for example, (see Fig. 2)

$$\gamma_q(0.9) \approx 0.55, \ \gamma_q(0.5) \approx 0.83;$$
$$\gamma_g(0.9) \approx 0.30, \ \gamma_g(0.5) \approx 0.54; \tag{3}$$

It follows from Eqs. (2), (3) and Fig. 2 that the energy collimation in a quark jet is stronger than in gluon one, and the collimation grows as energy increases.

The collimation of multiplicity N inside a jet behaves as

$$\frac{\theta_{\delta j}}{\theta_o} \ \alpha \ [N(E\theta_o, \Lambda)]^{-b/8N_c \ln 1/\delta j}, \tag{4}$$
$$b = \frac{11}{3} \cdot N_c - \frac{2nf}{3},$$

where $\Theta_{\delta j}$ is the cone, where the jet multiplicity fraction δj is concentrated.

Thus, the angle, say, $\theta_{1/2}$ decreases with the increase of the energy scale $Q = E\theta_o \sim E$ approximately as $N^{-1/4}(Q, \Lambda)$, *i.e.* parametrically much slower than in the case of the energy collimation of Eq.(2). Note, that an attempt to fix a jet axis with the accuracy better than the natural angular width of the corresponding energy flux seems to be physically unreasonable.

The average multiplicity of partons (gluons) in a gluon jet is given by the asymptotic limiting formula[5,6,20].

$$\ln(N^g) \cong \sqrt{\frac{32N_c\pi}{\alpha_s(Y)}} \ \frac{1}{b} + \left(\frac{B}{2} - \frac{1}{4}\right) \ln \alpha_s(Y) +$$
$$\tag{5}$$
$$+ O \ (1); \ B = \frac{a}{b}, \ a = \frac{11}{3}N_c + \frac{2nf}{3N_c^2}$$

(More elaborated analytical formulae may be found in Refs. 6, 10.)

Note that change of scale parameter Λ by a factor of O(1) would correspond to a correction of $O(\sqrt{\alpha_s})$ in multiplicity N. Such corrections are subleading to the MLLA and therefore this parameter cannot be compared directly with the values obtained from other experiments.

The first term in Eq. (5) is reduced by a factor $\frac{1}{\sqrt{2}}$, relative to the incoherent case. This reduction is a direct result of the destructive interference of soft gluon emission.

In the framework of the LPHD scenario the MLLA analytical expressions describe the hadron inclusive distributions with the values of parameter Λ and the normalization factors K determined phenomenologically from comparison with the data at the present energies.

When analyzing the distribution of charged particles one should take into account that it is, in fact, a mixture of different particles, spectra of which prove to be not exactly

$$\bar{D}^{ch} = \bar{D}^{\pi\pm} + \bar{D}^{K\pm} + \bar{D}^{p(\bar{p})} + ... \tag{6}$$

similar. This can, in principle, cause some $K^{ch}(Y)$ dependence at fixed values of $K^{\pi\pm}, K^{K\pm}, K^{p(\bar{p})}$ (see Ref. 21).

First comparison with the experimental data showed[10] that the relative yields of hadrons of various species can be reasonably simulated in PA by the conjecture $(Q_0)_{eff} = m_h$ (keeping the same value of Λ_{eff}). Such fits gave the following values of the conversion factors K^h and Λ_{eff}[10,21]:

$$K^{\pi\pm} \cong 1 \;,\; \frac{K^{K\pm}}{K^{\pi\pm}} = 0.22 \;,\; \frac{K^{p(\bar p)}}{K^{\pi\pm}} = 0.11$$

$$\Lambda_{eff} = 0.155 \text{ GeV}.$$

Fig. 3 displays the energy dependence of charged particle multiplicity in e+e- collisions together with the prediction of Eq. (5) (neglecting the possible weak energy dependence of the conversion factor). The PA result is seen[22] to agree well with the e+e- data and the HERWIG[23] Monte Carlo points, as shown in Fig. 3. Taking this agreement as a testing ground, one can predict the dynamics of the multiplicity increase, shown in Fig. 4.[24]

Assuming five flavours the ratio of particle multiplicities in a gluon jet and a quark jet with the accuracy to $O(\alpha_s)$ terms is given by[25].

$$N_g/N_q = \tfrac{9}{4}(1 - 0.27\sqrt{\alpha_s} - 0.07\alpha_s...) \tag{7}$$

The MLLA expression for the limiting spectrum of relativistic particles in the region $\ln(1/x) \gg 1$ can be presented in a form [6] (see also Ref. 12):

$$\bar{D}^{lim} \equiv [\bar{D}_g^g(l,Y)]MLLA = \frac{4N_c}{b}\Gamma(B)\int_{-\frac{\pi}{2}}^{\frac{\pi}{2}}\frac{d\tau}{\pi}e^{-B\alpha}\Big[\frac{ch\alpha + (1 - 2\zeta)sh\alpha}{\frac{4N}{b}Y\frac{\alpha}{sh\alpha}}\Big]^{\frac{B}{2}}$$

$$I_B\Big(\sqrt{\frac{16N}{b}Y\frac{\alpha}{sh\alpha}[ch\alpha + (1-2\zeta)sh\alpha]}\Big) \tag{8}$$

which is especially convenient for numerical integration. Here $\alpha = \alpha_0 + i\tau$, $\tanh\alpha_0 = 2\zeta - 1$, $\zeta = 1 - \frac{1}{Y}$, I_B is the modified Bessel function of order B.

This spectrum exhibits the above mentioned "hump-backed" structure with a maximum at particle energies $E_0 = x_0 E$ approaching asymptotically

$$\ln(\tfrac{1}{x_0}) = \tfrac{1}{2}Y(1+B\sqrt{\frac{b}{4N_cY}}) + 0(1) \tag{9}$$

In Ref. 26 a transparent expression for the $\bar{D}(l,y)$ has been obtained in a form of a distorted Gaussian.

Note that in the framework of the MLLA the l-distributions of q and g-jets differ only by the normalization factor $\frac{C_F}{N_c} = \frac{4}{9}$.

Account for the Next - to - MLLA effects, namely, $\sim O(\sqrt{\alpha_s})$ terms [21], shows that the distribution is a q-jet is pretty close to the limiting spectrum. The asymptotic position of the maximum is slightly shifted to higher $\delta_q(\ln(\frac{1}{x_0})) \approx -0.01$. Spectrum in a g-jet is shifted to lower x as compared to the limiting case: $\delta_q(\ln(\frac{1}{x_0})) \approx 0.04$, see Refs. 21,26.

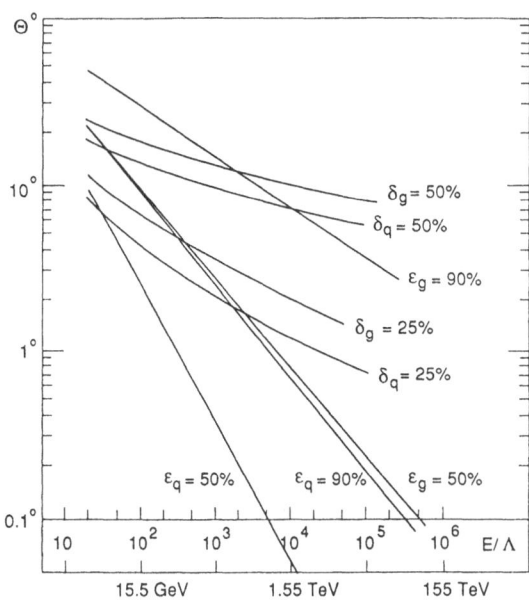

Fig. 2 Energy and multiplicity collimation in the QCD jets[12]. The lower scale corresponds to the case Λ=0.155 GeV. ε means energy fraction, δ-multiplicity fraction.

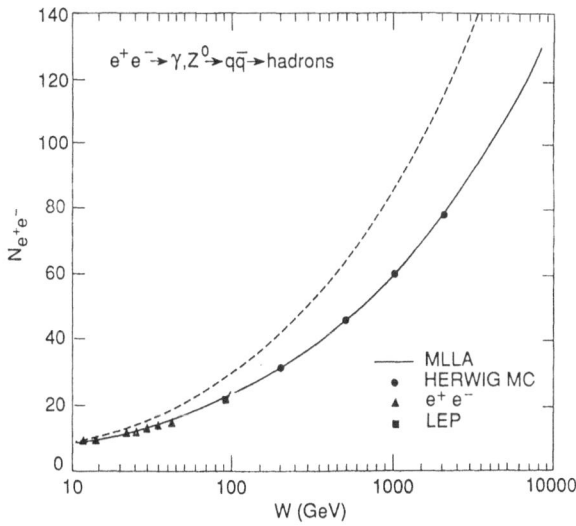

Fig. 3 The energy rise of charged particle multiplicity in e+e- collisions. The dashed line corresponds to the double logarithmic case.

A strong support to the PA ideas came recently from the OPAL data[19] on the inclusive momentum distribution of charged particles at the Z⁰. To observe the expected energy evolution these data were combined with the TASSO results[27].

Fig. 5 shows that the MLLA limiting formula (8) provides an excellent description of both the space and the energy evolution of the spectrum. Especially spectacular is the confirmation of the the energy evolution of the peak position (see Eq.(9)), which is independent of the normalization factors.

These results wonderfully confirm the QCD cascading picture of the multiple hadroproduction. One may expect even higher credibility of the MLLA expressions at larger jet energies where the subasymptotic corrections are less important.

Fig. 6 illustrates the dependence of the limiting distribution in a g-jet on its hardness[24] $Q=E\Theta_0$. In this figure one can clearly see the rise of the maximum $E_0 = x_0 E$ with the jet energy and an increase of the hump height, reflecting the rise of multiplicity.

Account for the specific QCD features of jets could be of importance for design of future experiments and the analysis of their data.

It is noteworthy that according to Eq. (9) E_0 in particle spectra grows rather slowly with the jet energy E; even at $E \approx 1.5$ TeV, its value reaches only $E_0 \approx 3$ GeV.

To explore the coherent origin of the hump-backed particle spectrum and in an attempt to study the depletion in its soft part for jets produced in hadronic collisions, it proves to be important to look at particles restricted to lie within a particular opening angle with respect to the jet [28]. For example, one might consider the energy distribution of particles accompanying the production of an energetic particle and lying within an opening angle Θ_0 about the direction of the trigger particle momentum.

Parton cascades in these situations will populate mainly the region

$$\frac{m_h}{\sin\frac{\theta_0}{2}} \langle E_h \langle E$$

with E_h and m_h being the energy and the mass of the observed particle, respectively ($m_h \geq \Lambda$). The maximum of the distribution, in E_h, is now forced to larger energies

$$\ln\frac{E_0}{m_h} = \frac{1}{2} \ln \frac{E}{m_h (\sin\frac{\theta_0}{2})} \tag{10}$$

$$-B \left(\sqrt{\frac{b}{16N_c} \ln \frac{E(\sin\frac{\theta_0}{2})}{\Lambda}} - \sqrt{\frac{b}{16N_c} \ln \frac{m_h}{\Lambda}} \right)$$

If one chooses Θ_0 moderately small and varies E, coherence will give a moving peak in accordance with the relation:

$$\frac{E}{E_0} \frac{dE_0}{dE} = \frac{1}{2} - B\sqrt{\frac{b}{64N_c Y(\theta_0)}} \tag{11}$$

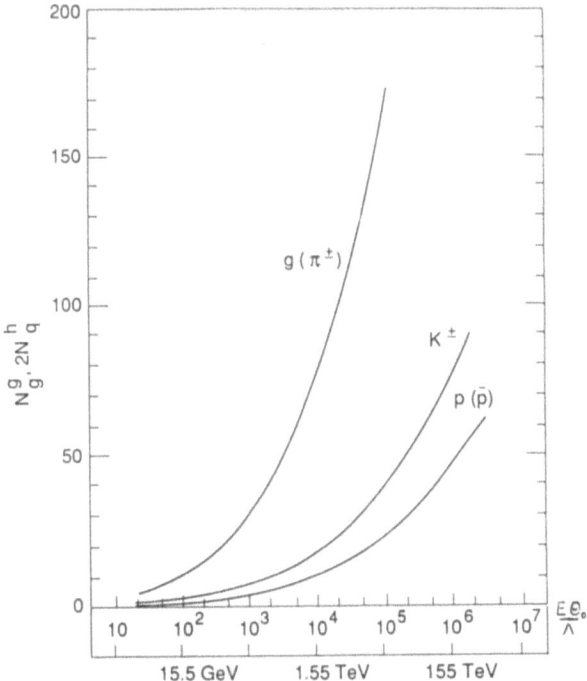

Fig. 4 LPHD expectations for the average particle multiplicities in the QCD jets[24]. The lower scale corresponds to the total jet opening angle and Λ=0.155 GeV.

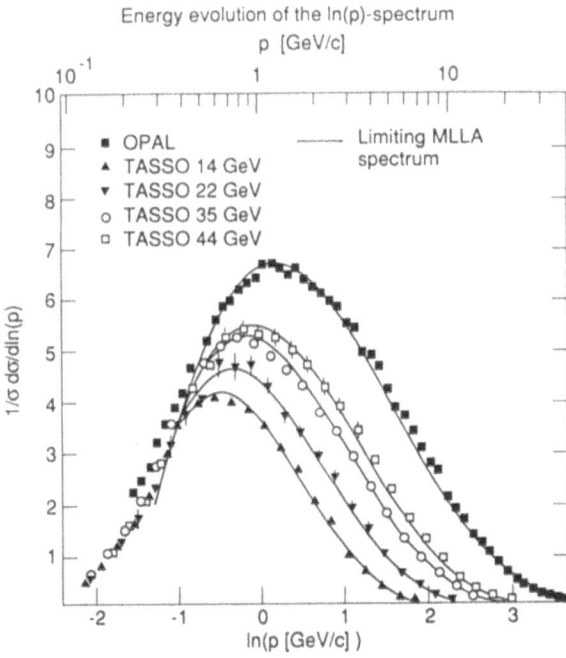

Fig. 5 Energy evolution of the ln(p) distribution of charged particles in the e+e- collisions[19] together with the predictions of the limiting MLLA formula (8), assuming $\overline{D}_q = \frac{4}{9}\overline{D}^{\lim}$

with

$$Y(\theta_o) = \ln \frac{E(\sin \frac{\theta_o}{2})}{\Lambda}$$

The angular cut Θ_0 is especially useful for jets produced in hadronic collisions, since one is able to eliminate much of the soft background.

Fig. 7 illustrates [12] the energy distribution of charged hadrons in different cones Θ around the jet axis at $\frac{E \, \theta^0}{\Lambda} = 10^4$.

Finally, let us make some comments on the multiplicity distribution in QCD jets. As was shown in Ref. 29, the predicted multiplicity moments, including the next-to-leading corrections, are very close to those of a negative binomial distribution

$$\frac{\sigma(n)}{\sigma_{tot}} = k \, (k+1)...(k+n-1) \frac{1}{n!} \left(\frac{\langle n \rangle}{k} \right)^n \left(1 + \frac{\langle n \rangle}{k} \right)^{-n-k}, \tag{12}$$

with

$$\frac{1}{k} \cong 0.4 - 0.9 \sqrt{\alpha_s} + O(\alpha_s) \tag{13}$$

It is not excluded that $O(\alpha_s)$ terms are rather important at nonasymptotic energies.

3. RADIOPHYSICS OF LARGE $p_\perp$ HADRONIC JETS

In the framework of PA, the source of multihadron production in QCD jets is gluon bremsstrahlung, so one should expect that the produced hadrons are a consequence of the colour dynamics at small distances. Therefore, the detailed features of the parton-shower system, such as the flow of colour quantum numbers, influence significantly the distribution of colour-singlet hadrons in the final state.

The drag[15] (or string[30]) effect in the $q\bar{q}g$ events of e+e- annihilation is one of the best example of the interjet collective phenomena revealing the wave nature of hadroproduction in HP.

It is instructive to note, that originally this effect was predicted as the result of a Lorentz boost exerted by a gluon on the nonperturbative string stretched between quarks[30]. But in the PA picture it is a result of interference among the gluon waves radiated from the $q\bar{q}g$ composite emitter[15].

The e+e- data (see Refs. 7,31 and references therein) have strongly supported the predicted drag of the interjet particles in the direction of the gluon jet (net destructive interference in the $q\bar{q}$ valley). They demonstrated that the wide angle particles really do not belong to any particular jet, but their emission properties depend on the overall jet ensemble.

Let us emphasize that the observation of the colour interference between soft hadrons from, say, q- and g-jets reveals the QCD wave properties of hadronic flows, Thus, it can be considered as an experimental proof of the common bremsstrahlung nature of the hadroproduction mechanisms for both jets. The properties of drag interference phenomena are deeply rooted in the basic structure of non-Abelian gauge theory.

Future detailed tests of QCD coherence dynamics require comprehensive studies of the total three dimensional pattern of particle flows in the multi-jet events. Of special interest here are the angular structure and the energy dependence of multiplicity flow.

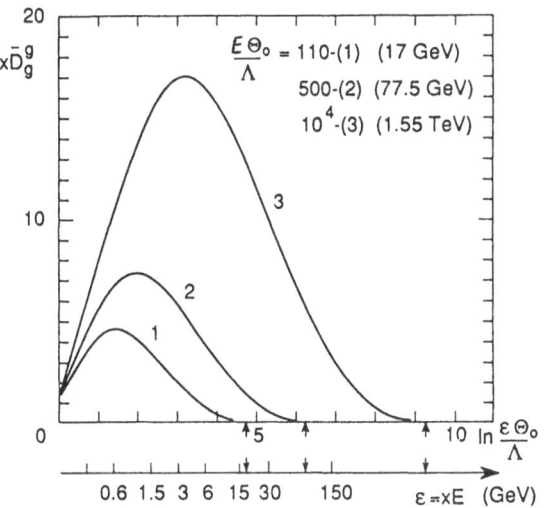

Fig. 6 Limiting energy spectrum of partons in a gluon jet with an opening angle θ_0 and energy E. The values in the brackets correspond to the total jet opening angle and Λ=0.155 GeV. The lower scale shows the energies of particles corresponding to such a case.

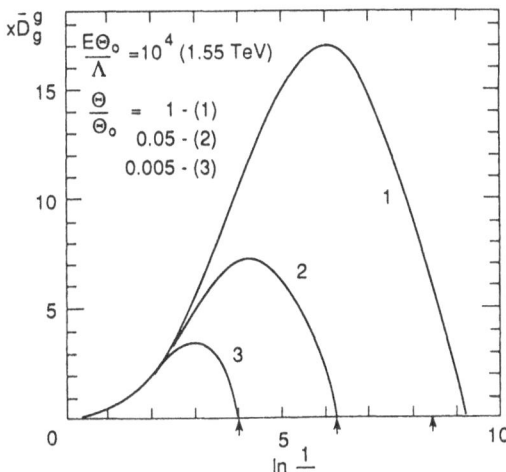

Fig. 7 Dependence of the energy spectrum of partons in a gluon jet with an opening angle θ_0 and $\frac{E\theta_0}{\Lambda} = 10^4$ on the aperture of the registered particle flow[12]. The narrower is the registration cone, the harder are the particles within this cone. The value of energy, given in brackets, corresponds to the total jet opening angle and Λ=0.155 GeV.

The studies of the interjet collective phenomena in hadronic collisions look very promising (see Refs. 1,4,12,16,32). Here the soft radiation pattern can be represented as an incoherent sum of contributions, subject to an angular ordering constraint.

Unfortunately so far the experimental study of the particle flow radiophysics at the hadronic colliders is still in its infancy. With the new experiments at the (super)colliders it is to be hoped that the interjet dynamics will reveal its strength both as a probe of perturbative and nonperturbative QCD and as a valuable tool, helping to study manifestations of new physical phenomena.

To illustrate, how the interjet coherence connects the structure of hadronic accompaniment in HP with the t-channel colour transfer, let us consider high $p_\perp$ scattering of partons $A+B \rightarrow A'B'$ ($E_A \sim E_B \sim \sqrt{s}$) at relatively small angles $\theta_s \approx \dfrac{p_\perp}{E} \ll 1$. The hardness of the process is determined by the momentum transfer $p_\perp \approx \sqrt{-\hat{t}}$ which naturally restricts the transverse momenta of the accompanying gluon bremsstrahlung $l_\perp < p_\perp$ and, so the development of partonic cascades.

In the structure of the final hadronic system three regions may be separated. Two of them adjoin the fragmentation regions of the colliding hadrons and occupy the intervals

$$\Delta\eta = \ln p_\perp / \Lambda , \tag{14}$$

where η is the pseudorapidity

$$\eta = \frac{1}{2} \ln \frac{1 + \cos\theta}{1 - \cos\theta}$$

The hadronic spectrum in each of these intervals is saturated with the particles from the bremsstrahlung cones of the incoming and the scattered partons, and so results from an incoherent sum to two jets with angular opening $\theta = \theta_s$. The height of the distribution is determined, roughly speaking, by the sum of parton 'colour charges' $C_A + C_{A'}$ and $C_B + C_{B'}$ respectively.

In the central region

$$|\eta| < \eta\ (\theta_s) \cong \ln \frac{\sqrt{s}}{p_\perp} \approx \ln \frac{E}{p_\perp} \tag{15}$$

(final particle angles larger than the scattering angle θ_s) the incoming and the scattered partons radiate coherently, and, as a result, the hadron density is determined by the colour charge C_t of the t-channel exchange. Since in the given kinematics ($-\hat{t} \ll \hat{s}$) gluon exchange dominates, we conclude that in the central region hadronic spectrum is determined by the 'colour strength' of the gluon current C_g, and what is of importance, it becomes universal, independent of the nature of scattered partons (A, B = q or g).

As was shown in Ref. 12, the resulting spectrum, accounting for the parton branching effects, is independent of the energy (pseudorapidity) of the particle registered at angle $\theta > \theta_s$ at fixed transverse momentum $k_\perp$:

$$\frac{d\sigma}{\sigma d\eta dk_\perp} \cong 4N_c \int_{k_\perp}^{p_\perp} \frac{dl_\perp}{l_\perp^2} \frac{\alpha_s(l_\perp^2)}{2\pi} \overline{D}_g\left(\frac{k_\perp}{l_\perp}, \ln \frac{l_\perp}{\Lambda}\right) \tag{16}$$

In this expression $x \cdot \overline{D}_g$ is the standard distribution of particles in a gluon jet, for which the product of energy and opening angle equals Q.

Integrating over $k_\perp$ of hadrons at fixed η, one obtains

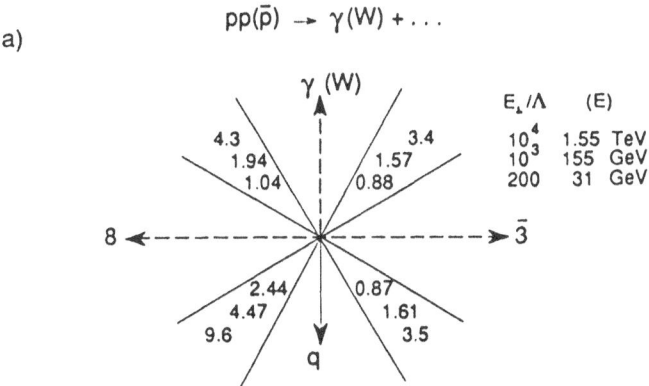

Fig. 8a). The energy rise of $\pi^\pm$ multiplicities in the $30°$ sectors in large $p_\perp$ prompt $\gamma(W)$ production [12]: $\dfrac{E\theta_0}{\Lambda} = 2 \cdot 10^2$, 10^3 and 10^4. The values of energy at $\Lambda = 0.155$ GeV are given for $90°$ in the parton-parton center of mass.

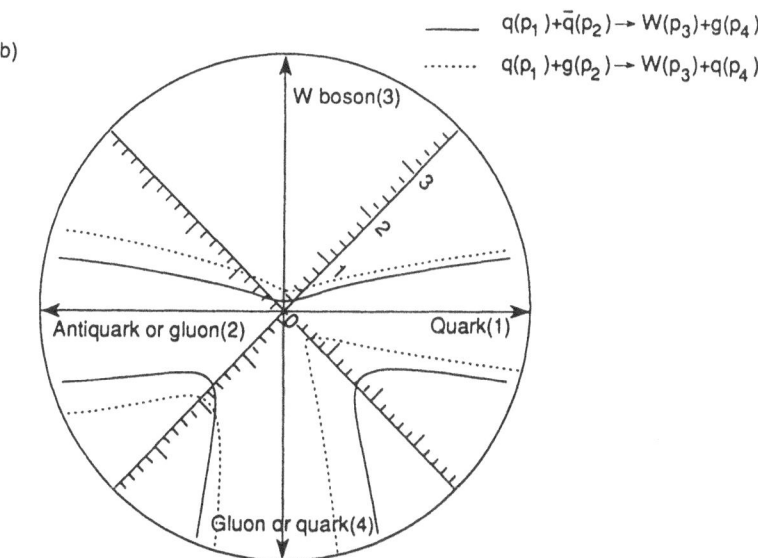

Fig. 8b). Polar plot showing QCD antenna pattern P for the large $P_\perp$ production of W-bosons of $90°$ scattering in the center of mass[32]. The value of P is given by the radial distance of the curves from the origin. The plots have been cut off at P=4, so the singularities in the collinear regional are not displayed.

$$f(\eta, \ln\frac{p_\perp}{\Lambda}) = \frac{d\sigma}{\sigma d\eta} = \int^{p_\perp} \frac{dl_\perp}{l_\perp} 4N_c \frac{\alpha_s(l_\perp^2)}{2\pi} \int_0^1 dx \overline{D}_g(x, \ln\frac{l_\perp}{\Lambda}) \tag{17}$$

Thus, the hadron yield in the interval $|\eta(\theta)| < \eta(\theta_s) \approx \ln\frac{E}{p_\perp}$ does not depend on the rapidity, so a flat distribution emerges, whose height is determined by the hardness of the scattering process.

The rich diversity of the collective drag phenomena has been studied for high $p_\perp$ hadronic collisions[2,16,24,32].

One of the simplest examples is the prompt production of the colourless object (γ, W, Z...) at large $p_\perp$, as shown in Figs. 8. Here the drag phenomena are very similar to 3-jet events of e+e- annihilation. The radiation pattern in Fig. 8a) was calculated[12,28] for final pions projected onto the scattering plane. The particle production proves to be largest between the directions for the incoming gluon and the outgoing quark, but approximately 2.8 times smaller between the directions of the incoming quark and the outgoing quark.

The drag effect here leads to an azimuthal asymmetry of particles, which can be seen by looking end on at the struck q-jet (see for details Ref. 12). The study of such an asymmetry may indicate which of the incoming particles has shaken off the hard gluon.

Fig. 8b, shows the QCD antenna pattern function P which is defined as the ratio of the two-to-three and two-to-two cross sections in the soft limit at 90° in the parton center of mass[32] (see also Refs. 12,26).

One can easily see from Fig. 8b, the essential qualitative difference between the two mechanisms of the production of W-bosons at the hadronic colliders.

(i) $q(p_1) + \bar{q}(p_2) \rightarrow W(p_3) + g(p_4)$, $\qquad$ (18)

(ii) $q(p_1) + g(p_2) \rightarrow W(p_3) + q(p_4)$,

Therefore it may be possible to distinguish whether a W is produced by process (i) or (ii) on an event by event basis (see Refs. 12,32 for details).

Figs. 9a,b, show the QCD antenna patterns for 90° scattering in the center of mass for two of the principal two-to-two scattering processes responsible for jet production in $p\bar{p}$ collisions, see Refs. 12,16,28,32.

$q(p_1) + \bar{q}(p_2) \rightarrow q(p_3) + \bar{q}(p_4)$ $\qquad$ (19)

$g(p_1) + q(p_2) \rightarrow g(p_3) + q(p_4)$

It can be seen from Fig. 9b, that there is an asymmetry of about factor four between the first and the third quadrant.

If detectable, this asymmetry would allow the separation of g and q-jets on an event by event basis. By moving to large rapidity one can select events in which the fraction of the longitudinal momentum carried by one of the partons is very large and the other is very small, leading to an enriched contribution from the qg-scattering, see Refs.32.

One of the interesting tasks for the future experiments seems to be the study of colour coherent phenomena in the structure of the final states in deep inelastic scattering (see Refs. 1,12,33). Of special interest here is the evolution of the distributions with q^2 and x_B.

Colour interference between jets seems to become a phenomenon of large potential value as a new additional tool for discriminating between HPs. For example, reconstruction

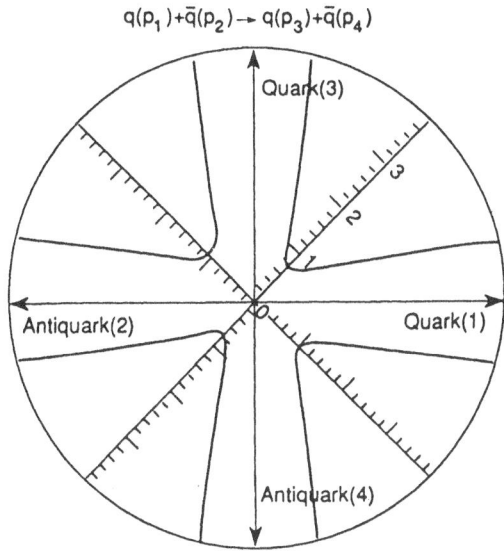

Fig. 9a). Polar plot showing QCD antenna pattern P for the identical quark process[32] $q\bar{q} \to q\bar{q}$.

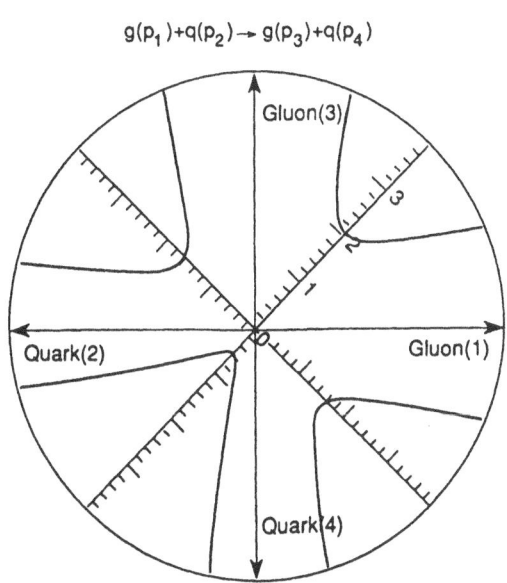

Fig. 9b). Polar plot showing QCD antenna pattern P for the process[32] $gq \to gq$.

of antenna pattern by the effects of particle drag may help to study the production of new colourless heavy objects - the Higgs boson H, new quarks and leptons, supersymmetric particles, and so on. Most of these objects produce hadronic jets, and the configurations of interjet particles should differ from familiar QCD processes like parton scattering.

An interesting example came from the study of the accompanying particle distribution in the heavy Higgs production[12]. Here the radiation pattern should be quite different depending on the production mechanism of a Higgs boson, see Figs 10,11. If H boson is produced via the g-g fusion of Fig. 10b), then the standard hadronic distribution $f(\eta, \ln \frac{M_H}{\Lambda})$ arises in hadronic spectrum. However in the case of the W-W mechanism of Fig. 10a, the central region (15) should be empty, and the process looks as the quasi-diffractive one (the gluon emission by initial and final partons at large angles cancel each other coherently because in t-channel colour is not transferred.

Monte Carlo simulation using PYTHIA programme for 500 GeV Higgs production at the LHC energy confirms this qualitative statement [34].

Another example is the comparison of the production of a colourless object via gg or $q\bar{q}$ collisions. Here, if the hard kinematics is the same, the heights of the accompanying plateau should differ approximately by a factor of two ($N_c/C_F = 9/4$).

4. PA AND SHOWER MONTE CARLO ALGORITHMIC SCHEMES

The well-elaborated shower Monte Carlo simulation programs for jet physics has been presented on the market (for reviews see, e.g., Refs. 1,35,36).

All of them are of a probabilistic and iterative nature. The models evolve steadily in time, in particular, by building in realistic fragmentation and proper QCD cascades. The QCD- inspired models describe very successfully the existing data, based on the concept of the well-developed parton shower, and prove to be very useful tools for experimentalists.

The model-builders aim to formulate the Monte Carlo algorithms to be as faithful as possible representatives of the QCD dynamics of jets. Over the last few years, one could observe an obvious convergence between the different classes of models and a growing tendency to reproduce the important ingredients of PA.

Note that the Monte Carlo simulation technique permits one to reproduce the flow of colour in the parton cascade, at least to leading order in N_c, since to this order the colours of all external lines may be traced continuously through the corresponding tree diagram. In a Monte Carlo simulation it is straightforward to intertwine the QCD parton shower with a nonperturbative fragmentation model for the conversion of parton configurations into hadrons at any desired scale of virtuality Q_0.

One of the aims for model improvement is the most adequate incorporation of the colour coherence, reflecting the quantum mechanics of QCD.

It is important to notice, however, that the description of multipartonic system development in terms of classical Markov chains is of limited value in principle. Of special interest here are $1/N_c^2$ suppressed soft contributions [6,12].

Another example is connected with the interjet collective phenomena in the multijet ensembles. These effects could be reproduced in a classical probabilistic language only in the large N_c limit (see Ref. 12). The problem is that nowadays we simply do no know how to handle $1/N_c^2$ terms in Monte Carlo simulations. Normally, the neglect is not serious ($\leq 10\%$), but under special conditions these terms may even become dominant.

The azimuthal asymmetry of QCD jets is one of the simplest examples when the colour-suppressed terms are essential, see Ref. 12 for details. Here the qualitative difference

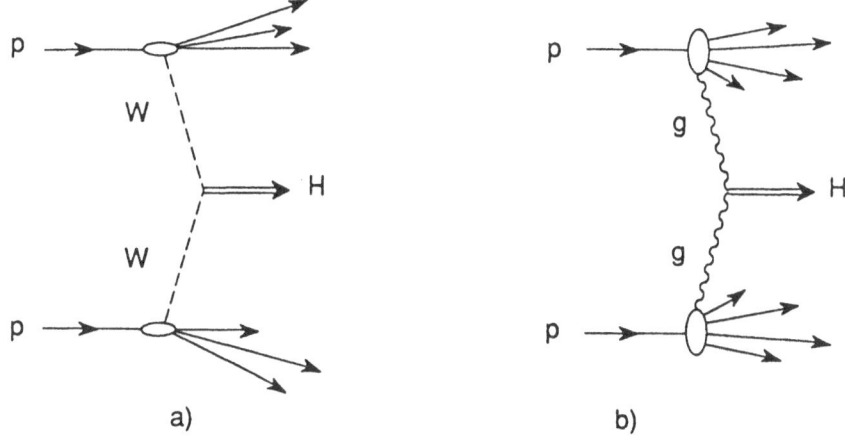

Fig. 10. Hadronic Higgs production via a) W-W fusion and b) g-g fusion.

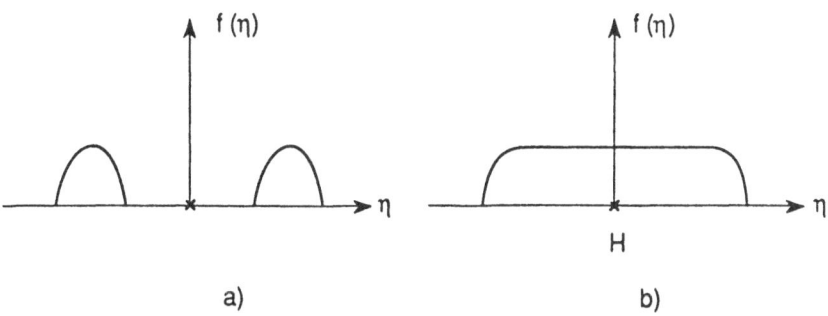

Fig. 11. Qualitative picture illustrating the difference in accompanying particle distributional for the Higgs production mechanism of Figs. 10a), b).

between the predictions of QCD and its large -N_c- limit proves to be the most spectacular in the case of high $p_\perp$ $p\bar{p}$ scattering with the identification of the scattered q-jet.

Concluding this section, let us emphasize that the relative smallness of the nonclassical effects does not diminish their fundamental importance. These consequences of QCD radiophysics are a serious warning against the traditional ideas on independently evolving partonic subsystems.

5. CONCLUSION

I would like to deliver this talk as my entrance fee to the 200 TeV Club. The main aim of the talk was twofold: On the one hand, I tried to demonstrate the present level of reliability and maturity of the QCD PA to the description of the jetty final states in HP. On the other hand, the idea was to show that the QCD studies within the PA are far from being exhausted. For example, it looks very promising to use the spectacular collective QCD effects as a valuable tool for studying the manifestations of new physical phenomena. Therefore, the future experiments of hadronic supercolliders definitely need PA-like activity (hopefully by the QCD chapter of the 200 TeV Club).

Acknowledgements

It is a pleasure for me to thank Professor A. Zichichi and L. Cifarelli for their kind invitation to this excellent workshop and the enjoyable discussions.

I would like to thank I. Bigi, Yu. Dokshitzer, G. Marchesini, A. Mueller, T. Sjöstrand, S. Troyan and B. Webber for fruitful discussions.

REFERENCES

1. G. Marchesini, invited talk at this workshop.

2. For reviews see, A. Bassetto, M. Ciafaloni, and G. Marchesini, Phys. Rep. C100 (1983) 201; Yu. L. Dokshitzer, V. A. Khoze, A.H. Mueller, and S.I. Troyan, Rev. Mod. Phys. 60 (1988) 373; Yu. L. Dokshitzer, V.A. Khoze, and S.I. Troyan, in "Perturbative QCD", ed. A. H. Mueller (World Scientific, Singapore, 1989), p.241.

3. Yu. L. Dokshitzer, Talks given at the International Schools of Subnuclear Physics,Erice, 1989, 1990.

4. V.A. Khoze, proc. of the Int. Symp. on Lepton and Photon Interactions at High Energies, Stanford, Aug. 1989 (World Scientific) p. 387.

5. A. H. Mueller, Nucl. Phys. B 213 (1983) 85 and erratum quoted in Nucl. Phys. B241 (1984) 141.

6. Yu. L. Dokshitzer and S. I. Troyan, "Proc. XIX Winter School of the LNPI", V.I. 144 (1984); preprint LNPI-922(1984).

7. W. Hofmann, Ann. Rev. Nucl. Part. Sci. 38 (1988) 27; P. Mättig, Phys. Rep. C177(1989) 141; K. Sugano, Int. Journ. Mod. Phys. A3 (1988) 2249.

8. R. Cashmore, lecture given at the Cargese Summer Institute, Aug., 1990.

9. M. Jacob, Talk at the 25th Int. Conf. on High Energy Physics, Singapore, Aug. 1990.

10. Ya. I. Azimov et al., Z. Phys. C27 (1985) 65; Z. Phys. C31 (1986) 213.

11. D. Amati and G. Veneziano, Phys. Lett. 83B (1979) 87; G. Marchesini, L. Trentadue, and G. Veneziano, Nucl. Phys. B181 (1981) 335.

12. Yu. L. Dokshitzer, V. A. Khoze and S.I. Troyan in Ref. 2.

13. B.I. Ermolayev and V.S. Fadin, JETP Lett. 33 (1981) 285; A. H. Mueller, Phys. Lett. B104 (1981) 161.

14. Yu.L. Dokshitzer, V.S. Fadin and V. A. Khoze, Phys. Lett B115 (1982) 242; A. Bassetto et. al., Nucl. Phys. B207 (1982) 189.

15. Ya I. Azimov et. al., Phys. Lett B165 (1985) 147, Yad. Fiz 43 (1986) 149.

16. R.K. Ellis, G. Marchesini and B. R. Webber, Nucl. Phys.B286 (1987) 643.

17. Yu. L. Dokshitzer et.al., Phys. Lett B245 (1990) 243.

18. Yu. L. Dokshitzer et.al., preprint CERN TH 5738/90.

19. OPAL Collaboration, M.Z.Akrawy et.al., CERN EP/90-94.

20. B.R. Webber, Phys. Lett. B143 (1984) 501.

21. Yu. L. Dokshitzer, V.A. Khoze and S.I. Troyan LU TP 90-12 (1990).

22. B.R. Webber, talk given at the 25th Rencontre de Moriond, March 1990.

23. G. Marchesini and B.R. Webber, Nucl. Phys. B310 (1988) 461.

24. Yu. L. Dokshitzer, V.A. Khoze and S.I. Troyan Preprint LNPI-1230 (1986).

25. J.B. Gaffney and A.H. Mueller, Nucl. Phys. B250 (1985) 109.

26. C.P. Fong and B.R. Webber, Phys. Lett B229 (1989) 289.

27. TASSO Collaboration., W. Braunschweig et.al., DESY 90-013.

28. Yu. L, Dokshitzer, V.A. Khoze, A.H. Mueller and S.I. Troyan in Ref.[2].

29. E.D. Malaza and B.R. Webber, Nucl. Phys. B267 (1986) 702.

30. B. Andersson, G. Gustafson, and T. Sjöstrand, Phys. Lett. 94B (1980) 211.
B. Andersson, B. Gustafson, G. Ingelman, and T. Sjöstrand, Phys. Rep. 97 (1983) 31.

31. DELPHI Collaboration, Contribution to the 25th Int. Conf. on High Energy Physics, Singapore, Aug. 1990.

32. R.K. Ellis, FERMILAB-Conf.-87/108-T (1987).

33. M. Ciafaloni, Nucl. Phys. B296 (1987) 249. L.V. Gribov et. al., Phys. Lett. 202B (1988) 276; JETP 34 (1988) 12.

34. T. Sjöstrand, private communication.

35. B. Andersson, preprint LU TP 88-2 (1988).

36. T. Sjöstrand, lecture given at the Cargese summer Institute, Aug. 1990.

HIGHER-ORDER CORRECTIONS IN QCD EVENT SIMULATION[1]

G. Marchesini

Dipartimento di Fisica, Università di Parma,
INFN, Gruppo Collegato di Parma, ITALY

Abstract

In this talk I review recent results on Monte Carlo simulations including important next-to-leading contributions in the resummed perturbatice QCD. In particular I shall analyze quantities related to soft emission, heavy flavour production, initial state radiation for large and small x.

In hard hadronic processes the QCD perturbative evaluation of most physical quantities is characterized by the presence of large double logarithmic corrections [1] due to the emission of both soft and collinear gluons. In order to reliably compute such quantities it is necessary to resum these large corrections.

Collinear and soft logarithms correspond to a Markov branching process and can be resummed by Monte Carlo methods [see Ref. 2]. In particular the soft logarithms are correctly resummed provided one takes into account the coherent properties of soft gluon emission by means of a *coherent branching algorithm* [3-5].

Theoretical understanding of resummed perturbative QCD nowadays goes far beyond the leading double logarithmic order. It is interesting that not only the leading contributions, but also some important next-to-leading correction can be resummed by Monte Carlo methods.

In this talk I recall some improvement in the QCD simulation which allows us to include these important subleading contributions. In particular I shell discuss improvements in the resummations of the the following quantities:

---

[1] Research supported in part by the Ministero della Università e della Ricerca Scientifica.

Talk given at the 12th Eloisatron Workshop: "New Technologies for Supercolliders", Erice, September 1990

New Technologies for Supercolliders, Edited by L. Cifarelli and
T. Ypsilantis, Plenum Press, New York, 1991

1. Next-to-leading corrections in the soft final state emission: mean multiplicity, multiplicity moments, and single- and double-particle inclusive distributions [1,6,7].

2. Soft corrections in heavy flavour production processes: coherence in heavy flavour emission [8-10].

3. Next-to-leading corrections to Deep Inelastic Scattering and Drell-Yan hard cross sections in the semi-inclusive region of large x [11-14].

4. All-loop corrections [15-17] to Deep Inelastic Scattering and Drell-Yan hard cross sections in the region of small x.

Before discussing these points let me briefly recall the connection between QCD event simulations of hadron emission and perturbative fields theory. Perturbative study does not provide information on the confinement mechanism which converts partons into hadrons. However, provided this mechanism does actually exists, perturbative QCD predicts [19,20] that for large Q^2 confinement of partons is *local in colour* and *independent* on the hard scale Q^2.

This property is due to the Sudakov form factor, which sums infrared virtual corrections. In QED this form factor depletes the cross section for a single charge radiated without the accompanying cloud of photons, within a given resolution. Similarly, in QCD the Sudakov form factor inhibits the separation of the colour charges forming a singlet. In the jet emission of partons one finds [19] in perturbative QCD, as well as in the Monte Carlo simulations [3], that the mass distribution of two partons forming a colour singlet is concentrated around values of the order of Q_0 and independent of Q^2 for large Q^2.

This QCD property is well confirmed by the phenomenological analysis done by the Leningrad group [20]. In this analysis one compares the fragmentation functions for π K and p obtained at e^+e^- accelerators at various high energies Q with the fragmentations function for partons computed by resumming leading perturbative contributions. The result is that the π-, K-, and p-fragmentation functions are proportional to the parton-fragmetation functions. The proportionality constants K_π, K_K and K_p are independent of Q and correspond to the hadronization conversion from partons to hadrons. Thus partons convert into hadrons locally in the phase space and this property has been named LPHD (Local Parton Hadron Duality).

Monte Carlo simulations for hadronic processes have two different components, a perturbative and a non-perturbative one. The perturbative component describes partons interactions according to perturbative QCD, with typical momentum transfers $q_t^2 \gg Q_0^2$ such that $\alpha_S(Q_0^2) \lesssim 1$. The non-perturbative component is parameterized by structure functions and a hadronization model below the momentum scale Q_0. The preconfinement property is used in the Monte Carlo programs [3-5] by assuming hadronization models which are local in colour and independent of the hard process and the energy.

1. Next-to-leading corrections in soft emission

An important prediction of perturbative QCD is the coherence of soft gluon radiation, which is associated with gluon interference in the soft region (see the talk by V. Khoze [18]). All quantities related to the soft particles such as the multiplicity, multiplicity moments, the shape of the single or double inclusive distributions in the region of small momentum fraction $x \ll 1$ are obtained by the time-like anomalous dimension $\gamma_N(\alpha_S)$ for the moment index $N \sim 1$. To leading order this is well known [1] as given by

$$\gamma_N(\alpha_S) = -\frac{N-1}{4} + \sqrt{(\frac{N-1}{4})^2 + \bar{\alpha}_S} \quad \underset{N \to 1}{\longrightarrow} \quad \sqrt{\bar{\alpha}_S} \qquad \bar{\alpha}_S = \frac{C_A \alpha_S}{2\pi}. \tag{1}$$

For a recent review of QCD predictions for some of these quantities at LEP see Ref. [10,21].

Although each perturbative contribution is singular as $N \to 1$, the limit of $\gamma_N(\alpha_S)$ for $N \to 1$ is finite but proportional to $\sqrt{\alpha_S}$. In general one finds [1,6] that for $N = 1$ the perturbative expansion becomes an expansion in $\sqrt{\alpha_S}$. For these infrared-sensitivity quantities, the next-to-leading corrections are especially important and need to be re-summed in order to obtain reliable asymptotic estimations.

The physical origin of the result (1) is the destructive interference of soft gluons which reduces the phase space QCD radiation to an angular ordered region. It has been shown [6] that one can correctly sum not only the leading but also the next-to-leading contribution by using the branching algorithm in the angular ordered phase space.

This branching algorithm has been implemented the Monte Carlo programs with coherence [3-5]. By using these simulations one correctly reproduces to this accuracy quantities related to the $N \sim 1$ moments. Here I briefly recall the main results concerning the hadron momentum distributions at small relative momenta.

One of the most interesting prediction [1,6] coming from (1) is that the momentum spectrum of relatively soft particles ($x \ll 1$) should exhibit the so-called "hump-backed plateau" in the shape of the single inclusive distribution. The maximum of the distribution is predicted [6,7] to be at a value of the momentum $P = P_{Max}$ given by

$$\ln \frac{P_{Max}}{\Lambda} = \tfrac{1}{2} \ln \frac{E}{\Lambda} \left(1 - \frac{\rho}{24} \sqrt{\frac{48}{\beta \ln(\frac{E}{\Lambda})}} \right) + O(1) , \tag{2}$$

where $\beta = 11 - 2N_f/3$ and $\rho = 11 + 2N_f/27$ for N_f flavours, Λ being the effective QCD scale, and $E = Q/2$ is the jet energy. The confirmation of this QCD prediction at Lep is shown in Fig. 1 (see [18] for a more detailed discussion on this point).

The two-particle correlation $R(x_1, x_2, E)$, defined by

$$R(x_1, x_2, E) \equiv D^{(2)}(x_1, x_2, E)/[D(x_1, E)D(x_2, E)] , \tag{3}$$

where $x_i = P_i/E$, has been computed [7] up to $\sqrt{\alpha_S}$. For e^+e^- annihilation with $N_f = 5$ flavours one has

$$R(x_1, x_2, E) = 1.375 - 1.125 \left[\frac{\ln(x_1/x_2)}{\ln(E/\Lambda)} \right]^2 - \left[1.262 + 0.877 \left(\frac{\ln(x_1 x_2)}{\ln(E/\Lambda)} \right) \right] \frac{1}{\sqrt{\ln(E/\Lambda)}} . \tag{4}$$

There is a preference for $x_1 \simeq x_2$ and for both to be small. The range of these correlations is of the order of $\ln(E/\Lambda)$, so they extend over long distances in the $(\ln x_1, \ln x_2)$ plane. Thus they should be easily distinguishable from correlations due to hadronization, which would be expected to have ranges of order $\ln(Q_h/\Lambda)$, where the hadronization scale Q_h is at most a few GeV.

289

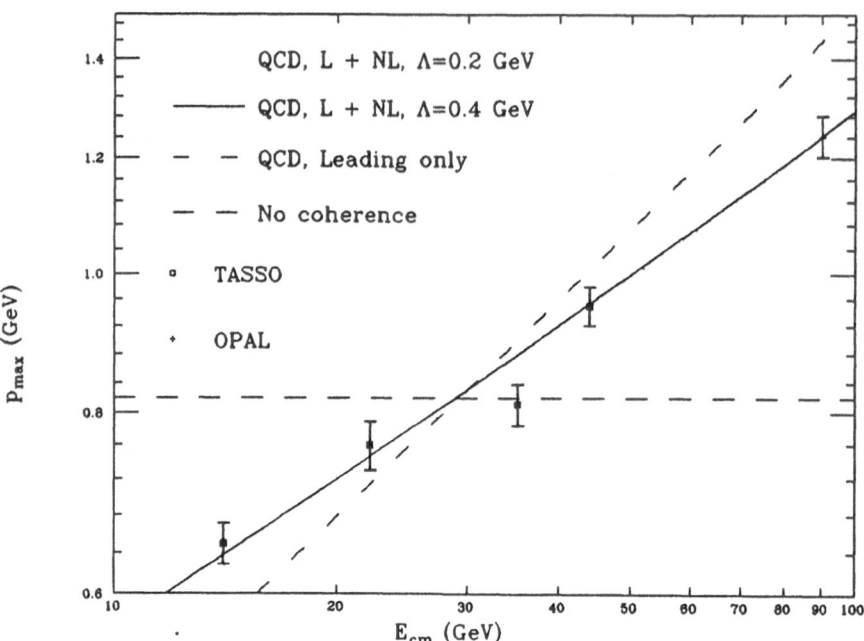

Fig. 1.   The maximum of the charged particle distribution compared with the prediction in (2).

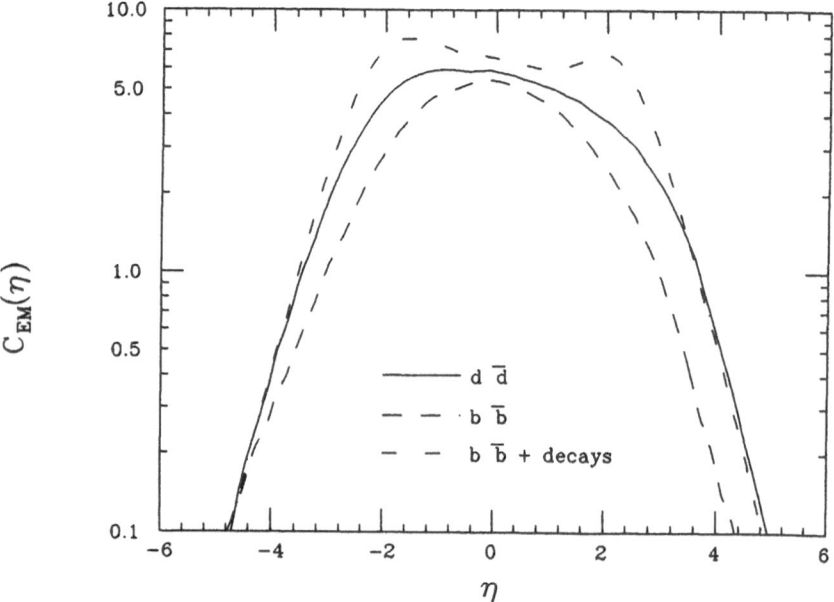

Fig. 2.   Energy-multiplicity correlations in $Z^0 \to d\bar{d}$ and $b\bar{b}$ as a function η (see Eq. (7)).

2. Soft corrections in heavy flavour production processes

In heavy flavour production processes [8-10], the main perturbative QCD properties are connected with the bremsstrahlung factor for the radiation of a soft gluon of transverse momentum k_t and energy ω from a heavy quark Q of mass M_Q and energy E_Q, given by

$$dW(k) = \frac{C_F \alpha_S}{\pi} \frac{d\omega}{\omega} \frac{k_t^2 dk_t^2}{[k_t^2 + \omega^2(M_Q/E_Q)^2]^2}, \qquad (5)$$

which leads to the depletion of soft gluon radiation at small angles θ

$$\theta \simeq \frac{k_t}{\omega} \gtrsim \frac{M}{E_Q}. \qquad (6)$$

This suppression of collinear radiation by a screening mass term has been included in HERWIG [5]. The heavy quarks undergo a QCD cascade similar to that for light quarks, but the phase space available for branching is now further reduced by the heavy quark masses, in accordance with (6). Inclusion of this effect in Monte Carlo programs is especially important in order to obtain a reliable description of heavy quark physics.

To investigate the influence of the collinear singularity screening by heavy quark masses one can compare [8-10] the shapes of light and heavy quark jets. Particularly interesting is the study of the energy-energy and energy-multiplicity correlations defined by

$$C_{EE}(\eta) = \int \frac{E_1 dE_1}{E} \frac{E_2 dE_2}{E} d\theta_{12} \, \delta\left(\ln\tan\frac{\theta_{12}}{2} + \eta\right) \frac{d\sigma}{dE_1 dE_2 d\theta_{12}},$$
$$C_{EM}(\eta) = \int \frac{E_1 dE_1}{E} \quad dE_2 \; d\theta_{12} \, \delta\left(\ln\tan\frac{\theta_{12}}{2} + \eta\right) \frac{d\sigma}{dE_1 dE_2 d\theta_{12}}, \qquad (7)$$

with η the relative pseudo-rapidity of the two observed hadrons. The energy-energy correlation is an infrared-finite quantity which requires careful treatment of collinear singularities in the small-angle region. The energy-multiplicity correlation is sensitive in addition to soft coherence effects and is given asymptotically by summing the leading Feynman diagrams, as is done in Monte Carlo programs with coherence [3-5].

Fig. 2 shows the comparison between d and b jets in Z^0 decays, for the C_{EM} correlations generated using HERWIG. We see that the b jets have a broader structure with less hadron flow at small angles (large η), even when heavy flavour decay products are included. This Figure illustrates the advantage of measuring correlations rather than plotting η relative to some jet axis: the correlations are not forward-backward symmetric and indeed give different dynamical information in the two hemispheres.

The screening mass term in (5) could prove to be a signature for jets initiated by heavy quarks. As an example in Fig. 4 is plotted the distributions in the quantity $\bar{C}_{EM}$ obtained by integrating the distribution $C_{EM}(\eta)$ over the rapidity region $1 < \eta < 2$. The two curves for light and heavy quark jet are clearly separated.

3. QCD Coherent branching and semi-inclusive processes at large x

We discuss now the theoretical reliability of a Monte Carlo simulations with coherence [3-5] for semi-inclusive quantities in deep inelastic scattering (DIS) and Drell-Yan (DY) processes. The analysis has been done in Refs. [11,13] for the transverse momentum distributions, and in Ref. [14] for the distributions near the boundary of phase space, namely for $x = |Q^2|/2p \cdot Q \to 1$ in DIS and for $\tau = Q^2/s \to 1$ in DY.

As recently discussed in Ref. [14], a Monte Carlo program with coherence using only the one-loop splitting functions can indeed correctly compute these semi-inclusive distributions to next-to-leading order. To achieve this accuracy one needs to use the two-loop expression for α_S and the following universal relation between the scale parameter Λ_{MC} used in the Monte Carlo simulation and $\Lambda_{\overline{\text{MS}}}$, the QCD scale in the $\overline{\text{MS}}$ subtraction scheme

$$\Lambda_{\text{MC}} = \exp\left(\frac{67 - 3\pi^2 - 10N_f/3}{2(33 - 2N_f)}\right) \Lambda_{\overline{\text{MS}}} \simeq 1.569\,\Lambda_{\overline{\text{MS}}} \quad \text{for } N_f = 5 \,. \tag{8}$$

Therefore a Monte Carlo with next-to-leading accuracy can be used to determine $\Lambda_{\overline{\text{MS}}}$ from semi-inclusive data.

This result is obtained by computing directly the mentioned semi-inclusive distributions by using the coherent branching algorithm and by comparing these distributions with the analytical results to next-to-leading order. One shows [11,12,14] that the two results agree provided that in the branching algorithm with coherence one uses the two-loop expression both for running α_S and for the two-loop Altarelli-Parisi quark or gluon splitting function at $z \to 1$:

$$P_i(z, \alpha_S) = \frac{A_i(\alpha_S)}{1 - z} \quad \text{where} \quad A_i(\alpha_S) = \frac{C_i}{\pi}\alpha_S^{(\overline{\text{MS}})}\left(1 + K\frac{\alpha_S^{(\overline{\text{MS}})}}{2\pi}\right) , \tag{9}$$

with $C_i = C_F$ or C_A for a quark or a gluon respectively and K given by

$$K = C_A\left(\frac{67}{18} - \frac{\pi^2}{6}\right) - \frac{5}{9}N_f \tag{10}$$

From Eq. (9) we see that the next-to-leading correction to the splitting functions for $z \to 1$ is a universal factor associated with soft gluon emission. Therefore it can be absorbed into the one-loop splitting functions used in existing Monte Carlo simulations with coherence and with the two-loop definition of α_S. This is simply obtained by rescaling the value of Λ_{MC}, used in the simulation, in such a way that

$$\alpha_S^{(\text{MC})} = \alpha_S^{(\overline{\text{MS}})}\left(1 + K\frac{\alpha_S^{(\overline{\text{MS}})}}{2\pi}\right) . \tag{11}$$

From this we finally obtain the announced result in (8).

It should be emphasized that the relation (8) has been derived only in the semi-inclusive region, and therefore $\Lambda_{\overline{\text{MS}}}$ can be determined by existing Monte Carlo programs with coherence only by fitting data in this restricted kinematical region. Away from the boundaries of the phase space, Λ_{MC} cannot be related to $\Lambda_{\overline{\text{MS}}}$ without a full treatment which includes the non-singular parts of higher-order QCD corrections.

4. Simulation of QCD initial state radiation at small x

For hadron-hadron and hadron-lepton collisions the range of applicability of perturbative QCD has been recently extended [15] to the semi-hard phase space region of small x, with $x \simeq Q^2/s$. In this region the multi-parton distributions still have a factorized structure and can be described as a branching process. Such a branching process is very different from the one valid in the finite x region, which is the basis of all present Monte Carlo generators. Indeed for small x the structure function is given by the Lipatov equation [22], while the standard branching leads to the Altarelli-Parisi equation. The main properties of the emission in the small x regions are essentially the followings (we consider the DIS process).

(i) *Coherence and phase space for the branching.* Destructive interference among soft gluons depletes the emission phase space. Both for $x \to 0$ and for $x \to 1$, one finds that the emission takes place in the angular ordered region

$$\theta_{i+1} > \theta_i \quad \Rightarrow \quad q_{t\,i+1} > \left(\frac{1 - z_{i+1}}{1 - z_i} \right) z_i q_{t\,i} , \tag{12}$$

where θ_i and q_{ti} are the angle and the transverse momentum of the emitted gluon q_i with respect to the incoming parton p and $(1 - z_i)$ is the energy fraction of gluon q_i.

(ii) *Non-Sudakov form factor.* In the region $x \to 0$ some of the gluons q_i have $z_i \to 0$. The corresponding virtual corrections contain $\ln z_i$-singular contributions, which are not of Sudakov type. These new virtual corrections factorize and exponentiate to give the following non-Sudakov form factor

$$\Delta_{ns}(q_{ti}, z_i, Q_{ti}) = \exp\left[-\frac{C_A}{\pi} \alpha_S(Q_{ti}) \ln\left(\frac{1}{z_i}\right) \ln\left(\frac{Q_{ti}^2}{z_i q_{ti}^2}\right) \right] , \tag{13}$$

where Q_{ti} is the total transverse momentum of the system formed by all partons emitted within a cone of aperture θ_i.

We describe the main features [15] of this new branching and present the result of the Monte Carlo simulation [16,17].

1) *Comparison with the standard branching.* In the small x region, the standard coherent branching process for initial-state radiation [5], is obtained by generalizing the Altarelli-Parisi evolution to agree with the one loop singular contribution of the anomalous dimension for $N \to 1$. For this reason we shall call this the "one-loop branching". In this branching the phase space (for small x) is given by transverse momentum ordering and the only virtual corrections are of Sudakov type ($\Delta_{ns} = 1$). In order to clarify the main features of the all-loops branching, we have also constructed a Monte Carlo program based on the one-loop branching.

2) *Feature of the angular ordering.* Due to the rescaling factor z_i, the phase space (12) for $z_i \to 0$ is much larger than the transverse momentum ordering used in the one-loop branching

$$\{\theta_{i+1} > \theta_i\} \sim \{q_{t\,i+1} > z_i q_{ti}\} \quad \supset\supset \quad \{q_{t\,i+1} > q_{ti}\} , \quad x \to 0 . \tag{14}$$

Since the phase space for the QCD coherent branching (12) is the same in all regions of x, we have constructed [16] a Monte Carlo simulation which for $x \to 0$ takes into account the mentioned results to all-loops, and for x finite takes into account all the leading contributions and the next-to-leading corrections important in the large x region. This new Monte Carlo simulation program is not yet in the form of an event simulator but is suited to study purely perturbative features.

For $z_i \to 0$ the lower bound of $q_{t\,i+1}$ in (12) vanishes. This fact has two consequences for the branching. In inclusive quantities no singularities are generated [15] from the vanishing of this limit. However, taking into account the running of α_S, in the branching

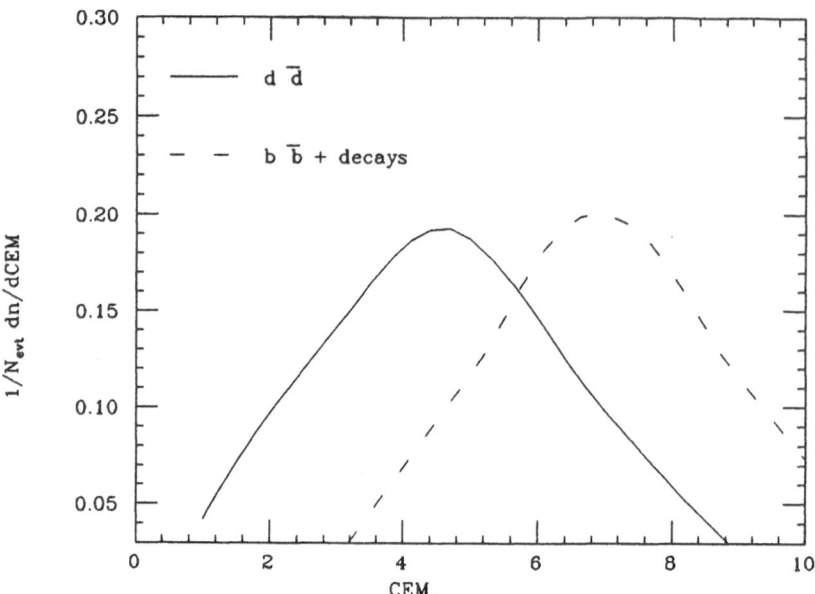

Fig. 3.   Energy-multiplicity correlations of Fig. 2 integrated over the region $1 < \eta < 2$.

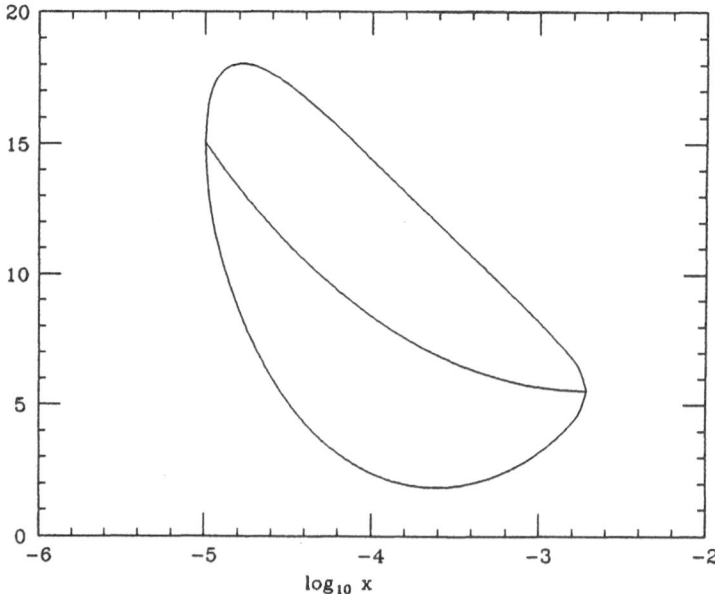

Fig. 4.   Average trajectory and dispersion in the variable $t = \ln(Q_t^2/\Lambda^2)$ versus x. See Ref. [17].

process we need anyway to include the cutoff $q_{t\,i+1} > Q_0$. The second consequence is that for $x \to 0$ there is no transverse momentum ordering. Therefore the transverse momenta generated during a small-x parton cascade perform a random walk.

We have analysed [17] the transverse momentum random walk for small x by considering for each cascade the trajectories in the variables Q_{ti} and x_i, the total emitted transverse momentum and the residual fraction of energy after the i-branching. The average values for trajectories and the dispersion are computed either by saddle point method or by Monte Carlo simulation. In Fig. 4 the result obtained by saddle point approximation are shown for typical initial and final values of transverse momentum and energy. At small values of $t = \ln Q_t^2/\Lambda^2$ we clearly see the effect of the transverse momentum disorder. The starting value of the evolution is at $\ln Q_s^2/\Lambda^2 = 5$, but the phase space (12) allows the cascade to generate lower values of intermediate t.

3) *Feature of the non-Sudakov form factor.* In this new branching the non-Sudakov form factor has the effect of screening the $1/z_i$ singularity of the gluon splitting function

$$\frac{\Delta_{ns}(q_{ti}, z_i, Q_{ti})}{z_i} \to 0 \qquad z_i \to 0 \,. \tag{15}$$

For finite x we have $\Delta_{ns} \simeq 1$, we can neglect the non-Sudakov corrections and the branching becomes the usual evolution process leading to the Altarelli-Parisi equation with coherence.

For $x \to 0$, the new form factor is important. The Q_{ti}-dependence in Δ_{ns} makes the branching *non local*, i.e. dependent on the development of part of the emission process. Because of this non local Q_{ti}-dependence, the $x \to 0$ branching does not reduce to the Altarelli-Parisi equation for small x.

At the inclusive level, the screening of the $z_i \to 0$ singularity in Eq. (15) is reflected in the cancellations of $\ln z_i$-singular terms between the virtual and real contributions coming from the the expansion of Δ_{ns} and from the transverse momentum integration in (12). These cancellations are crucial to reproduce the correct $\ln x$ powers in the expansion of $F(Q, x)$ and the $N \sim 1$ gluon anomalous dimension. For the structure function the iterative equation resulting from this new branching is, for small x, the Lipatov equation [22]. The corresponding $N \sim 1$ anomalous dimension is then given by

$$\gamma_N(\alpha_S) = \frac{\bar{\alpha}_S}{N-1} + 2\zeta_3 \left(\frac{\bar{\alpha}_S}{N-1}\right)^4 + \cdots , \qquad \bar{\alpha}_S \equiv \frac{C_A \alpha_S}{\pi} , \tag{16}$$

where ζ_i is the Riemann zeta function. Notice that the α_S^2 and α_S^3 terms are absent. This new branching, which generates the $N \sim 1$ anomalous dimension to all-loop, will be called the "all-loops branching process"

Each perturbative term has a singularity at $N = 1$, but the expansion (16) develops a square root singularity at $N = 1 + (4 \ln 2)\bar{\alpha}_S$. For fixed α_S, this singular anomalous dimension gives a structure function which, for small x, has the behaviour

$$xF^{(all)}(x, Q) \sim x^{-p} , \qquad p = (4 \ln 2)\bar{\alpha}_S , \tag{17}$$

leading to a violation of the Froissart bound. Ref. [23] discusses how this violation can be overcome by a consistent unitarization procedure.

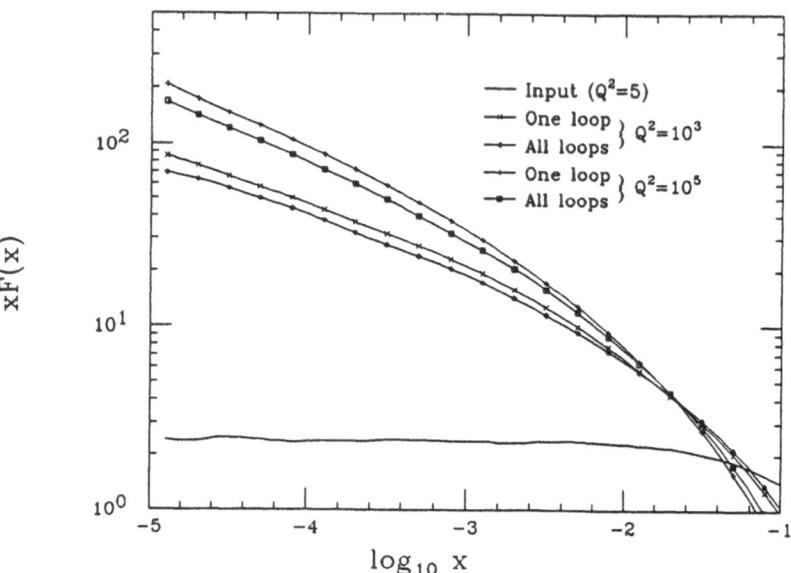

Fig. 5.   Structure functions in the two algorithms for the input condition in Eq. (19).

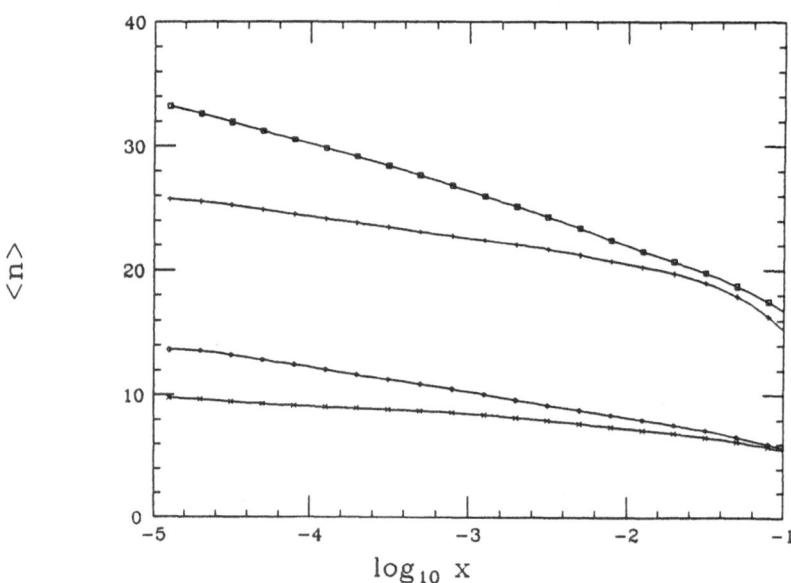

Fig. 6.   Associated distributions in the two algorithms for the input condition in Eq. (19).

If one takes into account only a fixed number of terms in the gluon anomalous dimension, the only singularities are at $N = 1$. The corresponding structure function is much less singular for $x \to 0$. In particular the structure function corresponding to the one-loop anomalous dimension $\gamma_N^{(1)}(\alpha_S) = \bar{\alpha}_S/(N-1)$ is given by

$$x F^{(1)}(x, Q) \sim \exp \sqrt{a \ln(1/x)}\,, \qquad a = 4\bar{\alpha}_S \ln(Q^2/Q_s^2)\,. \qquad (18)$$

4) *Structure functions.* The all-loops and one-loop simulation schemes were used [16] to study the evolution of a purely gluonic system over a wide range of momentum transfer scales Q^2, for a variety of input parameterizations of the structure function at low Q^2. Figure 5 shows the results for the flat input structure function

$$x F(x, Q_s) = A(1-x)^5\,, \qquad (19)$$

where the starting scale was set at $Q_s^2 = 5 \text{ GeV}^2$. The transverse momenta cutoff was chosen to be $Q_0 = 1 \text{ GeV}$ and $\Lambda = 0.2 \text{ GeV}$.

We see from Fig. 5 that the structure function evolves rapidly to a steeper form, indicating the generation of a stronger singularity as $x \to 0$. Even over such a wide range of x and Q^2, it is difficult to distinguish between a one-loop behaviour of the form (18) and the all-loops power type of behaviour (17). This can be explained on the bases of two effects:

a) Since the first correction to the one-loop expression in (16) is to the order α_S^4, we expect that the effective anomalous dimensions in the two formulations are nearly equal, leading to similar structure functions. For a more detailed analysis, see Ref. [17].
b) For x small but finite we have that the cutoff Q_0 needs to be taken into account even though the structure function is infrared finite. The effect of $Q_t > Q_0$ will be to shift the average trajectory upwards. Thus, over an intermediate range of x_i and Q_{ti}, we expect $F^{(all)} < F^{(1)}$.

We found a very strong dependence on the initial conditions, which persists until large values of Q^2. This underlines the importance of determining the input gluon structure function experimentally down to the lowest possible values of x. In this respect new data, especially from HERA, will play a crucial role.

5) *Associated multiplicity.* The similarity of the structure functions is due to a fortuitous cancellation among leading higher-order corrections which occurs in highly inclusive observables such as the anomalous dimension (16). In less inclusive quantities such as the associated multiplicity and distributions of emitted gluons, we have found clear differences (see Fig. 6) between the results of the two evolution algorithms, all of which are explicable in terms of the two new dynamical features of the all-loops algorithm, namely the enlarged phase space due to angular ordering and the screening of the $1/z$ singularity in the gluon splitting function by the non-Sudakov form factor.

6) *Rapidity gaps.* One of the most important feature of the new formulation is the suppression of large energy and rapidity gaps, and large neighboring pair masses, in the distributions of primary emitted gluons. When the full colour structure of final-state branching is taken into account, this will have the effect of suppressing the production

of high-mass colour-singlet combinations of partons. This in turn will produce a more local preconfinement of colour [19,20] and permit a more direct connection between the perturbative parton shower and observed hadron distributions.

Acknowledgements

Most of the results here reviewed here been obtained in recent works in collaboration with S. Catani, Yu.L. Dokshitzer, F. Fiorani, V.A. Khoze, E.M. Levin, M.G. Ryskin and B. Webber.

References

1. A. Bassetto, M. Ciafaloni and G. Marchesini, Phys. Rep. 100 (1983) 201.
   Yu.L. Dokshitzer, V.A. Khoze, S.I. Troyan and A.H. Mueller, Rev. Mod. Phys. 60 (1988) 373; Yu.L. Dokshitzer, V.A. Khoze and S.I. Troyan, in *Perturbative Quantum Chromodynamics*, ed. A.H. Mueller (World Scientific, Singapore, 1989), p. 241.

2. B.R. Webber, Annu. Rev. Nucl. Part. Sci. 36 (1986) 253.

3. G. Marchesini and B.R. Webber, Nucl. Phys. B238 (1984) 1; B.R. Webber, Nucl. Phys. B238 (1984) 492.

4. M. Bengtsson and T. Sjöstrand, Phys. Lett. 185B (1987) 435.

5. G. Marchesini and B.R. Webber, Nucl. Phys. B310 (1988) 461.

6. A.H. Mueller, Nucl. Phys. B213 (1983) 85, Nucl. Phys. B241 (1984) 141;
   E.D. Malaza and B.R. Webber, Phys. Lett. 149B (1984) 501, Nucl. Phys. B267 (1986) 702;
   Yu.L. Dokshitzer, V.A. Khoze and S.I. Troyan, in *Perturbative Quantum Chromodynamics*, ed. A.H. Mueller (World Scientific, Singapore, 1989), p. 241.

7. C.P. Fong and B.R. Webber, Phys. Lett. 229B (1989) 289, Phys. Lett. 241B (1990) 255.

8. Yu.L Dokshitzer, V.A. Khoze and S.I. Troyan, in *Proc. 6th Int. Conf. on Pysics in Collision*, ed. M. Derrick (World Scientific, Singapore, 1987).

9. G. Marchesini and B.R. Webber, Nucl. Phys. B330 (1990) 261.

10. Yu.L. Dokshitzer, V.A. Khoze G. Marchesini and B.R. Webber, in High Energy Hadronic Interaction, ed. J.Tran Than Van (Edition Frontières 1990).

11. G. Marchesini and L. Trentadue, Phys. Lett. 164B (1985) 150; L. Trentadue, Phys. Lett. 186B (1987) 405.

12. S. Catani, E. d'Emilio and L. Trentadue, Phys. Lett. 211B (1988) 335.

13. S. Catani and L. Trentadue, Nucl. Phys. B327 (1989) 353.

14. S. Catani, G. Marchesini and B.R. Webber, Nucl. Phys. B349 (1991) 635.

15. M. Ciafaloni, Nucl. Phys. B296 (1987) 249;
    S. Catani, F. Fiorani and G. Marchesini, Phys. Lett. 234B (1990) 339;
    S. Catani, F. Fiorani and G. Marchesini, Nucl. Phys. B336 (1990) 18.

16. G. Marchesini and B.R. Webber, Nucl. Phys. B349 (1991) 617.

17. E.M. Levin, G. Marchesini, M.G. Ryskin and B.R. Webber, Nucl. Phys. to be published.

18. V.A. Khoze, Talk at this meeting.

19. D. Amati and G. Veneziano, Phys. Lett. 83B (1979) 87; A. Bassetto, M. Ciafaloni and G. Marchesini, Phys. Lett. 83B (1979) 207; G. Marchesini, L. Trentadue and G. Veneziano, Nucl. Phys. B181 (1981) 335.

20. Yu.L. Dokshitzer and S.I. Troyan, Leningrad Nuclear Physics Institute preprint N922 (1984); Ya.I. Azimov, Yu.L. Dokshitzer, V.A. Khoze and S.I. Troyan, Phys. Lett. 165B (1985) 147, Zeit. Phys. C27 (1985) 65.

21. Z. Kunszt, P. Nason, G. Marchesini and B.R Webber, in *Z Physics at LEP*, CERN-Yellow Report 89-08, G. Altarelli, R. Kleiss and C. Verzegnassi eds.

22. L.N. Lipatov, Yad. Fiz. 23 (1976) 642 [Sov. J. Phys. 23 (1976) 338]; E.A. Kuraev, L.N. Lipatov and V.S. Fadin, Zh. Eksp. Teor. Fiz. 72 (1977) 373 [Sov. Phys. JETP 45 (1977) 199]; Ya. Balitskii and L.N. Lipatov, Yad. Fiz. 28 (1978) 1597 {Sov. J. Nucl. Phys. 28 (1978) 822]; J. Bartels, Nucl. Phys. B151 (1979) 293; T. Jaroszewicz, Acta Phys. Pol. B11 (1980) 965; Phys. Lett. 116B (1982) 291.

23. L.V. Gribov, E.M. Levin and M.G. Ryskin, Phys. Rep. 100 (1983) 1.

HIGH ENERGY PHENOMENON OF BARYON NUMBER VIOLATION

Ruibin Meng

High Energy Physics Division
Argonne National Laboratory, Argonne, IL 60439

ABSTRACT

We discuss the possibility of observing baryon number violation induced by electroweak instantons at high energy colliders. The event rate at the SSC is found about a few per hour if the naive-instanton amplitudes were valid for energy larger than 17 TeV.

INTRODUCTION

I am outlining work done in collaboration with G. Farrar[1]. My coverage here is eclectic, and I would refer readers to our paper for a more detailed discussion. Recently, Ringwald[2] and others[3, 4] have attempted to calculate the electroweak-instanton-induced amplitudes for scattering such as

$$u + u \to \bar{d}e^+\bar{c}\bar{s}\bar{s}\bar{\nu}_\mu\bar{t}\bar{t}\bar{b}\tau^+ + n_H H + n_W W$$

with large numbers of W and Higgs bosons, using the naive instanton approximation. Their results, if used at very high energy and coupled with approximate treatments of the final state phase space, suggested that the total cross section for B+L-violation might become large enough to be observed at SSC energies. This use of the naive instanton approximation for energies much greater than M_W has been challenged by many authors [5, 6, 7, 8, 9, 10, 11, 12, 13, 14]. In particular refs. [11] and [12] show that a better treatment leads to a *suppression* relative to the naive result, at least in related models. Nonetheless, it is interesting to consider the phenomenology of this system, since the extent of the suppression is not quantitatively known for the interesting case of electroweak gauge theory, and the difference between a factor of 1 or 4π multiplying $\frac{-1}{g^2}$ in the exponential can make the difference between an observable or unobservable effect. Therefore we take the naive instanton results for the amplitudes at face value, and investigate their phenomenological consequences.

New Technologies for Supercolliders, Edited by L. Cifarelli and
T. Ypsilantis, Plenum Press, New York, 1991

PHASE SPACE

The most important contribution to these processes in pp scattering arises from the case in which the initial particles are u or d quarks. We considered the case that an arbitrary number, n_H, of Higgs bosons accompany the fermions. We calculated n-body phase space[15], for $m = 81$ GeV and $E = 16$ TeV, by using the extreme relativistic and non-relativistic approximations. For comparison, we also computed the "true" phase space, by using either the monte carlo phase space generator RAMBO[16] or the analytic method described in ref.[15]. In the analytic approach, the delta function imposing overall energy-momentum conservation is replaced by a Laplace transform and the final expression for the phase space is a one dimensional integral of the nth power of a function of the Laplace transform variable. In the cases of interest to us, this function can be evaluated analytically. The final integration can be done by saddle point approximation, and yields the usual extreme relativistic and non-relativistic expressions for small and large n, respectively. For intermediate n, the equation determining the saddle point needs to be solved numerically. Since the analytic method is in excellent agreement with RAMBO, we use it in preference to RAMBO which is rather time consuming for large n. For $n \sim 120$, the relevant value according to Ringwald's calculation, we found that the extreme relativistic approximation overestimates the cross section by a factor of $\sim 10^{60}$, while the non-relativistic approximation underestimates the cross section by more than a factor of $\sim 10^{25}$, except for $n \gg 150$. This exercise teaches us the importance of doing n-body phase space quantitatively when n is large.

CROSS SECTION

We now turn to the determination of the cross section for B+L violation initiated by the collision of two u and/or d quarks, using Ringwald's amplitudes and the analytic method to evaluate phase space, and adding up the contribution of each σ_n. We find that B+L violation accompanied only by Higgs production is negligible at the SSC, for any value of M_H which is not excluded experimentally. Since the original work of Ringwald, it has been pointed out[9, 3, 17, 18] that the amplitude for production of gauge bosons has a different energy dependence from that for Higgs production. In particular Ringwald's eqn (66) should be multiplied by $\frac{2(3\omega^2 + \vec{k}^2)}{3m_W^2}$ for each gauge boson[19] We have therefore computed the cross section for arbitrary n_W and n_H, evaluating the phase space integrals using the analytic method, now with a matrix element having non-trivial dependence on the momenta of the gauge bosons and fermions. We find that SU(2) gauge boson production dominates Higgs production, and Higgs production can be ignored altogether to excellent accuracy. Figure 1 shows the cross sections for production of n_W $W^{\pm}$'s for $E_{qq} = 16, 17$, and 18 TeV, using the ref. [17] coeficient for the gauge boson factor. The cross section peaks for $n_W \sim 85$, a smaller number than for pure Higgs production, which is not surprising since that increases the momentum-dependent factor in the amplitude. As can be seen from Fig. 1, there is an extremely abrupt threshold in the total cross section at ~ 17 TeV.

We next turn to the computation of the cross section for pp collisions, to see if such a parton-level cross-section would be observable at presently proposed accelerators. In order to model the parton-level cross section, we take advantage of the sharp threshold and the fact that unitarity must be respected in a physical process. The naive-instanton-approximation

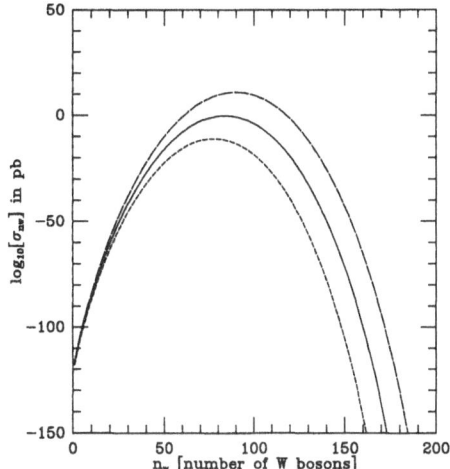

1. Quark-level cross section as a function of the number of gauge bosons produced: $\mathrm{Log}_{10}\sigma_{n_W}$ in pb vs. n_W for $\sqrt{s} = 17$ TeV (solid line), $\sqrt{s} = 18$ TeV (long-dashed line), and $\sqrt{s} = 16$ TeV (short-dashed line).

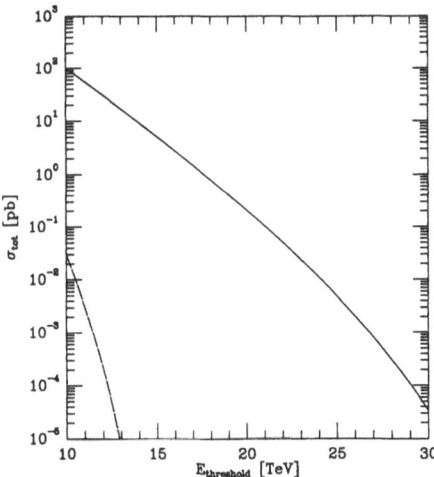

2. Total cross section at the SSC (solid line, c.m. energy $= 40$ TeV) and LHC (dashed line, c.m. energy $= 16$ TeV) as a function of the threshold energy.

amplitudes are pointlike, i.e., purely s-wave, so that their unitarity limit is $\sigma_{tot} \leq \frac{16\pi}{s}$ ($=70$ pb at $\sqrt{s} = 17$ TeV). We thus model the quark level cross section as $\theta(\sqrt{\hat{s}} - E_{thresh})\frac{16\pi}{\hat{s}}$. Using the structure functions of Ref.[20] and $E_{thresh} = 17$ TeV, we find a cross section of ~ 1.5 pb for pp collisions at 40 TeV c.m. energy, corresponding to a few events per hour at the SSC, while at $E_{thresh} = 21$ TeV the cross section at the SSC is about an order of magnitude lower. Anticipating that future theoretical developments will result in modifications to the amplitudes, we give in Fig. 2 the cross section at the SSC and LHC, as a function of the energy of the threshold.

EVENT CHARACTERISTICS

We now argue that the events, if they are actually produced, will be easily distinguished from conventional physics, and furthermore will provide a signature of their B and L violating nature, even though it is not easy to distinguish between quark and antiquark jets. A typical event would have $E_{qq} \sim 17$ TeV and would contain ~ 85 SU(2) gauge bosons. These events are very distinctive. The gauge particles decay to jets and leptons, such that on average each event contains ~ 120 jets, ~ 10 charged leptons (not including τ's or their decay products), and 30 ν's, each carrying about 95 GeV, with, typically, two $e^\pm$ and/or $\mu^\pm$ pairs reconstructing to a Z. Even though such a large number of jets will tend to overlap and thus not be completely separable, the large multiplicity of charged particles ($\sim 10^3$) and large total energy (~ 15 TeV) in the central detector, with unbalanced transverse momentum of several hundred GeV (from the ν's), is unlike any ordinary event. Although we cannot reliably estimate the QCD background, it is clearly negligible, since the cross section for producing events with ~ 120 jets of average energy ~ 95 GeV would (barring total breakdown of our understanding of perturbative QCD) be far below the unitarity limit.

Another characteristic feature of this process is the abrupt threshold in the cross section as a function of the total energy of the subprocess. Even if some non-"instanton" mechansim were discovered to produce large numbers of electroweak gauge bosons, it would be unlikely to imitate this stucture, since phase space alone would not produce the abrupt threshold, as can be appreciated by studying Figs. 1 and 2 in the relevant region of n. See ref. [21] for a discussion of "strong" weak interactions.

In principle, at least, it should be possible to determine that B+L has been violated, and at what rate. As noted above, besides the $W^\pm$'s and Z^0's which decay to lepton pairs or quark jets, with particles and antiparticles evenly represented in their decays, there are 10 primary antifermions when B+L is violated. Since each member of each fermion weak isodoublet is equally likely to be produced, 3/4 of the B+L violating events will have either a primary e^+ or μ^+, and 1/4 will have a primary e^+ and μ^+. Besides the excess in the number of antileptons as compared to leptons, which would require excellent solid angle coverage and high statistics to establish, there is an asymmetry in the mean energy of the fastest antilepton as compared to that of the fastest lepton. To illustrate its origin, first neglect the difference in the dependence of the matrix element on the energy of the primary antifermions and gauge bosons. In this case the most energetic e^+ or μ^+ would have on average twice as much energy as the most energetic e^- or μ^-, which is always the decay product of a gauge boson and never "primary". In the naive instanton approximation, however, the matrix element has a factor $\sim E_f$ for each fermion and a factor $\sim (4E_{gb}^2 - m_{gb}^2)$ for each gauge boson, reducing the asymmetry. In this case, for $E = 17$ TeV and $n_{gb} = 85$, the average energy carried by a primary antifermion is 116 GeV, while $< E_{gb} >= 186$ GeV. Thus the most energetic antilepton typically carries about 30% more energy than the most energetic lepton. Just as the magnitude of the B+L violating cross section cannot be reliably obtained from the naive instanton approximation, the form of the matrix element is also uncertain and we must await better theoretical tools to predict this asymmetry with confidence. Nonetheless these two simple cases illustrate that a measurable asymmetry can be expected. Furthermore, if events such as these are produced, the gauge boson spectra can (at least with an ideal detector) be *measured*. One can then compare the observed spectrum of highest-energy $e^{\pm}$ and $\mu^{\pm}$'s with that expected from decays of the gauge-bosons. If the fraction of events violating B+L is large enough, the excess of high energy anti-leptons will be observable. In principle, the spectra of the primary antifermions could be disentangled from that of the gauge bosons, leading to detailed dynamical information about the process, with which to confront theory.

CONCLUSION

We have seen that use of the naive instanton amplitudes for electroweak-instanton-induced baryon number violation in high energy collisions, combined with a correct treatment of the massive, multiparticle phase space, leads to a predicted cross section at the SSC which is large enough for unambiguous identification of the phenomenon. We must caution, however, that it is highly doubtful that the naive instanton approximation is valid at the relevant energies. Refs.[11] and [12] have shown that as the energy is increased from low to intermediate relative to the barrier height, corrections to the naive instanton approximation *reduce* the amplitude. Moreover we have found that even using the naive instanton approximation, an energy *higher* than the sphaleron energy (7-13 TeV[22]) is necessary in order to have an experimentally observable effect, so that the question of the experimental observability of baryon number violation in high energy collsions cannot be convincingly addressed theoretically until a reliable computational scheme for energies *higher than* the sphaleron is developed.

ACKNOWLEDGEMENT

The authers wish to thank the organizers of this workshop for arranging such an interesting meeting. We have benefitted from discussions with G. Altarelli, J. Ellis, S. Errede, I. Hinchcliffe, J. Huth, V. Khoze, and A. Ringwald.

References

[1] G. Farrar, R. Meng. *Phys. Rev. lett.*, 65, 1990.

[2] A. Ringwald. *Nucl. Phys.*, B330, 1990.

[3] O. Espinosa. Technical Report CALT-68-1586, Caltech, 1989.

[4] L. Mclerran, A. Vainshtein, and M. Voloshin. *Phys. Lett.*, 155B, 1985.

[5] R. Peccei. Technical Report UCLA/89/TEP/65, UCLA, 1989.

[6] K. Aoki and P. Mazur. Technical Report UCLA/89/TEP/67, UCLA, 1989.

[7] M. Cornwall. Technical Report UCLA/90/TEP/2, UCLA, 1989.

[8] K. Aoki. Technical Report UCLA/90/TEP/7, UCLA, 1990.

[9] V. Zakharov. Technical Report TPI-MINN-90/7-T, Univ. of Minnesota, 1990.

[10] A. Mueller. *Phys. Lett.*, 240B, 1990.

[11] T. Banks, G. R. Farrar, M. Dine, D. Karabali, and B. Sakita. Technical Report RU/90/20, Rutgers, 1990.

[12] A. Mueller. Technical Report LPTHE 90/19, Orsay, 1990.

[13] P. Arnold and M. Mattis. Technical Report LA-UR-90-1218, Los Alamos, 1990.

[14] L. Yaffe. Technical Report PT 90.7, Univ. of Washington, 1990.

[15] E. Byckling and K. Kajantie. *Particle Kinematics.* John Wiley and Sons, 1973.

[16] W. J. Stirling R. Kleiss and S. D. Ellis. *Comput. Phys. Commun.*, 40, 1986.

[17] V. Rubakov S. Khlebnikov and P. Tinyakov. Technical Report , 1990.

[18] M. Porrati. LBL Technical Report , LBL-28980, UCB90/17, 1990.

[19] A. Ringwald. Private communication.

[20] J. Morfin and W-K. Tung. Technical Report 90-74, Fermilab, 1990.

[21] A. Ringwald and C. Wetterich. Technical Report , DESY 90-067, 1990.

[22] F. Klinkhammer and N. Manton. *Phys. Rev.*, D30, 1984.

A 'SUPER MONTE CARLO' FOR MULTI-TeV PHYSICS SIMULATION

J. Alberty‡, F. Anselmo†(speaker), L. Cifarelli*†Δ, O. Di Rosa†,
B. van Eijk†, E. Eskut‡, D. Hatzifotiadou‡, G. La Commare†,
C. Maidantchik‡, M. Marino‡, G. Xexeo‡

† CERN/LAA Project, Geneva, Switzerland
‡ World Laboratory/HED Project, Lausanne, Switzerland
* University of Naples, Italy
Δ INFN Bologna, Italy

1. INTRODUCTION

The subject of this talk is if and how we can make predictions about cross sections and physical quantities distributions for multi-TeV hadronic collisions.

The outline of this presentation is as follows:
in Section 2 some problems and statistical and systematic uncertainties in simulating H.E.P. events for hadronic Supercolliders at multi-TeV are listed; in Section 3 the MSL solution, i.e. the COSMOS Project, is presented. The ideas and the status of COSMOS are briefly sketched.

2. PROBLEMS AND UNCERTAINTIES

H.E.P. event generators consist of two parts: one where perturbative QCD is applicable (Section 2.1) and one where it is not and hadronisation models have to be used (Section 2.2).

2.1   Problems and uncertainties in parton generation

One can state the problem of parton generation in the multi-TeV range as follows: at higher and higher energies one needs higher and higher order QCD corrections.

With respect to this problem, present event generators can be divided in two classes:
i) event generators that generate multi-parton states taking into account the correct higher and higher order subprocesses matrix elements (Matrix Element approach - hereafter called ME), e.g.: W + 1 jet, W + 2 jets, W + 3 jets, and
ii) those that first generate an elementary $2 \to 2$ or $2 \to 1$ subprocess and then add initial and final state QCD radiation, i.e., they let the two incoming hard interacting partons and, if any, the outgoing partons radiate gluons that, in turn, can split into q-q̄ or can radiate other gluons. The process stops at a certain scale that marks the end of the perturbative regime. This second approach is called Parton Shower (PS) and is based on the Altarelli-Parisi equation.

Examples of PS method are PYTHIA[1] and HERWIG[2] whereas EUROJET[3] is an example of the ME one.

The difference between the two approaches mainly shows up in the topology of the events like transverse momentum distributions or rapidity ones. An example of different transverse momentum slope is shown in fig.1 where the p_t distributions of a Higgs particle of 150 GeV mass, produced with q-q̄ and g-g initial states, using a ME generator[3] and a PS one[1] are superimposed. This picture deserves a bit of explanation. To try to compare Higgs + 1 jet with Higgs + parton showering we had to single out Higgs + 1 jet events among the ones generated by the PS Monte Carlo. Therefore, we defined a jet as a set of partons within a maximum distance of 1 in the pseudorapidity - polar angle plane with a minimum jet transverse energy of 10 GeV. Then, we rejected all events that had less or more than 1 jet.

In addition to this main difference it must be noticed that for careful quantitative signal versus background studies ME generators have to be used.

The best solution would be a kind of combination of the two approaches but, unfortunately, it is not a trivial problem.

2.1.1 <u>ME Approach</u>. We can summarize the problems showing up with the ME approach with the following formula giving the total cross section for hadronic collisions as a convolution of the parton level matrix element with the structure functions:

$$
\int_0^1 \int_0^1 dx_1 dx_2 \int_V P\frac{d^3p_i}{2E_i} \delta^4(p-p_1,...,p-p_n) |A|^2 G_1(x_1,Q^2)G_2(x_2,Q^2) \tag{1}
$$

where n is the number of final partons, p_i/E_i, is the momentum / energy ratio of the ith particle, V is the phase space volume of integration, the δ function provides the 4-momentum conservation and $|A|^2$, the matrix element squared, is a function of 3n - 4 independent kinematical variables. G_1 and G_2 are the structure functions, Q^2 is the scale and x_1 and x_2 are respectively the fractions of the first and second hadron energies taken by their hard-interacting parton. Eliminating the δ function we can rewrite the integral as follows:

$$
\int_0^1 \int_0^1 dx_1 dx_2 \int_V d\Phi \, \rho_n(\Phi) \, |A(\Phi)|^2 \, G_1(x_1,Q^2) \, G_2(x_2,Q^2) \tag{2}
$$

Now the hypercube of integration is no longer n-dimensional, but 3n-4 dimensional. Φ stands for the coordinates of a point in the 3n-4 dimensional phase space expressed in terms of any set of kinematical variables and ρ_n is the phase space density; it gets its shape from the Jacobian of the transformation of variables and from the integration over the δ function.

Every term in the above expression is a source of problems or uncertainties in actual Monte Carlo simulations.

a. Structure Functions

Many structure functions sets exist on the market. They are derived from experimental data and therefore their validity is restricted to certain x and Q^2 ranges. Outside these ranges various ways are used to get numbers: usually either they extrapolate or they take the value at the edge of the interval. Therefore, for instance, the lower integration limit for the x in the above expression should be replaced by a certain non-zero x_{min} value.

The influence of changing structure function set in computing cross sections can be seen in fig. 2 where the b-b̄ + X cross section with a transverse momentum cut of 10 GeV, a pseudorapidity cut of 3 and a minimum distance between the two bottom quarks of 0.7 in the pseudorapidity-polar angle plane is shown up to 200 TeV. The difference among the various sets results can be even a factor two if the energy is high enough.

308

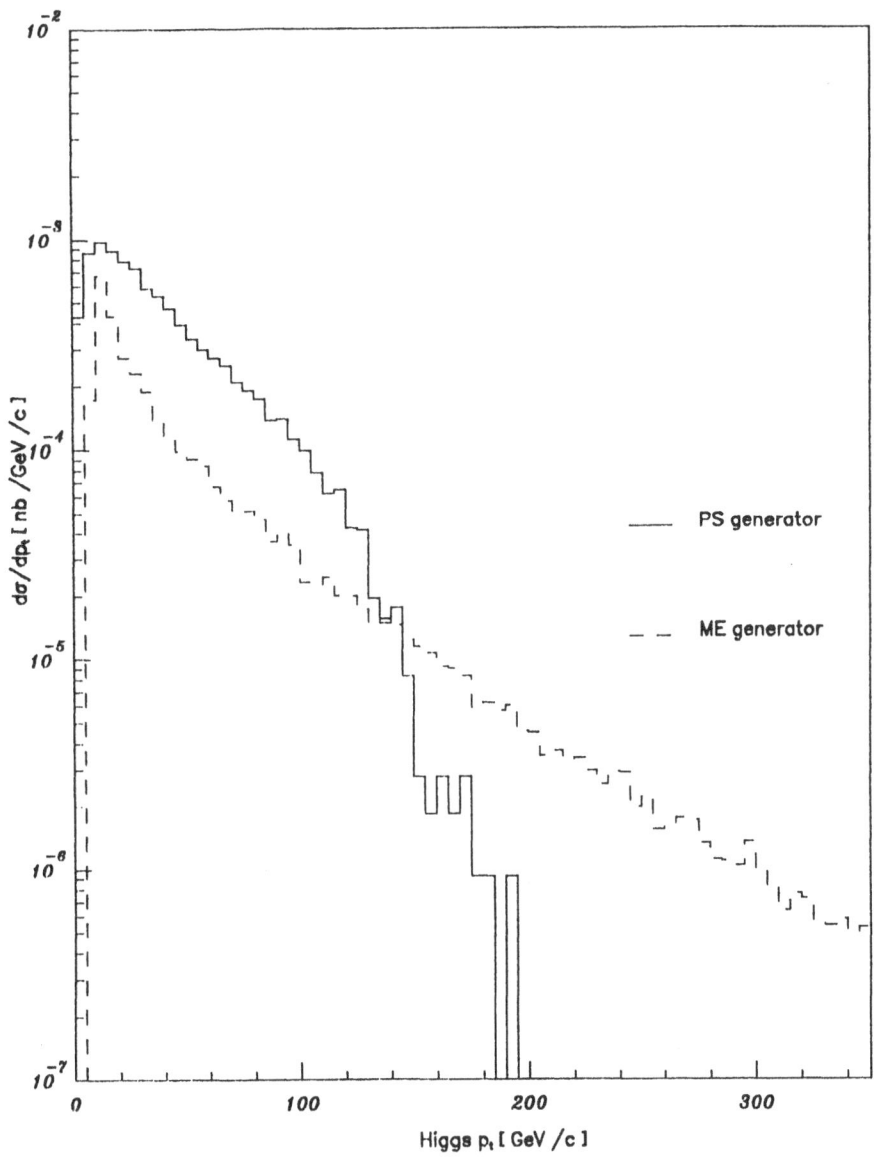

Fig. 1

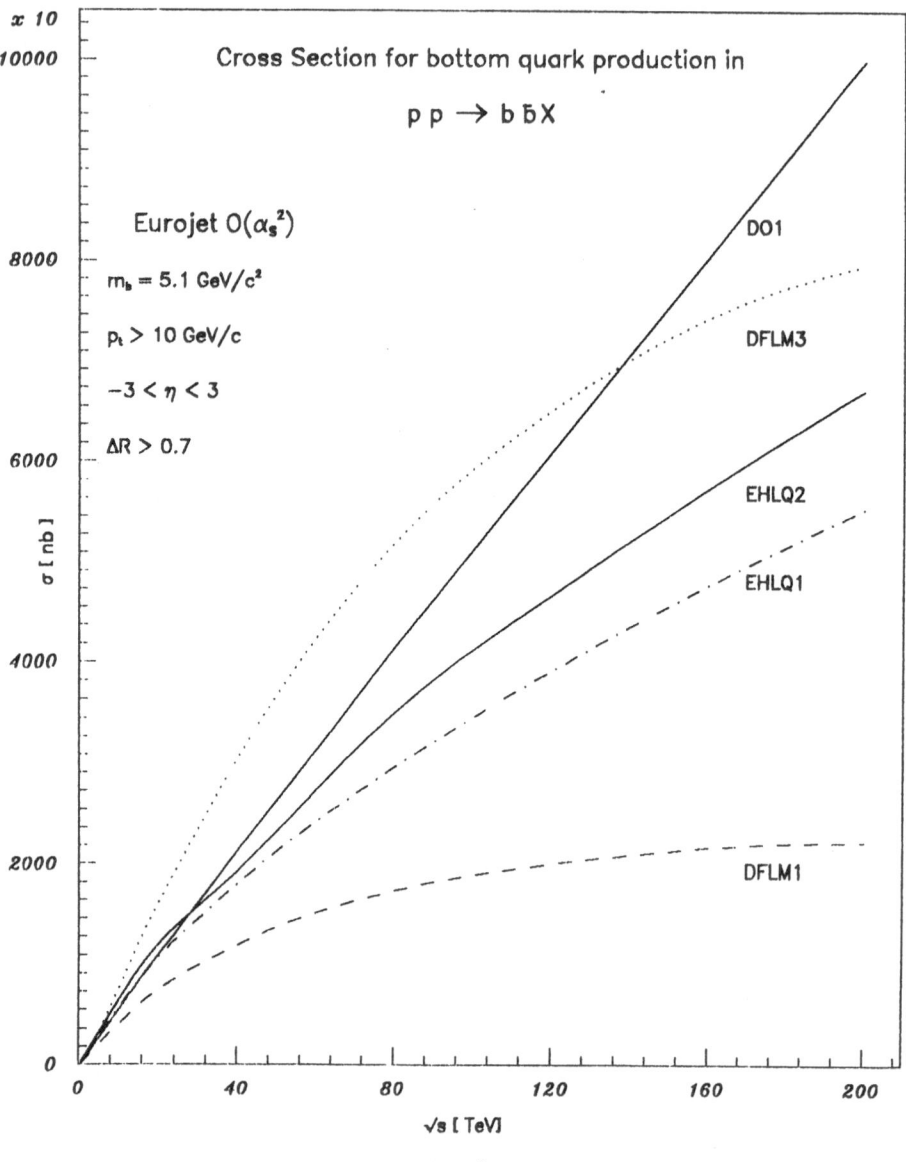

Fig. 2

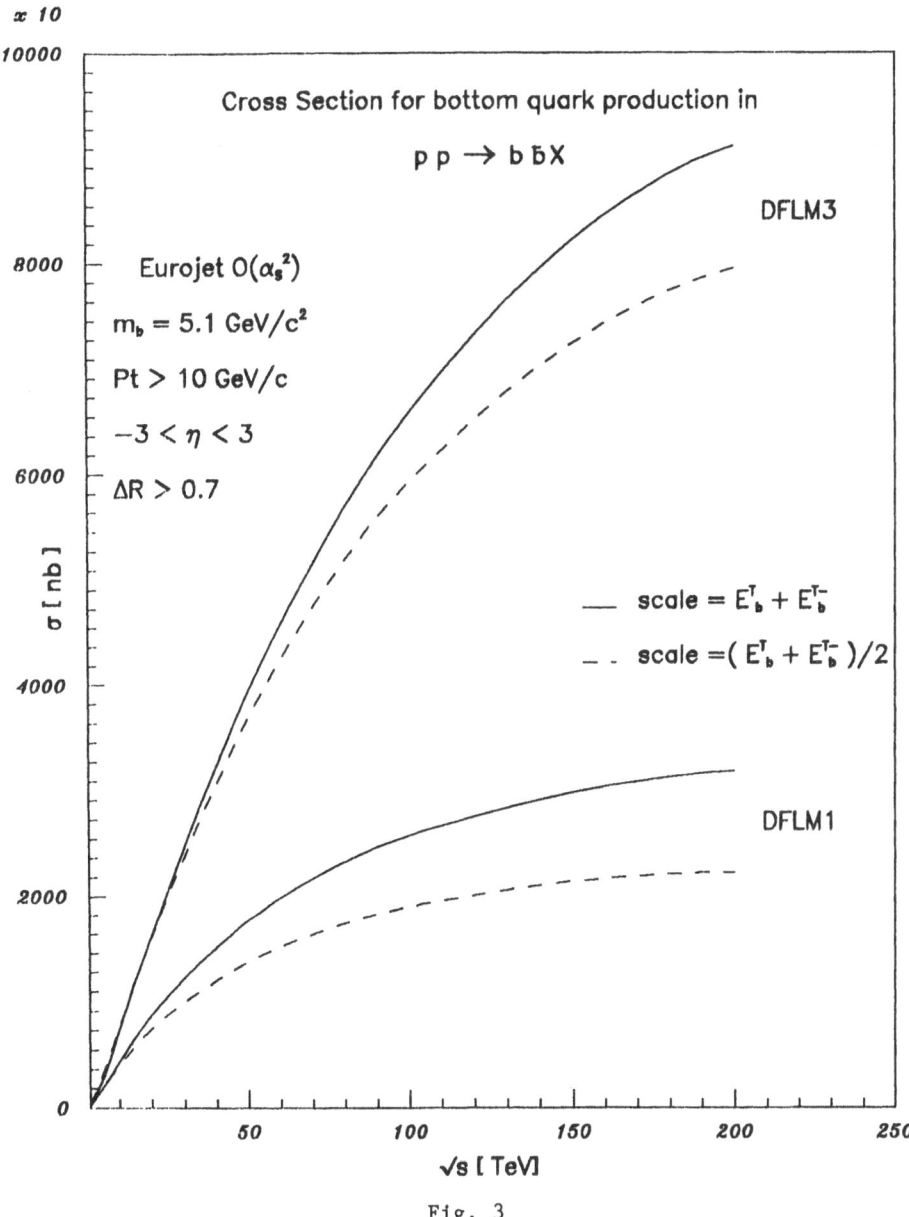

Fig. 3

Another problem is the well known one of the arbitrary choice of the QCD scale. An idea of how much a cross section can vary with the scale can be deducted from fig.3 where the same cross section as before, with a fixed set of structure function, is first calculated taking as scale the total transverse energy of the b-b̄ system and then the average one, i.e. the total transverse energy divided by two.

Hopefully, data from HERA will upgrade our knowledge about structure functions.

b. Matrix Elements

The ambiguity in choosing the scale is also present in the matrix element calculation where the strong running coupling constant enters.

c. Phase Space Generation

In a parton generation the phase space points should, in principle, be weighted by the phase space density and the matrix elements. In practice, this would be too hard to implement when you have a multi-parton final state. Therefore, momenta are usually generated according to simple distributions, e.g. gaussian or even uniform ones, depending on the particular problem. The widely used RAMBO[5] programme tries to generate phase space points with a distribution as uniform as possible. From a physical standpoint, this is acceptable if the matrix elements do not fluctuate too much.

The usual phase space generation procedure is the so called "hierarchical" one, i.e. n points are generated as a sequence of two points generations. RAMBO instead first generates n four-vectors without the four-momentum conservation constraint and then it scales them down to get the real n four-momenta.

This "democratic" procedure leads to phase space distributions much better than those produced by the "hierarchical" ones if the original mass is greater than half the summation of the final particles masses (for details and formulas see ref.5).

A possible way to generate the four-momenta of the outgoing partons according to the differential cross section is to use a programme like RAMBO together with an importance sampling MC integration routine (as it will be discussed in the next session). This is, for example, the technique used by I. Hinchliffe in his PAPAGENO parton generator[6].

Another problem concerning the phase space generation arises from the fact that the usual way to generate the outgoing partons four-momenta inside the requested phase space region is to first generate them and then to reject them if they are not within the cuts. If the phase space region of integration is small a lot of computing time can be wasted. There are programmes, e.g. OCTOPUS[7], that are more efficient with this respect if particles are light.

d. Integration

The major problem of phase space integration is that it is multi-dimensional. Whenever the number of dimensions is greater than four or five, clever numerical integration methods are necessary. This is because the number of calculations to get a fixed precision grows exponentially with the number of dimensions of the function. There are essentially three kinds of algorithms for numerical integration called: "statistical", "Monte Carlo" and "direct" one (see ref.4, chap. IX). The Monte Carlo (MC) ones have the fastest rate of convergence.

Let us take for the sake of simplicity the example of the integration of a one-dimensional function in the [0,1] interval:

$$I = \int_0^1 f(x)dx \qquad (3)$$

A rough MC method estimation can be given by the following mean:

$$S = \frac{1}{N} \sum_{k=1}^{N} f(x_k)$$ (4)

where x_k is a set of N uniformly randomly selected points in the [0,1] integration interval.

The error of the MC estimation of an integral depends on two factors:
1. the variance of the function to be integrated (the smoother and slowly varying the function is the better the MC works);
2. the number of random points at which the function is sampled (the larger the number of function evaluations is the more the mean of eq. (4) is meaningful). But, it must be kept in mind that in a MC algorithm the error decreases slowly as N increases ($\sim 1/\sqrt{N}$) and N is limited by the computer time. Therefore, people have mostly concentrated in developing variance reducing methods. Usually, these are based on simple considerations: either a variable transformation is performed to get a smoother function or the distribution of the function sampling is no longer wholly random and uniform.

The so called "importance sampling" methods estimate the integral as the following mean:

$$S = \frac{1}{N} \sum_{k=1}^{N} \frac{f(x_k)}{p(x_k)}$$ (5)

where $p(x_k)$ is the density according to which the random points are chosen.

The density $p(x_k)$ is modified to reduce the variance, the optimal one being obtained when:

$$p(x) = \frac{|f(x)|}{\int_V |f(x)| dx}$$ (6)

This is formally equivalent to making a variable transformation x=g(y), so that:

$$I = \int_V f(x)dx = \int_V f(g(y)) \, g'(y)dy$$ (7)

and performing a MC integration with UNIFORMLY distributed points in the integration interval. For some integrantion functions it is possible to further reduce the variance by using other methods applied to the smoothed f(g(y))g'(y) function. Among them the "stratified sampling" one. That is, instead of choosing N random points uniformly distributed in the integration region, the sampling is stratified by selecting two, or more, points in each of N/2 subintervals of the integration interval. Explicitly, the importance sampling estimation would be the eq. (5) one and the stratified sampling one would be, if only two points per sub-interval are chosen:

$$S = \frac{1}{N} \sum_{k=1}^{N/2} \sum_{j=1}^{2} \frac{f(x_{kj})}{p(x_{kj})}$$ (8)

where j points to the evaluations within the sub-interval.

2.1.1 PS Approach. The problems with the structure functions and the QCD scale are also present in event generators using PS approach.

What is different from the ME case is that the Altarelli-Parisi (A-P) equation (8) on which the PS approach is based is in leading logarithm approximation. This has many consequences. First of all, wide angle emitted gluons are not treated properly. In addition, there are many uncertainties when you try to implement this equation into a Monte Carlo algorithm. The A-P equation used in the final state radiation reads as follows:

$$\frac{dP_{a\to bc}}{dt} = \int dz \frac{\alpha_s(Q^2)}{2\pi} K_{a\to bc}(z) \tag{9}$$

$dP_{a\to bc}$ is the probability that the parton a would branch into b+c in a small variation dt of the evolution parameter t. $K_{a\to bc}(z)$ are the Altarelli-Parisi kernels:

$$K_{q\to qg}(z) = \frac{4}{3}\frac{1+z^2}{1-z} \tag{10a}$$

$$K_{g\to gg}(z) = \frac{6[1-z(1-z)^2]}{z(1-z)} \tag{10b}$$

$$K_{g\to q\bar{q}}(z) = \frac{1}{2}[z^2+(1-z)^2] \tag{10c}$$

An example of how PS event generators compute higher order cross sections given the lowest order one follows:

$$\sigma(gg\to q\bar{q}g) = 2\,\sigma(gg\to gg) \int dt\, dz\, \frac{\alpha_s(Q^2)}{2\pi} K_{g\to q\bar{q}}(z) \tag{11}$$

In this way the multi-dimensional MC integration is someway split in several integrations.

Almost all the quantities are arbitrary. Q^2 is the scale: a common choice for it is m_a^2, but also $z(1-z)m_a^2$. z can be any quantity giving the sharing of energy, or some combination of E and p, of the initial parton a taken by the parton b. The evolution parameter t is also arbitrary. At least two different algorithms exist that differ on the choice of this parameter: the LUND one [9] with $t=\ln(m_a^2/\Lambda^2)$ and the HERWIG one [2] where $t = E_a^2 \xi_a/\Lambda^2$ and $\xi_a = 1-\cos\theta_a$ with θ_a opening angle when $a\to bc$.

Moreover, the branching processes are to some extent independent. Some interference effect between the initial and final radiation for soft gluons is taken into account by the angular ordering trick. This is actually the reason why in the HERWIG PS algorithm the evolution parameter contains an angle (it is easier to impose angular ordering).

2.2 Problems and uncertainties in Hadronization

Nobody knows how to exactly handle the interactions between coloured partons at a scale of the order of Λ_{QCD}. Essentially three models exist with many possible variants: Independent Fragmentation[10], String Fragmentation[11] and Cluster Fragmentation[12] (for a good review see ref. 13).

314

Independent Fragmentation is not really a physics model but a simple parameterization; moreover, energy - momentum conservation is to be imposed. Therefore, it is considered not to be true even if it is able to reproduce experimental data when opportunely tuned. The way hadrons are formed from partons is via quark-antiquark formation from the vacuum polarization and subsequent arrangement of quarks and antiquarks. Several schemes have been proposed to take gluons into account[13].

According to the String Fragmentation model the tubelike colour field flux lines can break, by tunneling effect, into production of quark-anti-quark pairs. Gluons are kinks on the string, giving transverse momentum.

Both Independent and String Fragmentation models do need a fragmentation function to give longitudinal momentum to hadrons and a distribution function for transverse momenta, typically a gaussian one. Many fragmentation functions exist[13]; examples are: the Field-Feynman one, the Peterson one and the Lund one.

Cluster fragmentation has less free parameters. In the HERWIG version it is based on some "preconfinement" studies[14]. After the parton shower all gluons, if any, are forced to decay into quark-anti-quark pairs. The quarks that are closest in the phase space arrange themselves into "clusters" of more or less the same invariant mass. Clusters are then made decay in hadrons.

3. THE COSMOS PROJECT

From the above considerations it is clear that if we want to have meaningful distributions or cross sections, especially in the multi-TeV range, we have to cross check the various results obtained with different methods. We have to be able to estimate for instance what it changes using one hadronization model or another, the ME approach or the PS one, a PS algorithm or the other. Quantities whose values are very sensitive to the different PS algorithms or to the hadronization model cannot be considered as reliable or at least they must be taken with care.

Moreover, a common difficulty in using event generators is to understand how many parameters to set and what is their function.

3.1   The Basic Ideas

The COSMOS Project aim is to solve the problems above outlined providing the physicist with a work device able to perform a complete event generation with every possible combination of programmes performing one of the described tasks: integration, matrix element calculation, hadronization etc..

Therefore, COSMOS is built as a modular package, i.e. a software system organized in independent blocks that can be replaced with other blocks of the same functionality.

The complete event generation is subdivided in various steps shown in fig.4: parton generation, parton evolution, hadronization, decay and tracking into detectors.

Some auxiliary packages are needed: first of all we want all data concerning the generation to be organized in data bases. This because of the easier and clearer data manipulation (changing, querying, inserting..) we can get.

The user will also be able to give all the input information interactively driven by command menus and help facilities.

3.2   Status of COSMOS

The shaded blobs in fig.4 mean that some work has already been done about that item.

COSMOS is built as follows:
1. research and study of packages in collaboration with the MC authors and various groups;

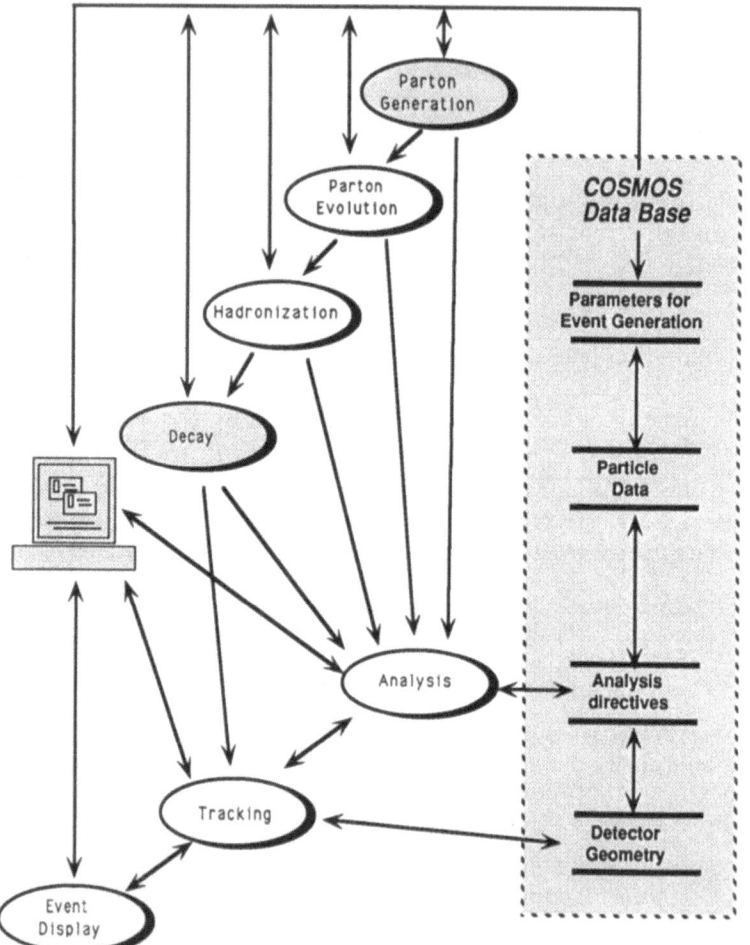

Fig. 4

2. all collected packages are put in standard FORTRAN 77;
3. comment lines of explanation are inserted in the code (40% of it);
4. testing the code;
5. documentation write up;
6. help facilities set up.

a. Parton Generation

Already inserted packages are:
SPHINX : package collecting many structure functions sets and allowing the user to choose what to do outside the grid of validity[15].

VEGAS : MC integration programme using Importance sampling and Stratified one[16].
CAESAR: it is a modified version of VEGAS only for Stratified sampling with a trick to get a given precision in less iterations[17].

RAMBO: phase space generator to get uniformly distributed four-momenta[5].

RAMBO modified by I. Hinchliffe: as RAMBO but used in combination with a MC importance sampling integration program like VEGAS or CAESAR to get four-momenta distributed according to differential cross sections[6].

Under study:
OCTOPUS: phase space generator for light particles the author claims to be more efficient than RAMBO when cuts are very restrictive[7].

Matrix elements already included :
- QCD jet production up to $O(\alpha_s^3)$
- heavy flavours production up to $O(\alpha_s^3)$
- Drell - Yan
- Direct photon production up to $O(\alpha\alpha_s^2)$ and $O(\alpha_s\alpha^2)$
- W/Z + n jets, up to n=3
- Higgs production up to $O(\alpha_s^3\alpha_W)$

b. Decay

The EURODEC package, the most complete on the market, has been already included.

c. Tracking Programme

It seems there is only one flexible programme for detector simulation i.e. the GEANT package[18] and therefore we shall restrict ourselves in using it.

d. Data Base

The Data Base Management System we used to handle data within COSMOS is the ADAMO system[19]. The modules implemented at this stage of the project are: PDKDB[20], containing information about particles and particle decay (see fig.6) and SUD[21] consisting of detector data modelled in such a way to be compatible with the GEANT tracking programme. Moreover, the COSMOS Application Builder (CAB)[22], is under development to constitute the frame to describe and implement a Monte Carlo Application.

e. User Interface

An interactive access to the above mentioned data bases is obtained using KUIP (Kit for User Interface Package)[23]. Examples of user interaction with the system are shown in figures 5,6,7.

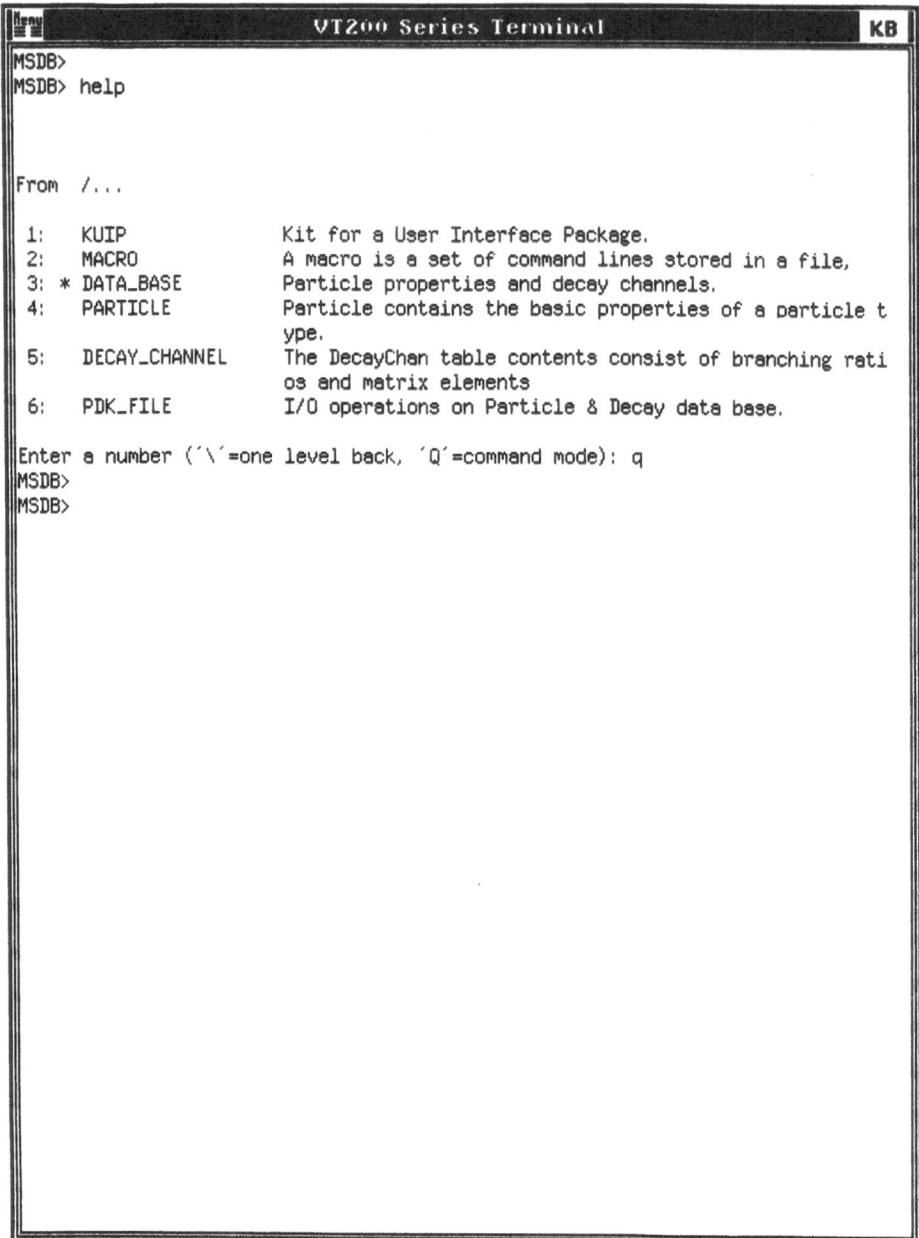

```
┌─────────────────────────────────────────────────────────────────────┐
║ Menu                    VT200 Series Terminal                    KB  ║
╟─────────────────────────────────────────────────────────────────────╢
║ MSDB>                                                                 ║
║ MSDB> help                                                            ║
║                                                                       ║
║                                                                       ║
║                                                                       ║
║ From /...                                                             ║
║                                                                       ║
║   1:   KUIP              Kit for a User Interface Package.            ║
║   2:   MACRO             A macro is a set of command lines stored in a file, ║
║   3: * DATA_BASE         Particle properties and decay channels.      ║
║   4:   PARTICLE          Particle contains the basic properties of a particle t ║
║                          ype.                                         ║
║   5:   DECAY_CHANNEL     The DecayChan table contents consist of branching rati ║
║                          os and matrix elements                      ║
║   6:   PDK_FILE          I/O operations on Particle & Decay data base.║
║                                                                       ║
║ Enter a number ('\'=one level back, 'Q'=command mode): q              ║
║ MSDB>                                                                 ║
║ MSDB>                                                                 ║
└─────────────────────────────────────────────────────────────────────┘
```

Fig. 5

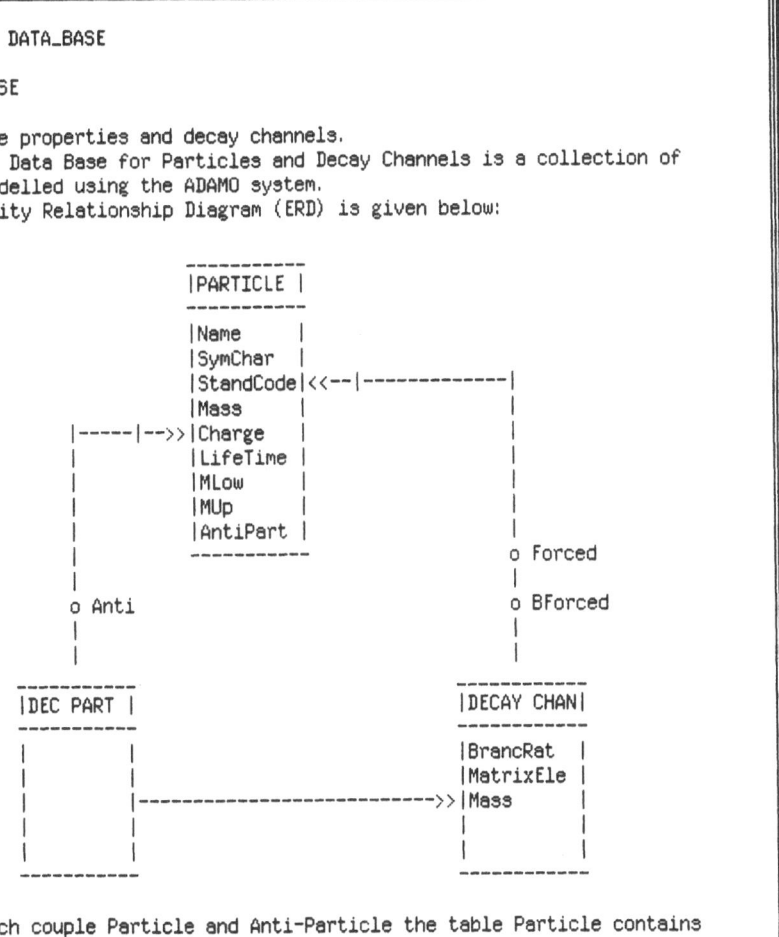

```
┌─────────────────────────────────────────────────────────────────────────┐
│ ≣Menu         VT200 Series Terminal                              KB       │
├─────────────────────────────────────────────────────────────────────────┤
│ MSDB>                                                                     │
│ MSDB> HELP DATA_BASE                                                      │
│                                                                           │
│  * DATA_BASE                                                              │
│                                                                           │
│    Particle properties and decay channels.                               │
│    The MSL Data Base for Particles and Decay Channels is a collection of  │
│    data modelled using the ADAMO system.                                  │
│    The Entity Relationship Diagram (ERD) is given below:                  │
│                                                                           │
│                            ------------                                   │
│                            |PARTICLE |                                    │
│                            ------------                                   │
│                            |Name     |                                    │
│                            |SymChar  |                                    │
│                            |StandCode|<<--|--------------|                │
│                            |Mass     |                   |                │
│                |-----|-->>|Charge    |                   |                │
│                |          |LifeTime  |                   |                │
│                |          |MLow      |                   |                │
│                |          |MUp       |                   |                │
│                |          |AntiPart  |                   |                │
│                |          ------------          o Forced                 │
│                |                                 |                        │
│              o Anti                            o BForced                  │
│                |                                 |                        │
│                |                                 |                        │
│          ------------                      ------------                   │
│          |DEC PART |                       |DECAY CHAN|                   │
│          ------------                      ------------                   │
│          |         |                       |BrancRat  |                   │
│          |         |                       |MatrixEle |                   │
│          |         |-------------------------->>|Mass  |                  │
│          |         |                       |          |                   │
│          |         |                       |          |                   │
│          ------------                      ------------                   │
│                                                                           │
│    For each couple Particle and Anti-Particle the table Particle contains │
│    the description of only one of them, data for the other one being      │
│    obtained through a charge conjugation.                                 │
│    The attributes of the entities Particle and DecayChan are self-        │
│    explained in the figure. DecPart is a dummy table used to implement a  │
│    'n' to 'm' relationship between the two tables. The attribute Anti     │
│    describes whether it is to be performed a charge conjugation on        │
│    pointed particle. That permits to obtain values for antiparticle too.  │
│    For each couple Particle and Anti-Particle only the decay channels of  │
│    one of them are stored. Channels for the other are obtained through a  │
│    charge conjugation.                                                    │
│    The relation between DecayChan and Particle is partial on Particle     │
│    since some of them are stable. The attributes Forced and BForced mark  │
│    out the forced channel for the particle and anti-particle             │
│    respectively.                                                          │
│                                                                           │
│ MSDB>                                                                     │
└─────────────────────────────────────────────────────────────────────────┘
```

Fig. 6

```
┌─────────────────────────────────────────────────────────────────┐
│ ▓▓▓           VT200 Series Terminal              KB  │
├─────────────────────────────────────────────────────────────────┤
│MSDB> HELP PARTICLE                                                │
│                                                                   │
│    PARTICLE                                                       │
│                                                                   │
│    Particle contains the basic properties of a particle type.     │
│    Each particle is identified by its Standard Code and name.     │
│    The information available are kept up-to-date conforming with the│
│    last experimental results.                                     │
│                                                                   │
│                                                                   │
│From  /PARTICLE/...                                                │
│                                                                   │
│  1: * INSERT      Insert a new particle in the Data Base.         │
│  2: * REMOVE      Remove a particle from the Data Base.           │
│  3: * MODIFY      Modify values of particle attributes.           │
│  4: * DISPLAY     Display values of particle attributes.          │
│  5: * LIST        List the Particle and PartiBuf tables contents. │
│                                                                   │
│Enter a number ( '\'=one level back, 'Q'=command mode): Q          │
│MSDB>                                                              │
│MSDB> HELP PARTICLE/INSERT                                         │
│                                                                   │
│  * PARTICLE/INSERT  Name SymChar StandCode Mass Charge LifeTime MLOW MUP AntiPar│
│t                                                                  │
│                                                                   │
│    Name       C 'The particle name'                               │
│    SymCha     C 'The particle symbolic charge'                    │
│    StandC     I 'PDG Standard Code' R=-99999:100000               │
│    Mass       R 'Particle mass in GeV/c**2' D=0 R=0:500           │
│    Charge     R 'Particle charge in elem. charge unit' D=0 R=-2:2 │
│    LifeTi     R 'Particle life time in seconds' D=0 R=0:          │
│    MLOW       R 'Mass lower cut in GeV/c**2' D=0 R=0:             │
│    MUP        R 'Mass upper cut in GeV/c**2' D=0 R=0:             │
│    AntiPa     C 'Antiparticle existence?'                         │
│                                                                   │
│    Insert a new particle in the Data Base.                        │
│    Each attribute has a default value that the user can modify. The boolean│
│    flag AntiPart has to be set to 'FALSE' if the particle is antiparticle of│
│    itself.                                                         │
│                                                                   │
│MSDB> PARTICLE/INSERT                                              │
│The particle name (<CR>=ELECTRON): TAU                             │
│The particle symbolic charge (<CR>= ): -                           │
│PDG Standard Code (<CR>=11): 15                                    │
│Particle mass in GeV/c**2 (<CR>=0): 1.784091                       │
│Particle charge in elem. charge unit (<CR>=0): -1                  │
│Particle life time in seconds (<CR>=0): 0.304E-12                  │
│Mass lower cut in GeV/c**2 (<CR>=0): 1.784091                      │
│Mass upper cut in GeV/c**2 (<CR>=0): 1.784091                      │
│Antiparticle existence? (<CR>=NO): YES                             │
│MSDB>                                                              │
│MSDB> █                                                            │
│                                                                   │
└─────────────────────────────────────────────────────────────────┘
```

Fig. 7

CONCLUSIONS

If one wants to use QCD Monte Carlo event generators to make predictions at multi-TeV ranges because of theoretical uncertainties and technical problems he has to be able to cross-check various kind of techniques, models, parameterizations etc..

A new instrument is born to solve at least some of these problems. Its name is COSMOS and it is a product of the Monte Carlo Simulation Laboratory of the LAA Project at CERN.

ACKNOWLEDGEMENTS

I would like to thank L. Cifarelli and T. Ypsilantis for their kind invitation and Professor Antonino Zichichi and the staff of the Ettore Majorana Centre for their friendly hospitality.

REFERENCES

1. H.U. Bengtsson, T. Sjöstrand, Computer Phys. Comm. 46 (1987) 43
2. G. Marchesini and B.R. Webber, Nucl. Phys. B130 (1988) 461
   I. Knowles, Nucl. Phys. B130 (1988) 571
3. A. Ali, B. van Eijk, in: "Proceedings of the 5th Topical Workshop on Proton -Antiproton Coll. Physics", St. Vincent, Aosta Italy (1985).
   B. van Eijk, in: "New Particles 1985", Conference Proceedings, University of Wisconsin, Madison, U.S.A. (1985).
   A. Ali, B. van Eijk, in: "Proceedings of the International Conference on Proton-Antiproton Physics", Aachen FRG (1986).
   B. van Eijk, CERN-EP/85-121, Preprint.
   A. Ali, B. van Eijk and I. ten Have, CERN-TH.4523/86, Nucl. Phys. B292 (1987) 1.
   B. van Eijk and R. Kinnunen, NIKHEF-H/187-15
   B. van Eijk and R. Kinnunen, Z. Phys. C41 (1988) 489
4. E. Byckling, K. Kajantie,"Particle Kinematics", Wiley, New York (1973)
5. S.D. Ellis, R. Kleiss,W.J. Stirling, preprint CERN-TH.4299/85
6. I. Hinchliffe, PAPAGENO, no documentation.
7. D. Kosower, Fermilab-Pub-90/58-T
8. G. Altarelli and G. Parisi, Nucl. Phys. B126 (1977) 298
9. M. Bengtsson and T. Sjöstrand, Phys. Lett. B185 (1987) 435
10. R.D. Field and R.P. Feynman, Phys. Rev. D15 (1977) 2590
    R.D. Field and R.P. Feynman, Nucl. Phys. B138 (1978) 1
    extensions:
    P. Hoyer et al., Nucl. Phys. B161 (1979) 349
    A. Ali, E. Pietarinen, G. Kramer and J. Willrodt, Phys. Lett. 93B (1980) 155
11. X. Artru and G. Mennessier, Nucl. Phys. B70 (1971) 93
    B. Andersson, G. Gustafson, G. Ingelman and T. Sjöstrand, Phys. Reports 97 (1983) 31 and references therein
12. R.D. Field and S. Wolfram, Nucl. Phys. B213 (1983) 65
    B.R. Webber, Talk at XVIII Rencontre de Moriond (1983) CERN Preprint CERN-TH-3569 (1983)
13. T.D. Gottschalk, CERN-TH 3810, CALT-68-1075
14. D'Amati and G. Veneziano, Phys. Lett. B83 (1979) 87
    G. Marchesini, L. Trentadue, G. Veneziano, Nucl. Phys. B181 (1981) 335.
15. B. van Eijk, preprint in preparation.
16. G.P.Lepage, Jou. of Comp. Phys. 27, (1978) 192-203
17. R. Laateever, J. Vermaseren, CAESAR, no documentation
18. R. Brun, F. Bruyant, M. Maire, A.C. McPherson, P. Zanarini, GEANT 3, CERN DD/EE/84-1
19. R. Brazioli et al., ALEPH Collaboration, ADAMO Notes, March 28, 1988.
    M.G. Green, RHBNC 89-01 and CERN-DD/US/131, March 15, 1989
20. G. La Commare, M. Marino and B. van Eijk, PDKDB: Particle and Decay Channels Data Base, private communication (to be presented at "MC 91", Workshop on Detector and Event Simulation in High Energy Physics, 8-12 April 1991, Amsterdam)

21. G. Anzivino, F. Arzarello, G. Bari, M. Basile, A. Bassi, L. Bellagamba, D. Boscherini, G. Bruni, G. Cara Romeo, M. Chiarini, L. Cifarelli, F. Ciralli, F. Cindolo, A. Contin, M. Costa, S. D'Auria, C. Del Papa, S. De Pasquale, F. Fiori, F. Frasconi, P. Giusti, G. Iacobucci, G. La Commare, A. Margotti, M. Marino, G. Maccarrone, T. Massam, N. Mc Cubbin, R. Nania, S. Quian, V. O'Shea, F. Palmonari, G. Sartorelli, M. Schioppa, G.C. Susinno, R. Timellini and A. Zichichi, "SUD: a Set-Up Descriptor for Modern High Energy Physics Experiments", submitted to <u>Particle World Communications in Subatomic Physics</u>

22. G . Xexeo, G. La Commare and J. De Souza, CAB, the COSMOS Application Builder, to be presented at "Computing in High Energy Physics '91", Tsukuba City, Japan, March 11-15, 1991

23. R. Brun, P. Zanarini, KUIP Users Guide, CERN-DD I202, February 26 1988

COMPUTING NOW AND THEN

Emilio Pagiola

CERN
1211 Geneva 23, Switzerland

NOSTALGIA

Super-colliders are a decade away or more. We can easily estimate how much computing will be needed to convert into physical data. the information obtained there What we can not do is to guess which resources, which organization will be then available for computing. To look briefly to the shifts and drifts of computer usage in high energy physics, to where we were and where we are, that is of some consequence.

Twenty five years ago, CERN acquired the supercomputer of those days. It was a CDC 6600, commanding then about 25 Millions Swiss Francs. With an unusual word length of 60 bits, it had 1 MByte of memory, countless tape units, a huge vertical disk storing some 24 Mbytes of data. In today,ss units it would have rated less than one Mips.

The mechanical printers were unforgettably noisy, as was the rest of the peripheral equipment, liberally driven by vacuum pumps and electric motors, hissing and grinding, when not jamming. The CPU was considered too precious for it to deal directly with input/output, which was delegated to a dozen of integrated peripheral processors. The floating point unit occupied more than one cubic meter. CERN wasn't satisfied with the manufacturer's operating system, and wrote one.

This marvel and pride was housed in a barrack, where it was at times rained on - which worsened only slightly the dismal Mean Time Between Failures, a quantity not yet called MTBF.

Fortran was the unchallenged language, CERN Fortran a monastic version of its rules, bravely aimed to make code generated at CERN universally portable.People were moving around with trays of punched cards, and experimental data were rushed on bicycle to the CDC 6600 for sample runs (Bicycle On Line, BOL, for short).

In the intervening years, computer resources expanded, computer literacy grew, Fortran (and tapes) staid. Networks have grown into a pervasive infrastructure, Electronic Mail into a factor of daily life. Graphics has made inroads into the communication between humans and computers.

Whilst batch computing gave way to time shared computing, and to personal computing, the expenditure, and its distribution remained, on average, remarkably constant. The computer centre accumulated facilities, now in the form of major Cray, DEC, and IBM mainframes, plus a huge tape vault. The users accumulated computers, in the form of VAXes, microVAXes, VAXstations and other stations, personal computers. This race between more in accumulation and more in numbers is sometimes described as distributed computing.

New Technologies for Supercolliders, Edited by L. Cifarelli and
T. Ypsilantis, Plenum Press, New York, 1991

The ménagerie of equipment and the maze of communication infrastructures at CERN are the result of the technological and financial constraints prevailing in the last few years For desk-top equipment, the cost of CPU power and disk space were the determining factors. It made sense to use dumb terminals, or to be content with Personal Computers (Macintosh or IBM compatible). When investing on workstations proper, it made sense to give weight to compatibility with other equipment of the group, hence the favor for VAXstations. When disk space was expensive, attempts at building station farms were confronted with the problem of fast remote data serving .

Today, massive CPU power is available, either with RISC processors, or with new generations of conventional architectures (i486, 68040). Vast power per se, desirable as it may be, is just a factor.

The radical event of 1990 is the price tag attached to it.

In the last few months several manufacturers, notably SUN, the market leader, and NeXT, the last entrant in the workstation race, have introduced complete stations offering 12-15 Mips for $4995. Earlier, IBM had introduced the RS/6000 family of stations, with power in the range 30-50 Mips, but with a entry price of $13000.

For $4995, desk-top systems provide 12-15 Mips, 1-2 Mflops, 8 Meg of RAM, a high resolution 17' Black & White Monitor. Ethernet connectivity, TCP/IP network protocols, NFS, X-Window, terminal emulation, MS-DOS emulation et similar frills are included. NeXT throws in *Mathematica* for free, and a 105 Mbytes disk.

Consequence: *these stations are now in the consumer market.* It is not clear if the suppliers themselves have a full view of the potential implications.

Consequence: in HEP where power is paramount, the advent of low-cost stations raises questions on renewed investment in such equipment as terminals - dumb and X -, and conventional PC's. Networks of stations challenge the conventional role of mainframes.

The framework has changed radically.

HARDWARE OUTLOOK

Computer power is not easily measured. At CERN, we talk of CERN units, the throughput capacity of some variant of the IBM 370. The industry talks of Mips, million of instructions per second, for integer power, which is the capacity of a VAX 780. Few people know or remember today what an IBM 370 or a VAX 780 could do. A CERN unit can be quoted to be worth approximately 3 VAX 780's. The floating point power is measured in Megaflops, million of floating point operations per second.

In these days, the magics for power at low cost is parallelism. To deliver its potential, it depends heavily on the performance of compilers. One form is RISC (Reduced Instruction Set Computer). CPU functions are broken down into simple, inexpensive instruction units, which can be multiplied and made to execute in parallel to improve performance - but comparable performance is offered by conventional architectures by Intel and Motorola. Other forms of parallelism add up several complete processors or specialized units, between which are spread tasks, or data, under process.

For some conventional desk-top workstations, binary compatible "Super-stations" are available, for about a quarter million dollars. These are multi CPU stations, which deliver CPU power in the 100-250 Mips range, are equipped with adequate central memory and with substantial i/o throughput, usually via a VME interface. "Superstations" are good candidates to take over the CPU, I/O intensive functions provided by the mainframes for batch processing, although their less sophisticated operating systems lack checkpoints or disk quota control, with the inherent risks of wasting important amounts of computing time.

Various "supercomputer" architectures, using hundreds of parallel processors, promise power in the 1000 Mips range. The rule of thumb which states that at constant price the CPU power doubles each 18 months is likely to hold true, or improve. Among the driving forces behind the CPU race are cutthroat competition and the demands of the entertainment industry, of HDTV for integer performance, and of Hollywood for floating point performance.

All computers are general purpose systems. Differences between mainframes, stations, and personal computers are a matter of semantics. Workstations and PC's differ from mainframes on the capacity of handling substantial input output volumes directed to several distinct applications. Workstations tend to be oriented to users working in teams, to be networked together, to run specialized software. Personal computers tend to be stand alone systems, most often running commercial applications.

The differences are waning. Workstation applications strived to optimize the stations resources, personal computers indulged in some measure of user friendliness. One could say that the performance per $ of an application, if high on a workstation, would be low on a PC, and viceversa. Increased, inexpensive resources let the workstations dedicate more of their resources to the users welfare, and the PC tackle more demanding tasks.

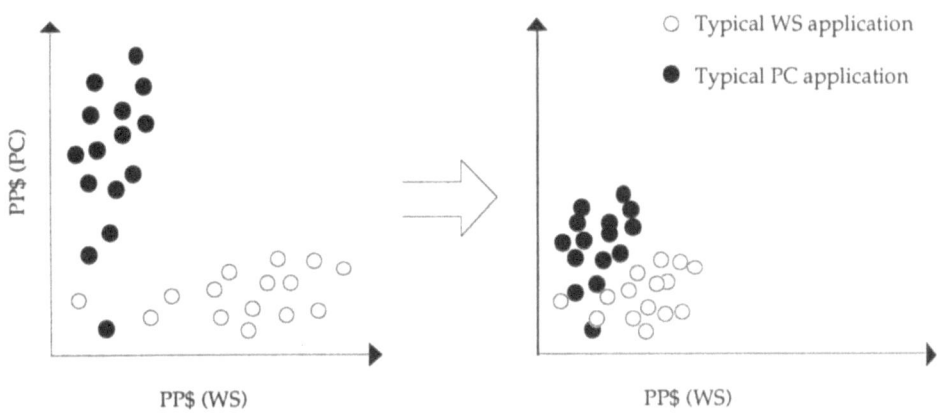

Fig.1   The waning difference between PC's and WS's. Applications had a high PC performance per $ and a low WS performance or viceversa. They are converging to the same performance per $ on both platforms.

The distribution of users among different kinds of computer systems is not easily assessed. Numbers and trends can be gleamed by looking at sales of computer equipment. From *Fig 2* we can infer that users of PC's are in the range of tens of millions, of mainframes and minis about a million, of stations probably half a million. The trend towards personal equipment, stations or PC's, is unmistakable.

This trend has deep consequences on the cost and quality of software. While software written for private use is used by the author and perhaps a few colleagues, and software written for mainframes or minis probably has a few hundred users, software written for personal computers has an audience measured in hundreds of thousands. Vast investment makes sense, and takes place. The quality of commercial software for personal computers is often without comparison with other products, and goes a long way in making good use of the most precious resource, the human being.

STORAGE UNITS

The second break-thru of 1990 has been in the cost and capacity of storage devices.

For disks connected via the SCSI (Small Computer System Interface) interface, the price per Mbyte is below $5, including enclosure and power supply. High-capacity configurations include one or two 780 (660 formatted) Mbyte disks in a single enclosure. or 1.2 Gbyte disks (1.0 formatted) Data rates are of the order of 1 Mbyte per second. To increase throughput, disks equipped with caches are entering the market.

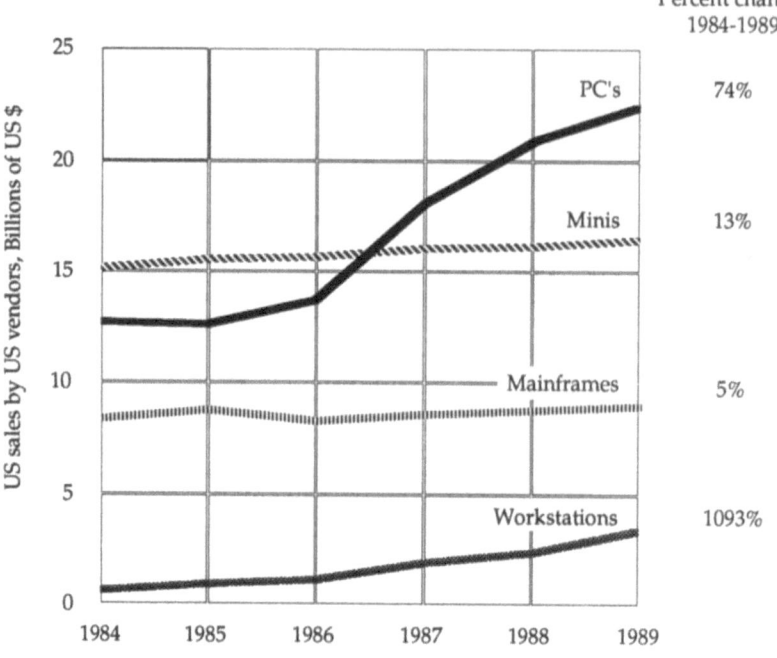

Fig.2   Evolution of the sales of computer equipment in the US, in the last five years. Minis are computer systems costing less than 1 million US $. Source: International Data Corporation.

Digital 8 mm Video Tapes and 4 mm Audio Tapes (DAT), with a SCSI interface, are commercially available, and are supported by all PC's and workstations. The capacity of 8 mm tapes is 2.3 Gbytes, soon to increase to 4.6, that of 4 mm DAT about one half. Data compression increases the raw capacity several times. The cost of the basic devices is few thousand dollars, the cost per stored MEGAbyte is about 2 cents. Robot handling of up to 10 cartridges is available. In short, tens of Gbytes can be accessed with a minimum of manual intervention

Conclusion: extensive capacity is available at accessible prices. It does not depend on a specific PC or workstation.

EQUIPMENT LIFETIME

There is a hard consequence of the rapid evolution of products that we can not ignore: *The lifetime against obsolescence of equipment does not exceed 3 years., at best*

This is not a source of worry. A low-cost workstation, with a 780 Mbytes external SCSI disk, a printer, and some third party software, add up to about $12000 (list prices). Spread on the lifetime of equipment, it amounts to $4000 per year.Quantity and academic discounts make the overall financial outlook still more favorable.

In this context, one can forgo the usual costly maintenance contracts During the first year the equipment is usually under the vendor's guarantee. The third year, the cost effectiveness of repairs is often questionable. It is usually sensible to carry a small stock of widely used systems. Pending repairs, systems which fail can be replaced with units taken from the stock, at no additional cost. The larger the number of systems involved, the better the procedure works. The difference in yearly cost for three·year no-maintenance systems versus six years systems with maintenance is small. It actually vanishes if the cost of upgrades, or the resale value of equipment are taken into account.

There are blessings in the short life span of equipment. Within reasonable levels of back compatibility, the freedom in the choice of equipment, to which we will not be committed for long. There is a chance to optimize the user, rather than arcane managerial criteria. We are led to take on the new without major ties to the old. Three years ago, it was good judgement that terminals, Macintoshes and VAXstations were the best desk-top equipment for the research sector at CERN. Today, attractive systems are low-cost stations, as offered by SUN and NeXT. In the near future different statements may hold true. Meanwhile, the equipment we selected on good grounds three years ago has given good service, and can die away peacefully.

Loss in capital investment because of rapid obsolescence is not a factor. The critical factors for acceptability of new systems are on network and software compatibility.

When possible, we should take advantage of the shifts in the market .

THE IMPACT OF LOW-COST STATIONS

Stations have computing power, fully integrated connectivity, and development tools. The new, inexpensive entries can help in establishing the fact that each individual may be correctly equipped. It is a perverse consequence of the distribution of wealth that crucial facilities are most scarce where they are most needed. Low-cost stations can help smoothing that for computing. The horrendous cost of mainframes requires a careful and painful social process of decision, selection, allocation. This can be put in perspective. The mutual obligation of providing a service and accepting it can be loosened.

The computer landscape is becoming one of local autonomy and remote services, instead than one of centrally provided and regulated functions.

Can stations do the job?

IN THE OFFICE

Data analysis, program development, job submission, writing letters, communications, preparing presentations are the daily activities Stations preserve the functionality of several other systems, because they emulate alphanumeric and T4000 terminals, as well MS DOS, and run X-windows.

The standard paraphernalia of peripherals for desk-top equipment are printers, scanners, FaxModems and back-up systems.

Most printers are Laser printers of some model. In the world of desk-top equipment, the technique for high quality printing (besides the Laser engine) is Postscript, a mark-up language which usually executes on a chip in the printer. The same procedure can be used to generate the characters and graphics displayed on the monitor. That is called DisplayPostscript. The printers then do not need the proprietary Postscript electronics, and are cheaper. Say that one, two printers are installed per corridor, or each 20-30 users.

FAXModems are PC or station interfaces to the telephone network, allowing to send Faxes from files stored on disk, or receive and store on disk incoming Faxes. Depending on the way the Fax was originated it is stored as a graphic image, or as an intelligible file. Fax stations capable of also handling conventional, *paper* Fax documents, require a printer and a scanner. Add a scanner per corridor. Note in passing that "public" FAXstations require, in principle, a FAX guardian, given the legally binding nature of FAXes received, a fact which is frequently ignored or overlooked.

Assign a station to drive the printers and operate a FaxModem, to act as an Electronic Mail server. Individual stations are not well suited for Electronic Mail, since there is no way to know when they are running.

In many cases, the combination of low-cost CPU's and of low-cost disks shifts the issue of serving to the issue of back-up, with shadows of management issues. Is it really a problem?. Allow one Mbyte of new stored data per user per day. Backing up over Ethernet represents about 10 seconds per user, which means that one hour per day provides back-up to 300-400 users, the occupants of a sizeable building. The back-up itself can be done by a station equipped of one 4 mm or 8 mm deck for back-up, one for restore. The time to fill a single cartridge is about one week: one trivial intervention per week per building is required. The cost of the contraption is the cost of one station and of two cartridge decks. Add $2000 of tapes (retention time one year): some $15000.

Folding everything together, to provide a complete infrastructure to a network of stations (or PC's) adds less than 10 percent to the cost of individual stations. Add that to the cost of stations as sold to the users, and we come out even.

The complete structure requires almost no supervision.

AT HOME

Statistics on home computing are inadequate to detect convincing patterns. The fact is that larger bills come from those users with the dumbest equipment. On the grounds that the cost of equipment is offset by savings on telephone fees, a number of users have a PC to work at home. They transport most of their data via diskettes, rather than via telephone connections.

Low-cost stations, at a price equal or lower than conventional PC's, offer interested users an opportunity to equip themselves at home with exactly the same system as in the office. Some imaginative possibilities: external SCSI disks (e.g. 780 Mbytes units) are easily disconnected, transported and installed; the 4 or 8 mm tape decks use ejectable cartridges smaller than a packet of cigarettes. One can conceive that part of the motivated staff carry these storage units home and back. Data intensive interactive sessions become possible at home. Wether this is desirable or not, we will not judge.

ON THE ROAD

More important, perhaps, is the ability of accessing computer resources when traveling. The industry's answer to this is the portable computer. Essentially all truly portable portables today are IBM PC compatible. All low-cost stations offer a PC (MS-DOS) emulation, an possible bridge between work in the office and work on the move.

DATA PROCESSING: STATION FARMS

Low-cost stations can be used in conjunction with low-cost, high capacity disks in farms. Here, advantage is taken of the disk capacity. A server fills in turn the disk of each stations with the whole content of one cassette, and returns to fetch the output from the station disks. This approach requires essentially no development. It is of interest when the time needed to process one record on a station is appreciably larger than the time needed to read it in and write it out across the network.

The optimal number, n, of stations in the farm is

$$n = 1 + \frac{T_{proc}}{T_{in} + T_{out}}$$

$$C = c \times (n\text{-}1)$$

with T_{proc} , T_{in} , T_{out} : Time to process, to transfer in & out one unit (event, cassette...), c : CPU rating of one station (e.g., in Mips), C : CPU rating of the farm (e.g., in Mips)

If the actual number of stations is larger then n, the stations idle. If it is smaller, the network idles.

The total user data rate per day of an Ethernet segment is about 100*3600*24~8.5 Gbytes/day. Because of other loads, probably only 2/3 of that, i.e. 5.6 Gbytes/day are available to feed user farms, corresponding roughly to 28 cassettes/day, or to about 5000 cassettes in, 5000 out per year.

The number of cassettes to be processed per year may be several times that, in fact. To multiply farms implies an increase in the overall data rate, i.e a multiplication of Ethernet segments connected to the backbone, and that the bandwidth of the backbone itself is larger than Ethernet's, for example, using FDDI (100 Mbps instead than Ethernet's 10), from the computer centre to each major experimental collaboration. The limiting factor may then be the handling rate of cassettes at the computer centre.

An alternative is to feed the farms via a local server equipped with a pair of 4 or 8 mm Robots. say. These robot units typically contain 10 cartridges and one read/write unit. A server thus equipped would keep alive a farm for about a week, for more than a month with data compression.

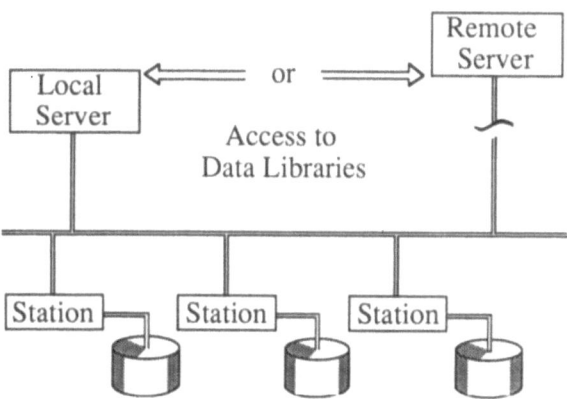

Fig. 3   A simple station farm. Data are transferred to a station disk and retrieved once the station completes processing.

ALTERNATIVES

The case of VAXstations

VAXstations are a case apart. Totally compatible with the full VAX spectrum, they immediately accept applications running on any VAX system. VAXstations are an extension on the user's desk of the VAX manifold, which can be considered as a single, distributed computer system. Direct compatibility makes VAXstations invaluable in counting rooms and function-specific clusters However, as of today, the price/performance ratio is not such to make them the mainstay of end-user equipment. The VMS operating system is popular but it is proprietary.

The issue of X-terminals

X-terminals are two things in one: a support for X-Window (a candidate to become the standard Graphic User Interface), and an extrapolation of the conventional terminal (no active intelligence on the desk) to the world of bit-mapped graphics, multi-window display. They connect directly to Ethernet.

X-terminals have an advantage in cost, lower than a workstation's, and in the overall management of central resources, which still deal with passive terminals. User autonomy is contained. On the negative side, response depends on network traffic, which may increase substantially. Central resources need upgrades. User autonomy is contained

Without affecting arguments for or against X-Window, low-cost stations, which can also run X-Window, challenge the concept of X-terminals. The difference in price between a well configured X-terminal and a low-cost station, both in percentage and in absolute terms, does not justify the sacrifice of the autonomous power of the workstation (presently, more than 10 Mips), neither the "no return" commitment to X-Window, neither losing the chance of reducing the load on central resources, generated by word processing, some areas of graphics, program development, etc. Low-cost stations can be used both as office and home equipment. X-terminals are not well suited as home equipment, once performance and telecom costs are taken into account.

NETWORKS

A fundamental aspect of computing is data transmission. Between computers, between computers and humans, between humans, via computers.

Networks let computers swap data among themselves. To move the data around, a variety of carriers, topologies and conventions are used. The most popular carriers are Ethernet and telephone lines. The use of Fiber Optics technologies, which offer faster data rates, is growing.

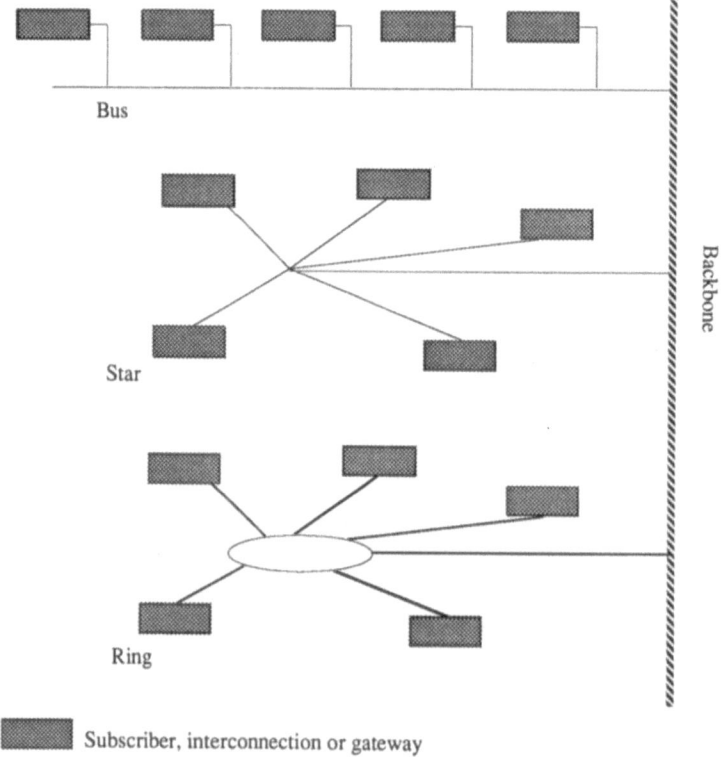

Fig. 4   Basic network topologies. Actual networks are a combination of these topologies. Like in Russian Dolls, frequently at the end of a network connection sits just another network.

Not all applications demand the same network performance. In the counting room we need to ship data around short distances at very fast data rates. For such backplane networks, Gigabits per second could be an appropriate bandwidth. Local Area Networks (LANs), covering a corridor or a building, let users share local resources within short response time. Bandwidth requirements are high. Backbone networks feed and connect different centres of activity on a site. Wide area networks establish world wide connectivity. In all cases, bandwidth is the Holy Grail.

Bandwidth and response are different requirements, however. Bandwidth describes the amount of data which can be transmitted in a time unit, response the delay after which a transaction is completed. In several instances, a truck or a jumbojet provides the best bandwidth, although the response would take a couple of days.

The telecommunications infrastructure is the lifeblood of modern computing. It allows subscribers, or clients, to communicate between themselves, or to access servers, providing such functions as Electronic Mail, printing, remote log-in. Two kind of servers have a vital role. Data, or file, servers bring data to the client's CPU. CPU servers bring user's data to a remote CPU.

The distribution of resources and services in the network infrastructure is not a neutral issue. It is in fact caught in the endless debate between horizontal and vertical social organizations, between power in more and more in power, between centre and periphery. As always, it can be expected that a combination of the two approaches will emerge.

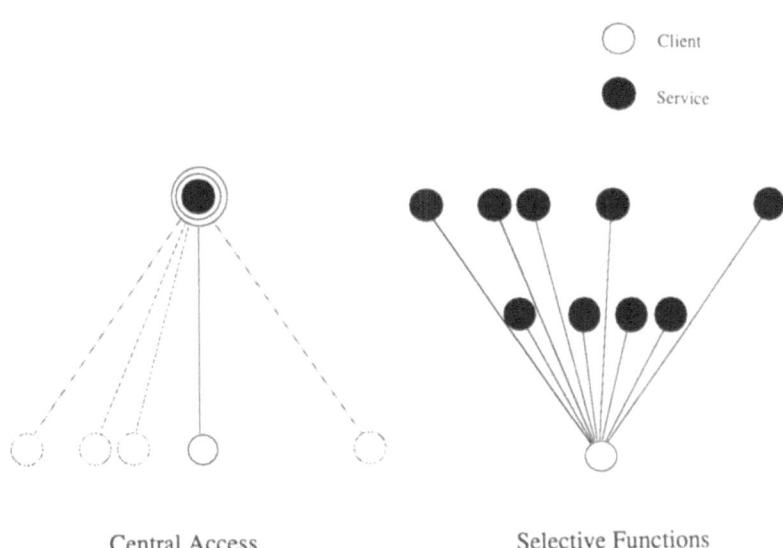

Central Access   Selective Functions

Fig. 5   The basic infrastructure choice. In Central access, users compete for services, in distributed access services compete for users.

High energy physics is notorious for the amount of data generated, processed and transmitted. It has determined for years the hunt for CPU power. It is now poised to lead the hunt for networks and bandwidth.

Today, at CERN, the infrastructure is a mixture of fiber optics, of cables at high (10 Mbps), medium (256 Kbps) and low speed (9600 bauds), and of interconnection electronics (servers, gateways, bridges, etc.). The network deals with a variety of protocols, of in-site interconnections, of transactions. Several dedicated connections exist to remote sites, others are accessible via the public (X25) PTT networks.

Members of experimental groups connect to the computer centre, to their "basement" computers, to the counting room computers, possibly to lab computers. They are equipped with a pot-pourri of terminals, Macintoshes and PC's, and stations (VAXstations, Apollos), roughly in the ratio 4 : 2 : 1. The in-house public network is Ethernet. Macintoshes and PC's are connected to LANs, with gateways to the public network. Dumb terminals are connected to terminal servers. Many stations are dedicated to specific functions, and organized into private clusters, others are connected to the public Ethernet. There are public terminal clusters and local printers for the computer centre's output.

The management and operation of the infrastructure, and the ability to effectively plan ahead are central issues. become intractable with its heterogeneity.

With stations in place of terminals (dumb or X), many users' transactions - such as printing, serving and back-up - may be local. Remote log-ins for electronic mail, text editing, word processing, graphic presentation, etc, are reduced. Traffic directed to remote resources can be expected to be dominated by asynchronous transactions, such as file transfers, rather that by response sensitive transactions. The topology of the network can be more closely adjusted to a backbone/local networks configuration. On simplified cable and protocol configurations, diagnostic tools can be made effective, easing the crucial job of overseeing regional areas of the overall network.

SOFTWARE SUPPORT, THIRD PARTY SOFTWARE

A concept of support has evolved with time for users of mainframes. It does not translate well to end-user computing. This is also true in the case of a substantial unification of the equipment. The key word in personal computing is *personal*.

The major functions of conventional support are: support of the libraries, of the operating system, of the standard social facilities offered on the system, regulation of user access and allocations, and user guidance. With personal desk-top systems, users have (against money) their choice of Word Processors, Spreadsheets, graphic tools, other applications. Basic software tools, such as compilers, which are provided by vendors as proprietary elements of the operating system on mainframes, are frequently obtained from third parties. All these products are usually of excellent quality, with user interfaces and documentation unparalleled on mainframe computing. When software policies for personal equipment exist, they are often only a rationalization of user habits. Popular software has been adopted.

The amount and cost of third party software is a major consideration when contemplating the procurement of new equipment.

First, there is the direct cost of software purchase. Because of the smaller commercial base software for stations is more expensive, as a rule, than software for personal computers. However, with the exception of special applications, notably in engineering, where software is the leading cost component, in the research sector the amount, and the cost, of third party software is usually limited. At most one must consider compilers, a word processor, a graphic application, possibly a spreadsheet. Occasionally, additional applications, such as Mathematica.

There is the cost of compatibility, at the file level, of the new software with the old. A peculiarity of commercial applications is that any one is much more (file) compatible with the same application running on a different system, than with a similar application, coming from a different software company, running on the same system. *Software, not Hardware, guarantees file compatibility*. When using multi-system applications, files are usually compatible, often at the point that compound documents generated on one system can be ported via electronic mail to the same application on different systems. However, if the user changes his personal equipment to a different brand, the application has to be purchased again.

Finally, relearning is always a serious factor. Many people cannot stand the sheer futility of investing time to obtain the same results by different means. Software houses often play on this fact to win captive customers.

To contain software servitudes, it makes sense to adopt as reference those products with the largest world-wide acceptance, compatible with our requirements. We should try to obtain, by negotiating with the suppliers, the right to transport the licence to use from one system to another within acceptable terms.

Conclusion: whenever we think of standards, if possible *adopt as standards everybody else standards*

In the world of personal computing, the most effective forms of support are provided by the users themselves, Areas of interest grow. Users who rely on one application to address some of their computing needs band together. Within areas of interest, areas of competence develop. The more aggressive, capable and concerned users, pave the way for their colleagues. The knowledge accumulated in the areas of competence spreads soon to the full area of interest. Application-oriented support is best delegated to spontaneous areas of competence, which can well be temporary.

This mechanism, for the price of some lack of structure (and, at times, of being poorly known by the public), is driven by the direct motivation of the users, and is deeply convivial. It is remarkably successful. It usually lasts only as long as it is needed. It should be recognized, encouraged to take whatever formal steps appear necessary, such as training, seminars, negotiations for site licences or quantity purchases, etc. Central support is limited to the continued availability of the libraries. In general, in the world of personal computing, the winning word is user's autonomy

OPERATING SYSTEMS, SOFTWARE ENGINEERING

A factor which helps keep down the cost of the new high performance stations is the adoption of UNIX as operating system. This means saving the considerable expenses in time and money to research and develop a proprietary operating system. It also means a chance that all stations, all computers, will run one day under the same operating system. UNIX is touted as THE open, standard system to come. In fact, most companies band together to defend this or that other major flavor of UNIX, although they end up promoting instead their own version of UNIX.

UNIX is part of the astonishing growth of intellectual advances in computing which took place in the 70's and gained fruition only in the second half of the 80's. The mouse, bit-mapped graphics, the desk-top metaphor, Smalltalk, were all worked out then. In a tortuous way they hit the computer market a decade later, and changed the shape and role of computing. So was UNIX.

There is actual enabling power behind the esoteric syntax of UNIX commands. Graphic User Interface are available with all vendors, to hide it to intimated users not charmed by the intricacies of the inner machinery. Thus we stand to gain if (one!) UNIX emerges as a standard. We will then have finally a single standard, that standard powerful, not constraining. A standard which can be accessed without the need to master all its threatening beauties.

A word of caution, however. UNIX is not, today, by far, the most widespread operating system, neither in numbers, nor in the value of equipment it drives. The dominating operating system is Microsoft DOS. There are literally tens of millions of users of that system. IBM itself can not really challenge - with OS/2 - the system to which it had earlier opened the door. True, DOS is a command line O.S., but UNIX is also a command lines O.S.. Graphic User Interfaces also exist for MS DOS. UNIX workstations are coming down in prices, PC's are going up in performances. The deciding factor can well be the emergence of truly lightweight, small enclosure portable computers, which sell strongly, in the millions. Those, as of today, run essentially under MS DOS.

The UNIX bandwagon has brought to preeminence the C compiler. Object Oriented Programming is the darling of the day in the software industries. Traditionally, HEP uses Fortran, and an algorithmic approach. We have slowly come to appreciate the merits of Software Engineering. We must realize that excessive reliance on accumulated investment and know how might be unwise. HEP is no longer a dominant force in computing. When joining High Energy Physics, young people are entitled to be in touch with state-of-the-art technology practices. Physicists are entitled to do physics, rather than computing by proxy.

This, in my eyes is the challenge in front of us. To provide a framework where the professional is in the front edge of the technology, where the physicist can use, combine and design tools which help him to solve the problem at hand, without either becoming a computer expert, or having to feel persistently unsecure in what remains an alien technology.

DETECTOR SIMULATION FOR COMPLEX MULTI-TeV EXPERIMENTS

A. Bassi$^\Delta$, G. Bruni$^\Delta$(speaker), L. Cifarelli$^{*\dagger\Delta}$, G. Iacobucci$^\Delta$,
G. La Commare†, M. Marino‡, and R. Nania$^\Delta$

† CERN/LAA Project, Geneva, Switzerland
‡ World Laboratory/HED Project, Lausanne, Switzerland
* University of Naples, Italy
Δ INFN-Bologna, Italy

ABSTRACT

We have defined a new Entity-Relationship database which contains all the information needed to describe any kind of High Energy Physics apparatus. The database, called SUD (Set-Up Descriptor), has been implemented using the ADAMO database management system. Different programs can access the data and their mutual relationships in a user-friendly way, via standard FORTRAN calls. As an example of application, the database has been interfaced with the GEANT Monte Carlo program.

1. INTRODUCTION

The ever-increasing complexity of events and detectors in modern High Energy Physics (HEP) experiments leads to a complicated software, both for the apparatus description and the physical event analysis or simulation.

Most of this complication comes from a poor use of data structures. For instance, the detector parameters (materials, dimensions, positions, rotation angles, hits, digits, etc.), and the track properties (momenta, directions, vertices) are often inserted into ad hoc common blocks with different organizations in different parts of the programs. As a consequence, the relationships among different parameters are difficult to handle, and lead to a coding which easily expands in a rather confusing way. The situation will be even worse in future Super-Collider experiments.

A great help to the HEP user comes from modern computer science research, which is going towards the creation of packages capable of easily handling huge amounts of data, still preserving all their essential mutual relationships.

These packages generally assume a certain data model to which the user has to conform. Of course more time has to be devoted to the design phase, but a careful organization of the data structures is of great advantage: programs can be written in a much clearer way, which in turn translates into a much simpler coding. The natural consequences of this approach are an easier debugging phase and a better control on future program developments. Thinking along these lines, it is possible to foresee a general scheme to handle the different steps needed to develop a full Monte Carlo (MC) simulation program for a HEP experiment.

New Technologies for Supercolliders, Edited by L. Cifarelli and
T. Ypsilantis, Plenum Press, New York, 1991

This can be done i) at the level of the interface with the Event Generators (EG) which provide the Physics input, ii) at the level of the geometrical description of the detector itself and iii) at the level of the simulation of the detector responses. The development of such a complete program chain started some time ago[1,2]. In this work, we will focus on the part dealing with the description of the experimental set-up.

The aim is to model a database general enough to be used for the description of different kinds of detectors (a calorimeter, a vertex detector, a muon spectrometer, etc.) by the different software packages normally needed in an experiment (MC simulation, reconstruction, graphics, etc.). The model should also be flexible enough to permit an easy accommodation of new data structures and relationships which are not yet included and may become useful.

In order to develop a database along these lines, we were faced with two needs: a) to identify the common building blocks of different apparata; b) to choose a data model with an easy FORTRAN interface. For point a) we adopted as a starting point, the main ideas already developed in the well-known GEANT Monte Carlo simulation program[3]. Concerning point b), we made extensive use of the ADAMO package[4], a FORTRAN implementation of the Entity-Relationship (ER) model[5].

The final result of this work is the so-called Set-Up Descriptor (SUD), which is already currently used within the ZEUS collaboration at HERA and in the LAA/MSL project at CERN[6,7,8].

The organization of the present work is as follows. In section 2, we briefly recall the main features of the ER modelling techniques and of their implementation in the ADAMO package, referring for details to the original literature. The general guide lines of the chosen SUD data model are presented in section 3, while a detailed description is given in section 4. Section 5 shows, as an example of application, how the SUD database has been interfaced to the GEANT program itself. The conclusions are given in section 6.

2. THE ER MODEL AND THE ADAMO PACKAGE

In the Entity-Relationship (ER)[5] model, proposed by P. Chen in 1976, the world is represented by a set of *"entities"* and their mutual *"relationships"*. An entity is specifyed by a list of *"attributes"*, which can assume different values.

The ADAMO system is a particular software implementation of the ER data model, using either ZEBRA[9] or BOS [10] as the underlying memory manager system.

In the ADAMO context, the model has a useful graphical representation, called the *"ER diagram"*, where the entities are associated with boxes and their mutual relationships with arrows.

The ADAMO package offers a set of tools to go from a formal description of the ER diagram (via the DDL, Data Definition Language) to a FORTRAN coding, where the data structures are mapped on to tables. In addition a FORTRAN package (the TAP, TAble Package)[11] provides a set of subroutines to:

- manipulate the data (count, remove, add, ...),
- make consistency checks,
- fetch/insert tables from/to database files.

The entities and their relationships can be accessed from a FORTRAN program via the so-called *"window"*common block. An example of these concepts is illustrated in Fig.1, showing a simple ER diagram: the entity "Position" is related to the entity "Rotation" (the double-head arrow means that many spatial Positions can be linked to the same Rotation); the attributes of Position are the cartesian x, y and z coordinates, while the attributes of Rotation are the three angles specifying the rotation. The figure also shows, in a simplified way, the tabular representation of the entities and their relationships, and what the window common block looks

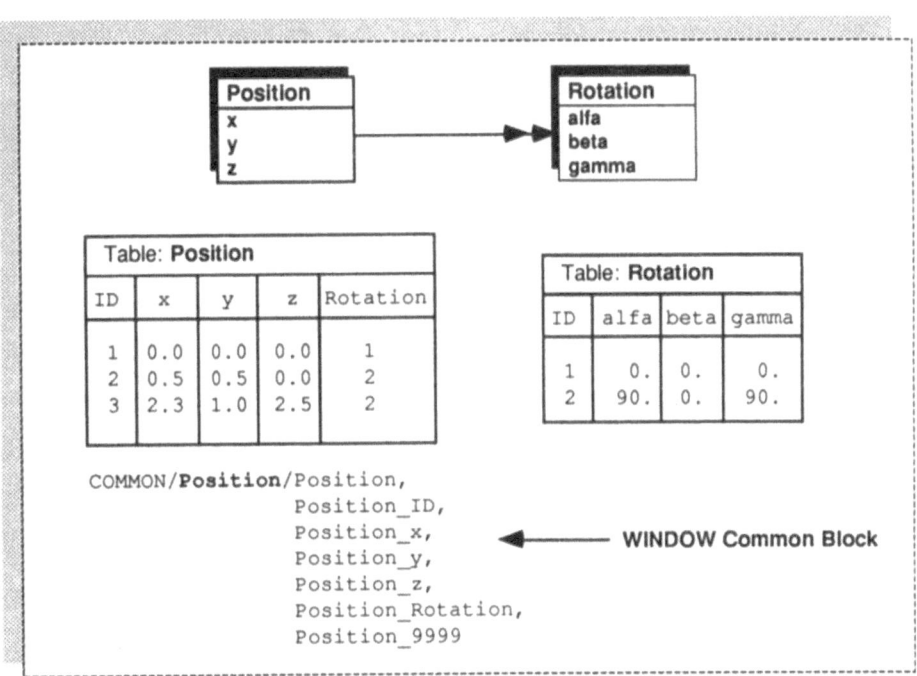

Fig. 1. Example of the tabular representation of entities and relationships in ADAMO

like. The column called "*ID*" is the row identifier, which is unique and different from row to row. The relationship between Position and Rotation is held in the column named Rotation of the Position table. The window common block can be "moved" along the lines of the corresponding table. Given the equivalence between an entity and its tabular representation, in the following we will give the same meaning to the words "*entity*" and "*table*".

3. THE SUD BASIC IDEAS

Among the many possible ways of describing a detector, the scheme already developed in the GEANT Monte Carlo program has been adopted as a starting point. This program basically consists of three parts:

(I)   description of the detector,
(II) definition of the detector response,
(III) tracking of particles through the detector.

We are mainly concerned with parts (I) and (II), which allow the set-up specification. These are the static parts of the program, as they belong to the initialization phase, and should be configured according to a well defined procedure.

In the GEANT scheme, the basic element to build a detector is represented by a volume (and its possible divisions), which is described by its shape and dimensions. The volume is then positioned and oriented in space with respect to a given reference frame. A tracking medium (which may be made of a combination of possibly different materials) is assigned to the volume. Among the volumes, the sensible volumes are those giving a signal when a particle traverses them. The response of a sensible volume in terms of hits, together with a set of digitization prescriptions, can be specified. In summary, we have the following list of actions:

(i)   define a set of volumes and their divisions (if any),
(ii) define their positions and orientations,
(iii) assign a tracking medium to each volume,
(iv) define the sensible volumes and their hits/digits.

This general way of describing a detector is suitable to be used in an ER model. The advantage of doing so relies on the fact that once we have a data model which includes all the entities and relationships required by the GEANT scheme, a general coding can be developed which is valid for an arbitrary apparatus. All what should be done is just fill the database with the pertinent data. The complexity of the coding will then be reduced because of an improved data organization.

4. THE ER MODELLING OF SUD

In this section we build the ER model of the set-up description, closely following the points (i) to (iv) of sect.3.

(i) Define the volumes and their divisions

A volume is specified by a name and a shape, which are the attributes of the entity named VOLU. It may be associated with a set of parameters (PARA-VOLU relationship). A set of different parameters is generally linked to the same volume, which explains the "*1:many*" relationship (double-head arrow, in Fig.2). A volume can be a division of another volume (VOLU-DIV relationship). Notice that in this case we have just a "*1:1*" relationship (single-head arrow, in Fig.2). The attributes of the DIV entity specify what is actually needed by GEANT concerning a division.

In Fig.2 a vertical bar across an arrow indicates that the relationship is "*partial*". For instance the VOLU-DIV relationship is partial on VOLU side, since there might be a volume having no divisions.

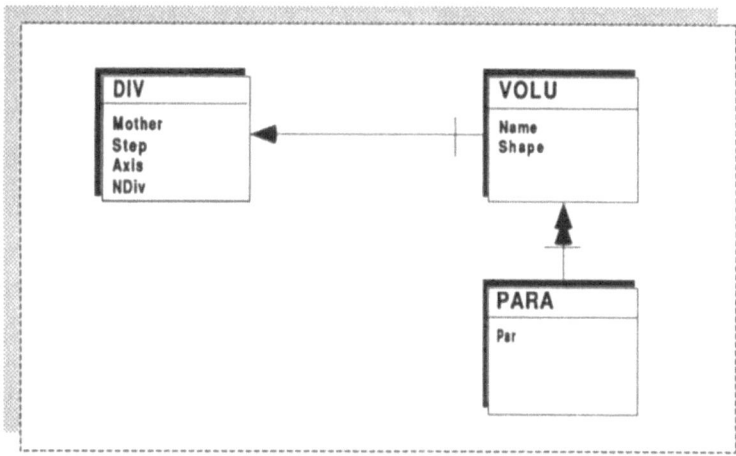

Fig. 2. ER diagram for the definition of volumes and divisions

(ii) Define the positions and orientations

A volume can be positioned in space, inside another "mother" volume, so that a relationship exists between the VOLU entity and an entity which is called POS. The attributes of POS are self-explanatory, "konly" being a GEANT flag which deals with overlapping volumes and whose possible values are "only/many" [3].

The orientation is specified by a set of rotation matrices (ROTM entity), which refer to a spatial position (POS-ROTM relationship). Many different positions can be linked to the same rotation matrix, and there might be rotation matrices not linked to any position. In the ROTM entity shown in Fig.3, the ID of the table is shown only to recall that, as a convention, the rotation matrix identification number is taken equal to the row number of the corresponding ADAMO table.

The relationship between PARA and VOLU (or POS) is "*generalized*", as shown in Fig.3 by a double-line arrow from PARA, which splits into two arrows pointing at VOLU and POS, respectively. This is so because in GEANT the volume parameters can be either defined altogether once at the beginning, or just a given shape can be first defined and the remaining parameters specified at positioning time (recall the two GSPOS and GSPOSP routines of GEANT)[3].

(iii) Assign tracking media to the volumes

To any volume a tracking medium (VOLU-TMED relationship) is associated, which can be the same for different volumes. A tracking medium can have a set of corresponding tracking parameters (TPAR-TMED relationship) and can be made of a set of materials (TMED-TMATE relationship). The TPAR entity allows the user to re-define some otherwise standard values of the tracking medium parameters. The materials which define a tracking medium can enter in different proportions (PROPOR-MATE relationship). The two relationships between PROPOR and MATE (named "in" and "of") specify which fractions of materials are contained in other materials.

(iv) Define the sensible volumes and their hits/digits

The attributes of the tables shown in Fig.5 are equal to the arguments of the GEANT routines which actually define the sensible volumes and/or their hits and digits.

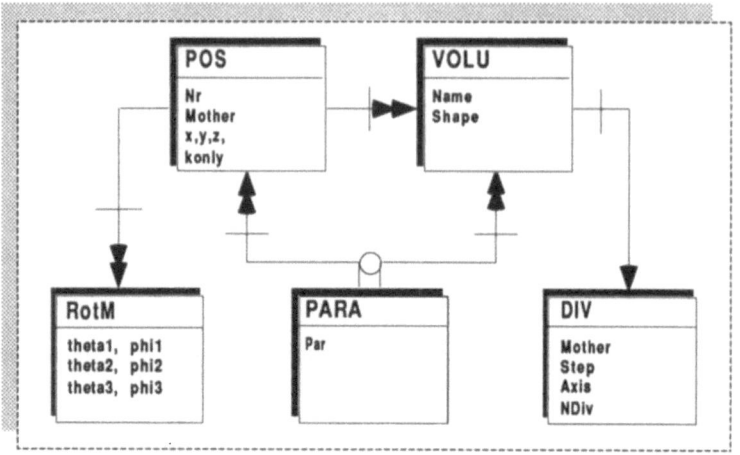

Fig 3. ER diagram for the definition of volume positions and orientations in space

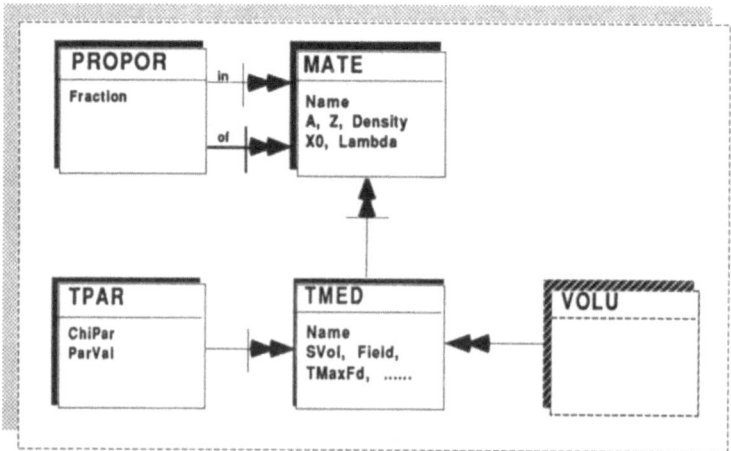

Fig. 4. ER diagram for the definition of materials and tracking media

A volume may be sensible (VOLU-SDET), and therefore endowed with a set of volume descriptors (SDETV-SEDT). A sensible volume can have many hits (SDETH-SDET) and digitization prescriptions (SDETD-SDET), and also a set of user-specified parameters (SDETU-SDET).

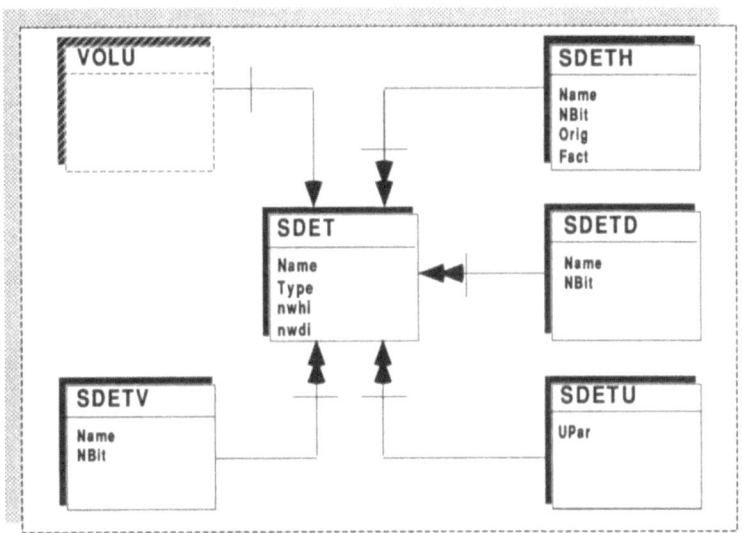

Fig. 5. ER diagram for the definition of hits and digits

(v) Generalizations of the ER diagrams

In this subsection we will discuss some generalizations of the ER diagrams presented so far, which have been achieved for the sake of completeness and flexibility of the whole SUD database. Let us start with a generalization within the ER diagram relative to the definition of hits/digits, shown in Fig.5. This comes from the fact that a sensible volume can have more that just one kind of hit response. In the GEANT program, the problem is solved via the introduction of the so-called "detector aliases". This solution can be implemented in the ER model by introducing a new entity, called SDETA in Fig.6, which corresponds to the alias-sensible-detector. There is thus a new relationship between SDET and SDETA which, when active, links a given sensible detector to one or more aliases. To complete the generalization picture, the simple relationships between either SDETH, or SDETD, or SDETU, and SDET (see Fig.5) become now generalized relationships with both SDET and SDETA, as shown in Fig.6.

A second generalization comes from the frequent need to deal with a simplified or partial set-up, with respect to the complete configuration stored in the database. For example, a certain detector or parts of it may not be wanted in the current version of the apparatus. Moreover, it could be that alternative detector descriptions are stored in the database, such as, for instance, a "fine-grained" and "porridge" version of a calorimeter, so that it might be convenient to switch from one description to the other. This need is not explicitly covered by the current version of the GEANT program, and we have therefore envisaged a solution whose scheme is illustrated in Fig.7.

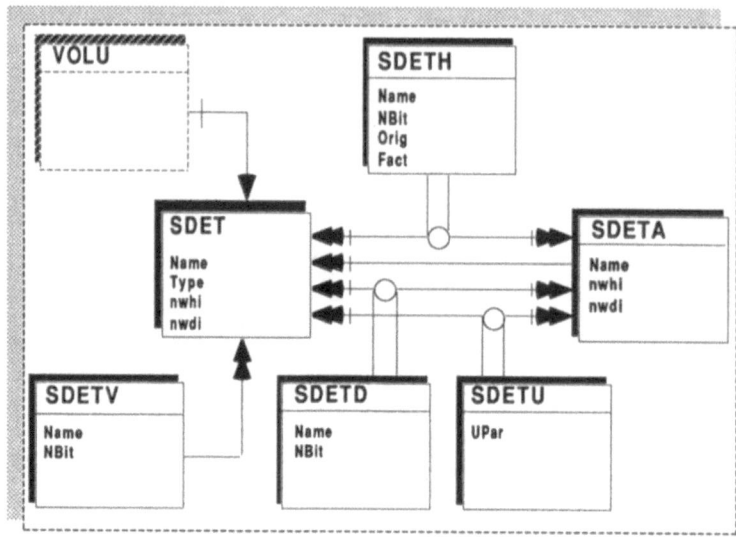

Fig.6. Generalized ER diagram for the definition of hits and digits

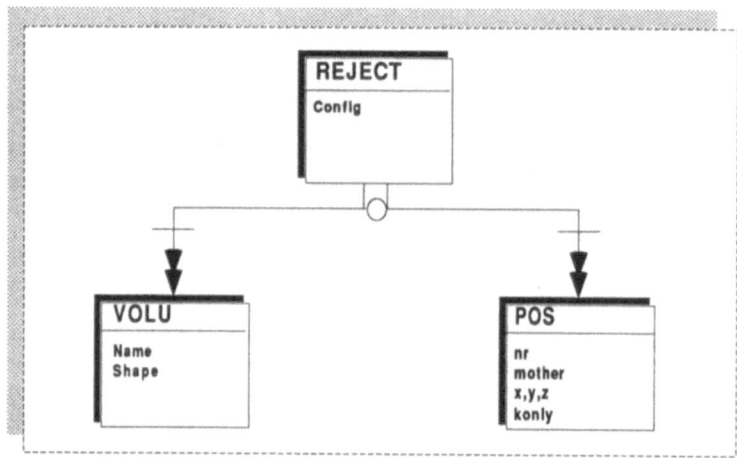

Fig.7. ER diagram relative to the detector selection scheme

The entity called REJECT has a unique attribute, "Config", and a generalized relationship with the VOLU and POS entities. For a given configuration, specified by the value of Config, there might be, in the relationship columns of the REJECT table, a list of volumes and/or positions which have to be temporary deleted from the database before calling the routines which manage the set-up definition. A clear advantage of this solution is that the chosen configuration flag (Config) can be easily set via data cards in the program.

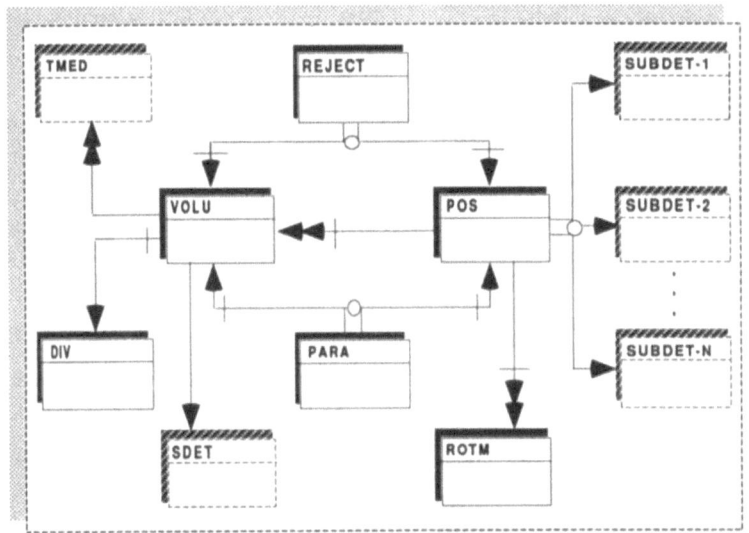

Fig. 8. Complete ER diagram of the SUD database

The third and last generalization is required to store the detailed sub-detector-dependent data which cannot be included in the general tables (strip widths, pitches, wire spacing, etc.). This is shown at the right hand side of Fig.8, which is the complete ER diagram of the SUD database. The POS entity is linked, via a generalized relationship, to the entities generically called SUBDET-1, ... SUBDET-N. This is the only detector-dependent part contained in the ER diagram.

5. THE SUD INTERFACE WITH THE GEANT PROGRAM

The SUD database can be interfaced to different programs. The interface to the GEANT package is of course straightforward as shown in the block diagram of Fig.9. The UGINIT routine of GEANT contains the initialization sequences, and the procedure to open the database files. Inside the UGEOM routine, the routines which manage the definition of materials, tracking media and rotation matrices, are called once for all. Then follows a call to the geometry definition routines (volumes, positions) which look at the database and transmit all the relevant information to GEANT.

Finally the closure of the database files is performed inside the UGLAST routine.

6. THE SUD INTERACTIVE VERSION

A SUD database can be created and filled using ADAMO tools such as the TAble Package (TAP)[11] and the Table Interaction and Plotting (TIP and TI)[12] one. However this requires the user to be familiar with these tools.

A more elegant and user-friendly approach is to provide an interactive system where the user can easily access the SUD database in a transparent way. To this purpose, we have developed the SUD Interactive System (SUD-IS)[13] based on the KUIP[14] command interpreter.

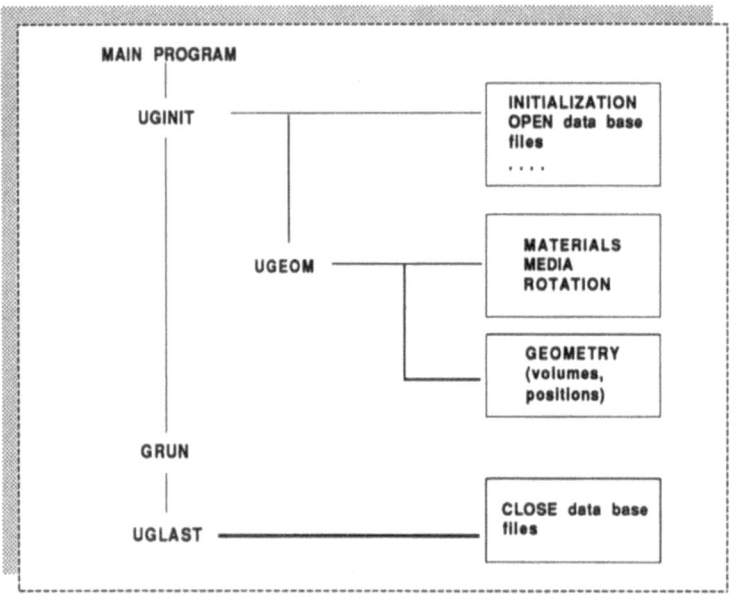

Fig. 9. Steering of the interface of SUD with GEANT

A self-explanatory dialogue method, together with an on-line help facility available for each command and operation, enables the user to define an entirely new detector and/or edit an old configuration. At any moment the session can be saved by updating an existing data file.

The commands are grouped in OBJECT/ACTION menus. The actions are intended to define, modify, remove, and list various objects, such as materials, tracking media, rotation matrices, and volumes, with their positions, divisions, and possibly their hits/digits.

Fig.10 shows an example of SUD-IS session. Two windows are shown: one relative to the definition of a volume, the other to an on-line help facility.

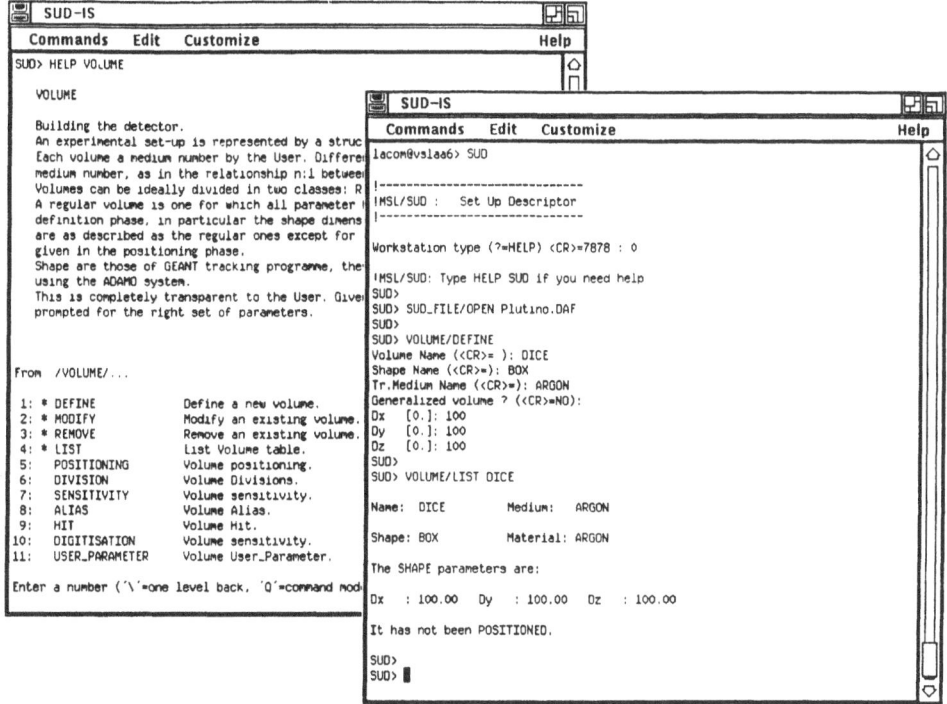

Fig 10. Example of SUD-IS session

7. CONCLUSIONS

A new ADAMO based data model has been developed to describe the detailed configuration of a HEP apparatus. It is general and simple, it applies to any arbitrary apparatus and leads to a short, easy-to-debug, user-friendly coding. A particular application of the model has been worked out, in connection with the GEANT Monte Carlo package, showing that is indeed possible, and even necessary, for future HEP experiments, to go towards a new way of handling complex data sets.

8. ACKNOWLEDGEMENT

The model presented herein has grown with time, from the very original ideas to the present status, essentially in the framework of the ZEUS collaboration at HERA. We acknowledge all our colleagues of ZEUS, and in particular N. McCubbin, for many useful discussions.

345

REFERENCES

1   A. Bassi et al., Z2 internal notes 88-1, 88-12 and ZEUS notes 88-7, 88-97
2   G. Anzivino et al., A General Definition for a HEP Apparatus Based on the ADAMO Package in the Framework of the GEANT Monte Carlo Programme, in Proceedings of the 1st International Workshop on "Software Engineering, Artificial Intelligence and Expert System in High Energy Physics", Lyon, March 1990
3   R. Brun et al., GEANT3 User Guide, CERN/DD/EE/84-1, Sept. 1987
4   Z. Qian et al., Computer Physics Comm. 45 (1987) 283
5   P.P. Chen, ACM Trans. on Database System 1 (1976) 9
6   R. Nania, Monte Carlo Simulation for the LAA Detector, in Proceedings of the 9th INFN Eloisatron Project on "Perspectives for New Detectors in Future Supercolliders", Erice, October 1989, World Scientific, Singapore
7   F. Anselmo et al., FMC: Full Monte Carlo Chain User Guide, v 1.2, CERN/LAA/MSL/90-01, February 1990
8   F. Anselmo et al., MEGA: Monte Carlo Event Generator Adaptor, CERN/LAA/MSL/91-08, June1991 (submitted to Particle World Communication in Subatomic Physics)
9   R. Brun and J. Zoll, Zebra User Guide, CERN Computer Centre Program Library Long Write Up (Q100), Jan. 1987
10 V. Blobel, DESY report F14-86-02, July 1986
11 L. Basadonna et al., The TAble Package Programmers reference manual, ADAMO Note 6, November 1987
12 A. Aimar et al., TIP User guide, Internal note, June 1989
13 G. La Commare and M. Marino, SUD - IS Users Manual, CERN/LAA/MSL/91-12, June1991
14 R. Brun and P. Zanarini, KUIP - Kit for an User Interface Package, CERN Computer Centre Program Library Long Write Up (I 202), February 1988

IMAGE PROCESSING AT SUPERCOLLIDERS

Werner Krischer

CERN, Geneva , Switzerland

ABSTRACT

For experiments on future hadron colliders event rates of 10^8 / sec and data volumes of 1 Megabytes / recorded event are expected. The speed-up of several orders of magnitude in processing power over today's on-line systems has to come mainly from progress in architectures. We have identified some representative triggering and data compaction algorithms, analyzed them as low level image processing tasks and have started to look into a few matching signal processing architectures that are commercially available. Fortunately we can benefit from the work of the very large international signal (image) processing and HDTV community and their commercial products. During the last decades they have solved many problems in the development of highly parallel signal (image) processing algorithms and architectures, e.g. systolic and/or wavefront array processors. The now (or in a predictable future) commercially available architectures seem to deliver the necessary computing power for future triggering and data compaction systems and will certainly have an important role to play in the design of such systems.

INTRODUCTION

Detectors at supercolliders will be faced with a formidable problem : The bunch separation will be 10-20 ns and we have to expect multiple collisions per bunch crossing ($\geq 10^8$ events/second). (The corresponding figures for LEP or UA1 are 22 or 3.8 μs and 0.3 or 50000 events/sec, respectively). After an analog trigger reduction of $\approx 1000 : 1$ the average decision frequency for the second level trigger is assumed 100 kHz. Per recorded event we expect ≥ 1 megabytes of data. This increased data volume is due partly to the larger dimensions of the future detectors, but mostly to the finer resolution in space and grayvalue. The detector output will look more like an "image". If executed today on commercial fully programmable processors, trigger algorithms are too slow by a large factor. The speed-up of several orders of magnitude over present-day on-line systems can not be expected only from improvements in IC fabrication, new technologies, algorithms and/or programming techniques. It has to come mainly from using adequate parallel architectures, properly integrated in the data flow. Our aim is therefore to understand the existing parallel architectures to be able to embed suitable ones later in future readout systems.

A judgement of architectures depends completely on the classes of algorithms one wants to run, and on finding a balance between total throughput and latency, important in real time. In [1] we have identified some typical algorithms, that we used as benchmarks for triggering and data compaction. The purpose of these tests was to find out which architecture matches best which algorithm and not to find one architecture that performs equally well on all algorithms. Some algorithms can run a million times faster than on a VAX 8300. Their analysis led us quite naturally to a certain algorithmically specific architecture, a pipelined image processing system, which we will describe here in some detail. We also tried out several Single Instruction Multiple Data [SIMD] processor arrays. The most successful SIMD architecture of our benchmark, the "Associative String Processor", became the subject of the massively parallel processing collaboration (MPPC) and has been described elsewhere [2]. We will restrict ourselves here to a description of our model of a pipelined image processing system. We will repeat the reasons that led us to this choice not only with the same algorithms as in [1], but we prefer to do this mainly with some recent work done for the lead / scintillating fibre calorimeter (SPACAL) [3].

TYPICAL ALGORITHMS AND IMPLEMENTATION ON PIPELINED IMAGE PROCESSORS

Peak Finder

A frequently recurring algorithm is the reduction of a graylevel image (e.g. 8 bits / pixel) to a binary one (1 bit / pixel) or to a list of x's, y's (and grayvalues). Simple thresholding is usually not sufficient. We have chosen a two-dimensional peak finder. A centre pixel is considered a peak if the 3 following conditions are fulfilled :

1. The centre pixel C is bigger than the average XAV   O
   of its 4-connected neighbours X   OXO
2. XAV is bigger than the average of the 8 next outer neighbours O.   OXCXO
                                                     OXO
3. The centre pixel is a maximum in a 3*3 neighbourhood.   O

The computation of the grayvalue difference of 2 groups of pixels like in the conditions 1 and 2 can be interpreted as a convolution. Testing if these differences exceed certain constants is a typical point operation on one image. Finding the maximum in a neighbourhood is one of the most basic morphological operations in image processing. Producing the x,y - list is the feature extraction task. Figure 1 shows a possible configuration for the peakfinder algorithm.

Track Finder

A list of x's and y's (< 256) of straight tracks coming from a known vertex is given. The tracks should be found by an angular histogram. This can be interpreted as the logical "AND" - operation between the binary image to which the x's and y's belong and a mask containing the histogram bin-numbers. The only non-zero pixels of the resulting image have a gray value equal to the bin-number at the original (x,y) - position. The tracks can then be found from a gray level histogram. This algorithm will take on a system with Region Of Interest (ROI) - capability and a pixel time of 100 ns 256*256*100 ns = 6.5 ms.

Another way of looking at this problem is to store the x - and y - addresses in separate images and perform a table lookup on the combined 16 - bit image. In this case the algorithm will only take 130*1*100 ns = 13 µs for the given 130 addresses, which we had to process.

Figure 2 shows configurations for these two cases, the so called "iconic" and the "symbolic" one.

Cluster Analysis in a Calorimeter

We have recently worked on the development of cluster finding algorithms for the SPACAL calorimeter [4]. Our test data came from beam exposures in December 1989 and June 1990. We mapped the hexagonal structure of these 20 or 155 tower modules onto a rectangular pixel memory in such a way, that a space invariant operation on the hexagonal array remains space invariant after the mapping on the rectangular array. For interactive algorithm development, we mapped several thousands of test events into big images like in Figure 3 (top right). Figure 3 (bottom right) shows zoomed parts of it. One notices the qualitative difference between electron (e) triggers on the bottom (occupying few pixels), and pion (π) triggers at the top (much more extended clusters). Our goal was to express this difference in lateral spread into an e/π - discrimination algorithm. Instead of developing

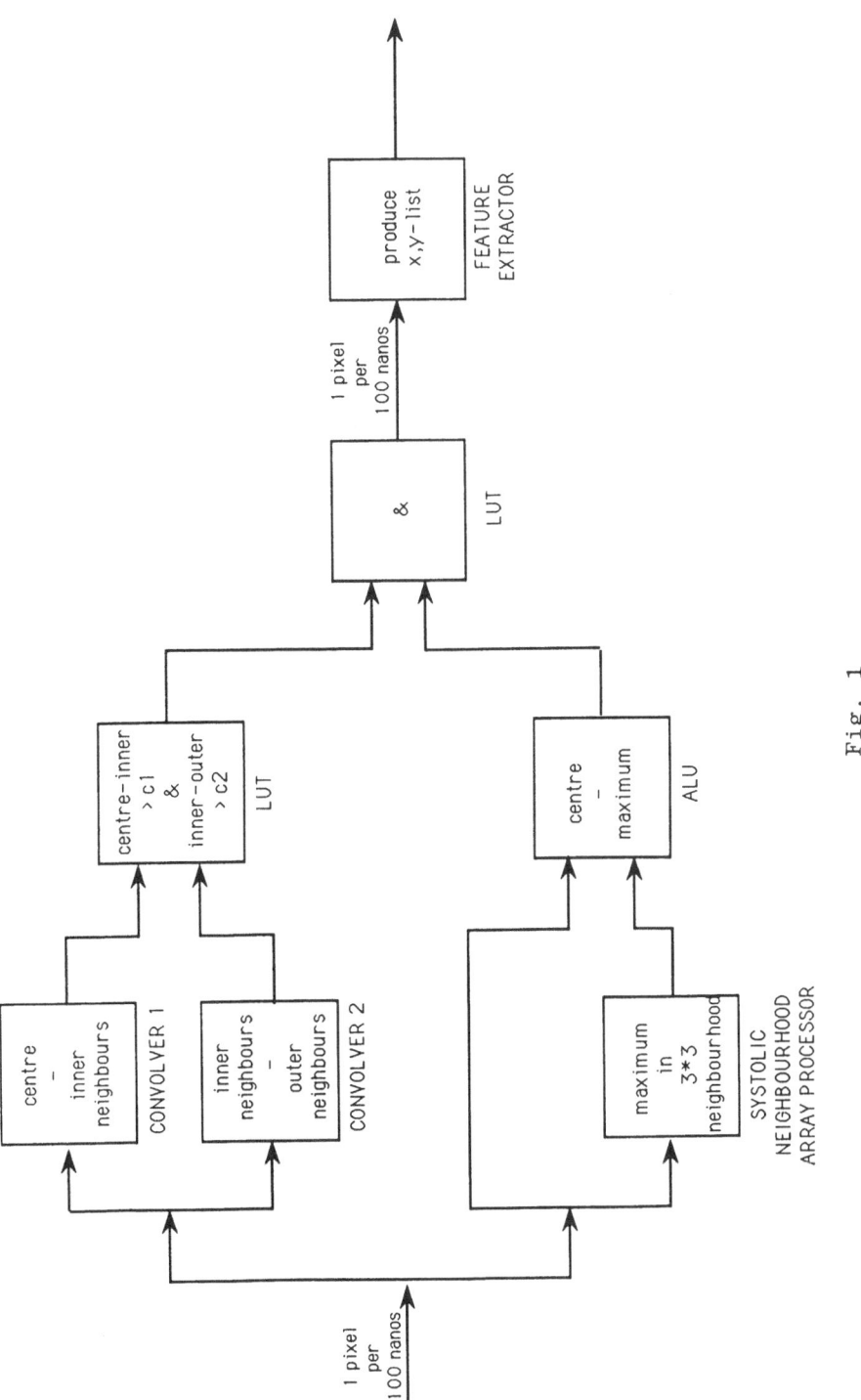

Fig. 1

349

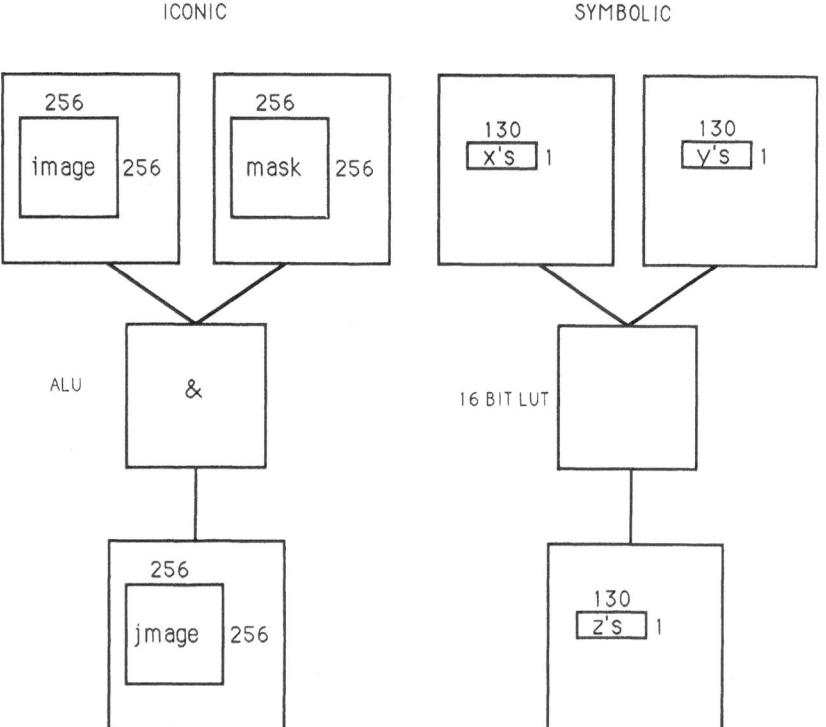

Fig. 2

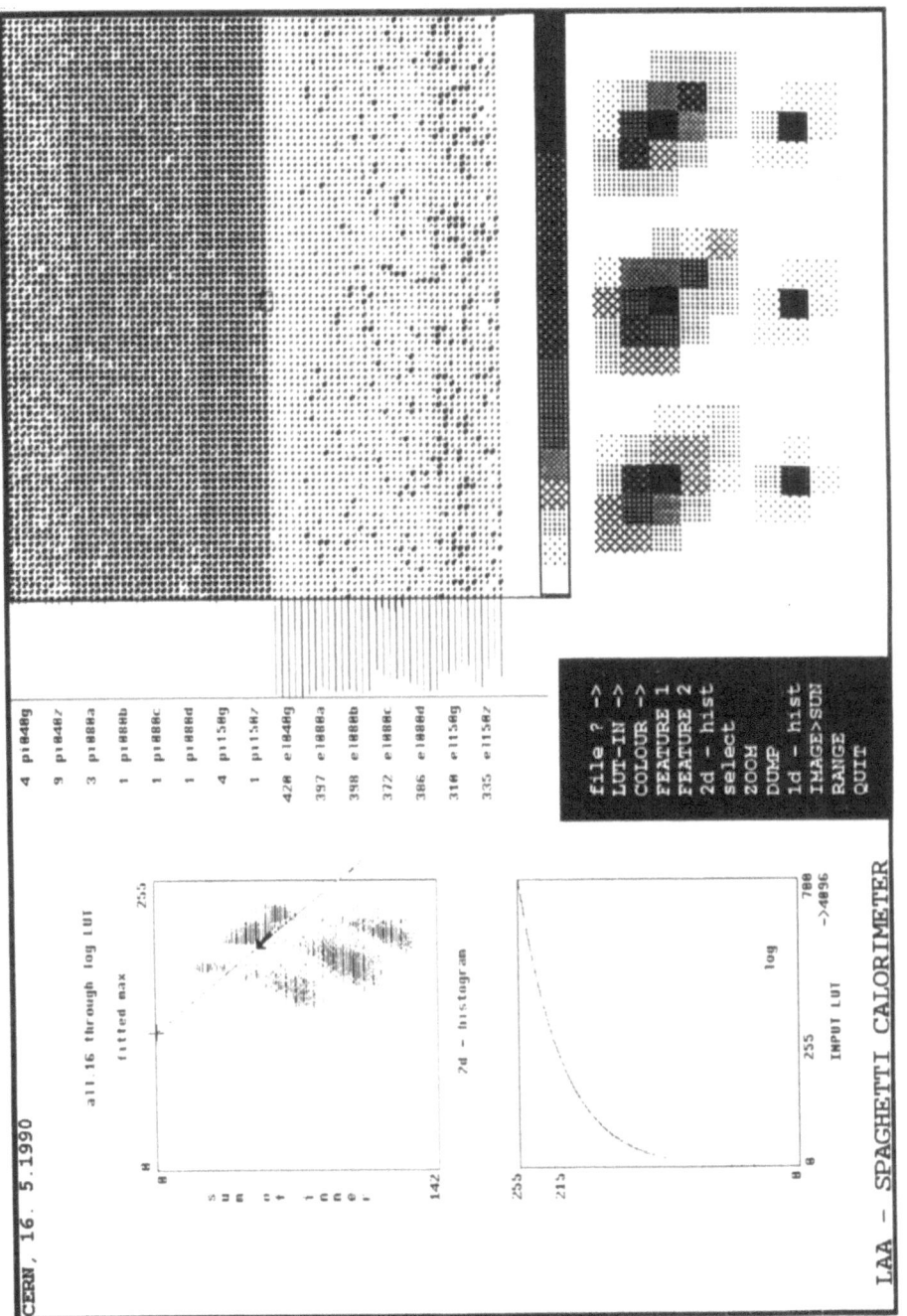

Fig. 3

algorithms for the isolated grayvalues given, we approximated the real trigger situation by using no a priori information about cluster position, and developed algorithms for such big images, built up of many small events. The detection of the pixel with beam impact, i.e. of local maxima, therefore obviously had to be one of the processing steps (this corresponds to no pointer information from a first level trigger!).

We extracted from the original image some different features, e.g. total energy , local maximum, sum of the inner or outer neighbours, the sum of the 3 biggest values, the peak of the paraboloid $z = a + b(x^2+y^2) + cx + dy$ calculated through the 2 biggest grayvalues and their closest neighbours, etc. We then made two-dimensional histograms from these features, and chose the pair of variables that gave the best cluster separation in this "feature space" and thus allowed to produce an optimal classification algorithm. Figure 3 (top left) shows such a two-dimensional histogram. One sees how one can interactively define a discriminant in the feature space. It should be noted here that simply comparing local maxima with inner and outer neighbours as in the peakfinder above did not work at all on this data because the resolution of this calorimeter was not fine enough. We got the best results by going to subpixel precision. The dynamic range in the original files extended from 0 to about 700. We reduced it to 0 - 255 in our simulation using several different compression scales, e.g. sinusoidal, piece-wise linear, logarithmic. The triggering algorithm's discrimination power can be substantially improved by choosing the appropriate transformation function. A logarithmic LUT turned, e.g. in the case of the plot of the fitted maximum against the inner neighbours, a completely useless histogram into an optimal one. It should be noted that a very local neighbourhood has been used here, which eases implementation and allows for minimal isolation criteria, i.e. is most robust against pileup in high-luminosity situations.

A highly simplified possible implementation of our algorithms on a pipelined image processing system could look like in Figure 4. The input image is first passed through a look-up table, which accomplishes the nonlinear transformation, apart from a dynamic range adjustment. The chosen features consist essentially of simple neighbourhood operations like convolutions or minimum/maximum determinations, and can be easily implemented by commercially available neighbourhood processors with several parallel data paths. Their outputs (the features) are then combined at the input of a big look-up table, in which all the decisions are coded. The at present available 16 - bit look-up tables allow 2-dimensional numerical classification in real time.

Our simulation program allows to see immediately the effect of different input-LUT's, features and discriminants in the feature space and it produces the contents of the input- and output- LUT.

AN ALGORITHMICALLY DEDICATED PIPELINED ARCHITECTURE

Among the systems on the market, that perform exactly the type of algorithms we have described earlier, we have chosen for our tests the MAXVIDEO - family from DATACUBE, Peabody(MA) (see [5]). This is sold as a family of (at present) about 40 very powerful VME - compatible boards. A flexible and expandable bus-system (MAXbus) allows to define the data paths between the different modules. A single output port can drive up to 10 input ports. Any input port can be physically connected to two output ports. Switching can be done on a pixel-by-pixel basis. We can thus design practically any single or multiple parallel processing pipelined architecture. At present data is transferred between I/O ports at a rate of 10 or 20 Mbytes/sec. Regions Of Interest (ROI) can be specified from 1 to 4096*4096 pixels and the processing time is proportional to the number of pixels. Examples of interesting modules are the triple ported Region Of Interest memories (ROI's) from 512*512 * 16 bits to 8 Megapixels, cascadable up to 16384*16384 * 8 bits. Convolvers exist from 3*3 up to 16*16 arbitrary coefficient kernels and a large kernel convolver for 2*256*64 with 2 coefficients only. This corresponds to a processing power of 180, 5120 and 320000 million arithmetic operations per second, respectively. A systolic neighbourhood processor can be used for nonlinear processing, e.g. morphological operations (erosion, dilation) or minimum/maximum finding in a 3*3 neighbourhood. Other examples are programmable crosspoint switches, pointwise ALU's, a module to count the occurences of many different events and to extract the coordinates of a specific event, MIMD - or SIMD - like configurable DSP's to fill the gap between special purpose hardware modules and more general software solutions. Many modules contain look-up tables (from 256 * 8 bits to

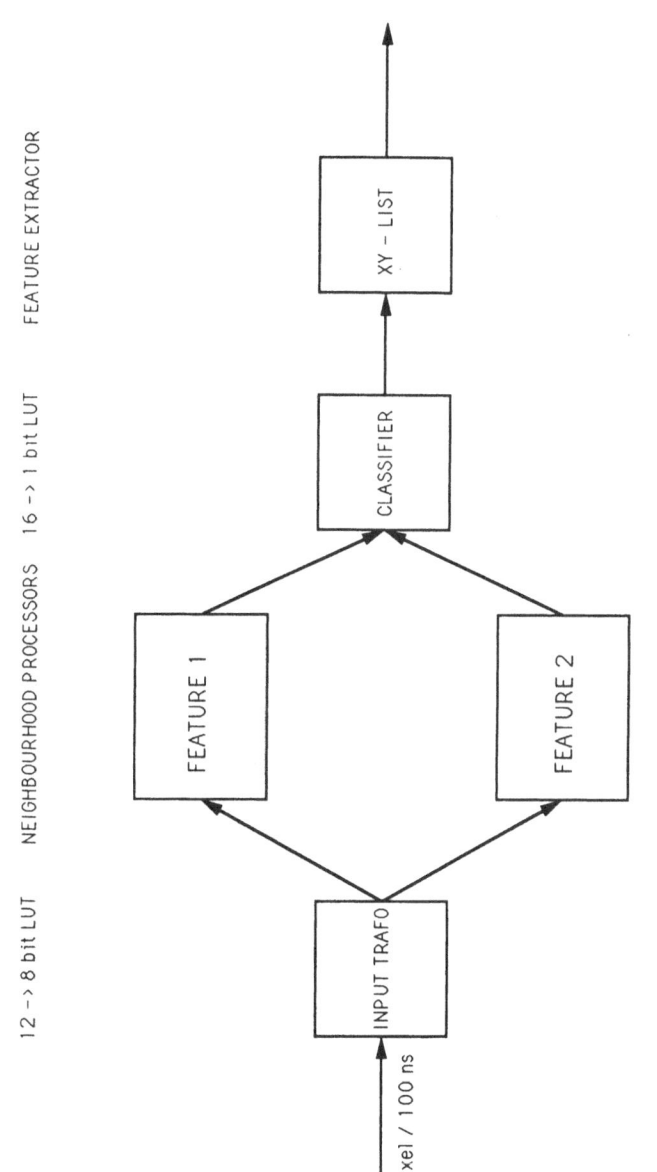

12 –> 8 bit LUT   NEIGHBOURHOOD PROCESSORS   16 –> 1 bit LUT   FEATURE EXTRACTOR

1 pixel / 100 ns

INPUT TRAFO

FEATURE 1

FEATURE 2

CLASSIFIER

XY – LIST

SIMPLIFIED BLOCK DIAGRAM

Fig. 4

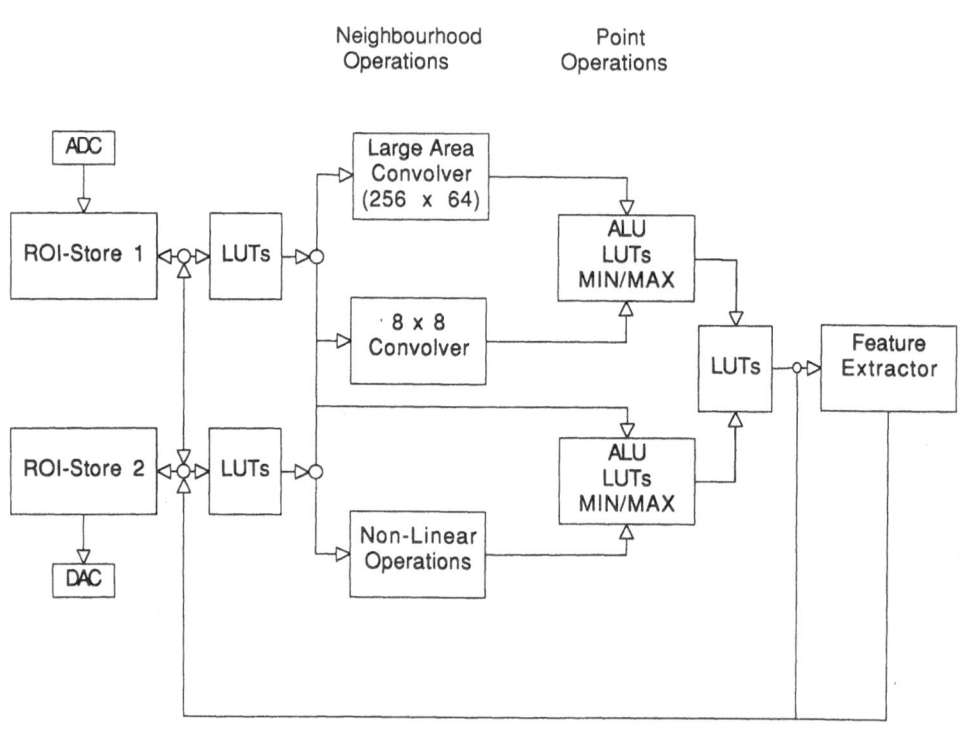

Fig. 5

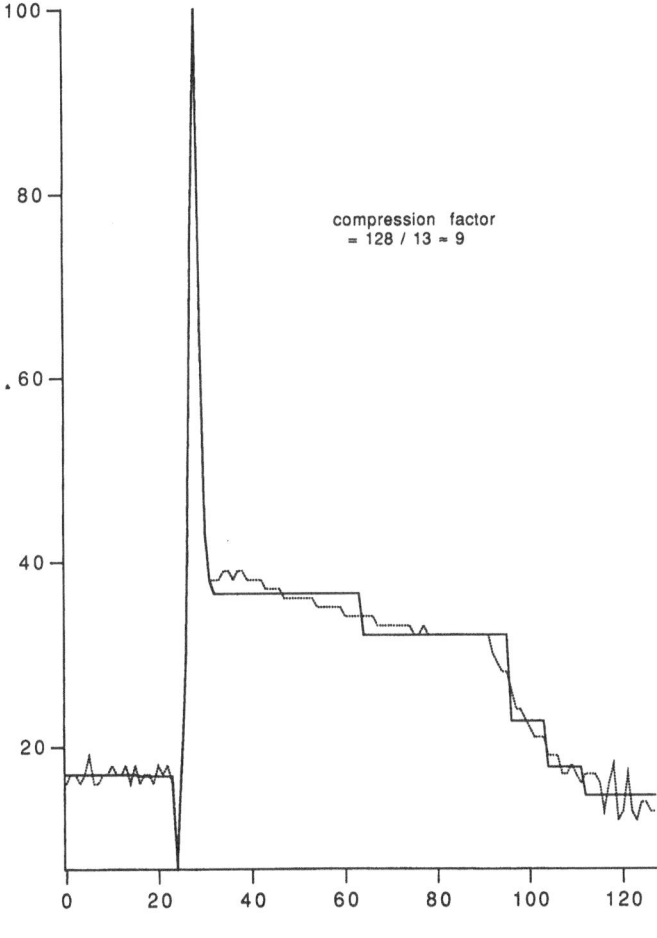

compression factor
= 128 / 13 ≈ 9

HAAR - transform

Fig. 6

65536 * 16 bits, i.e. an arbitrary function F(X,Y) can be performed on two 8- bit images X and Y).

Figure 5 shows in a simplified form the complete system we have put together for most of our algorithms. This configuration could today be replaced by a single board that works at a 20 MHz rate.

Software

Whereas general purpose computers fetch, process and store data under the tight control of a processor, pipeline processors operate on a continuous flow of pixel data. Their control is set up prior to the operations of the pipe. DATACUBE supports UNIX and OS/9 development systems. Another company has been founded to interface to non VME-bus and other operating systems. Some relatively high level software (MAXWARE) had been provided as C - source code. It allows to call functions like "read red LUT" , "convolve with coefficient bank nr. i", etc, but it leaves you e.g. the worry about the pipeline delays! A new object oriented "IMAGEFLOW" software is advertized to allow you to "program pipelines and not image processing hardware". An ASCII - configuration file defines the hardware set-up parameters, an Application Programmers Interface hides all the hardware details and there is no loss of real time performance.

FURTHER EXAMPLES OF SIGNAL PROCESSING ACTIVITIES

Apart from our belief that digital signal processing in general will play a more and more important role in our data acquisition systems (e.g. [3],chapter 3.2.3), there are certain domains of very active research in the international digital signal processing (in particular the HDTV) community that are of particular interest in the context of processing and/or storage of signals from our sensors.

Image coding

We usually threshold the signals from our detectors and record for every track hit their coordinates, maximum, integral, etc. If for some reason we cannot or do not want to define a threshold or if, because of real-time constraints, it is not possible to extract the whole information on-line, we may be forced to use standard signal coding techniques that have been developed for efficient storage and/or transmission of digital images (facsimile, digital television,...). Figure 6 shows one of the many techniques (see e.g. [6]), we have tried on our signals. One sees how perfectly this HAAR - transform "fits" our signals. The 13 most important transform samples preserve the essential information of the 128 time samples of this example, only smoothing the background. This transform is one of a new class of integral transforms ("Wavelet transforms" [7]), which have two important properties : they are particularly well suited to represent pulse-like signals (as from our electronic detectors) and they are the fastest of all known integral transforms.

Advanced signal processing algorithms and architectures ([8])

We have mentioned above the calculation of a paraboloid through 4 points in the SPACAL example. Although a least squares fit through more points was not superior for SPACAL we would like to ask the question if an on-line least squares fit of some function $z(u,v)$ is conceivable or not. A least squares fit of a paraboloid $z_i = a + b(u_i^2 + v_i^2) + cu_i + dv_i$ ($i = 1$, m) through the m points u_i ,v_i , z_i can be written as the general linear least squares problem $\| Ax - z \|_2 = $ minimum, where the matrix $A(m,n)$ and the vector $z(m)$ are known and usually $m > n$. In our case $n = 4$ and $x^T = (a,b,c,d)$. The method of normal equations $x = (A^TA)^{-1}A^Tz$ is not stable [9]. If we used algorithms like the orthogonal triangularization $A = QR$ with $x = R^{-1}A^Tz$ we could use the "systolic" architecture of Figure 7 (derived in [10]) and a linear one for the backsubstitution. Here the circular boxes compute the coefficients of the Givens rotation and the square ones perform the rotation. In [11] a "systolic" architecture is described that produces immediately the residuals of such a fit. Several different processors have been used to implement this type of algorithm. The DATAWAVE processor ([12]) would be ideally suited for this type of application. This is

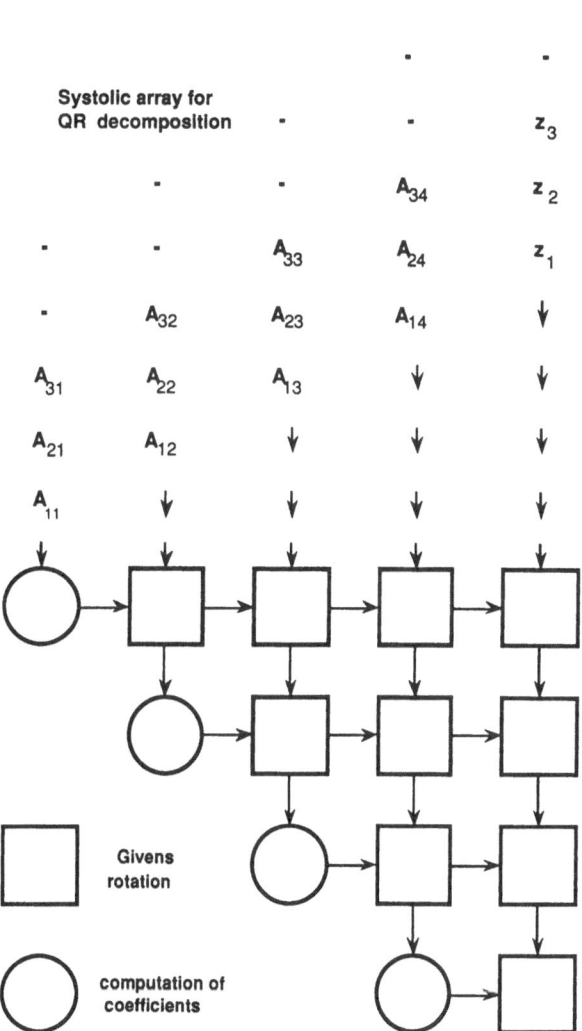

Systolic array for QR decomposition

A_{34} z_2

A_{33} A_{24} z_1

A_{32} A_{23} A_{14}

A_{31} A_{22} A_{13}

A_{21} A_{12}

A_{11}

□ Givens rotation

○ computation of coefficients

Fig. 7

just one example of the type of problems that have been solved - theoretically and in practice - by the signal processing and/or HDTV community.

TECHNOLOGY OUTLOOK

Our present model of a pipelined system is the MAXVIDEO system. It represents the state of the art in image processing. Parallelism is used at all levels : at the task, image, neighbourhood and the pixel-bits level. For control and communication the industry standard VME - bus is used. The MAXbus used for data has an expandable bandwidth of at present 10 or 20 Mbytes/sec per pipeline. The open architecture, based on modularity and communication, allows to improve the functionality (bandwidth and processing power) at any moment by adding more hardware of existing or novel architectures and technologies, off-the-shelf or home-built. DATACUBE announces for ≈ 1992 a 100 MHz version. Other companies sell 100 MHz versions today, but without a ROI feature and with none of the extremely powerful neighbourhood processors (a convolution with a n^2 mask needs n^2 passes through the system). Intermetall GmbH , Freiburg im Breisgau , from the ITT semiconductor group announced recently a completely programmable single chip video processor [12] with a 125 MHz clock rate and a peak performance of 4 Gops. The sustained throughput rate is supposed to be 750 Mbytes/sec. The price of the chip for large quantities could be in the range of 30 - 40 \$. The chip is executed in 0.8 μ technology. They extrapolate this for 1993 in 0.4 μ technology to a 250 Mhz clockrate and a peak performance of 32 Gops. This processor has been designed for HDTV applications e.g. to save silicon area by having only one programmable processor for the treatment of all television standards.

Together with the possibility to process regions of interest defined by the first level trigger all these systems are very hot candidates for second level triggering.

CONCLUSION

A speed-up of several orders of magnitude in intelligent computing over today's on-line systems will be needed to make use of fine-grain detectors in future hadron colliders. This speed-up will have to come mainly from progress in architectures. We have analyzed representative triggering and data acquisition algorithms as low level image processing tasks. Sequential von Neumann computers can not exploit the parallelism in these algorithms. Fortunately we can benefit from the work of the very large international signal (image) processing and HDTV community and their commercial products. During the last decades they have solved many problems in the development of highly parallel signal (image) processing algorithms and architectures. Some of these systems are now (or in the near future) commercially available. They seem to deliver the necessary computing power and will certainly have an important role to play in the design of future triggering and data compaction systems.

ACKNOWLEDGEMENTS

I want to acknowledge the continuous support of A. Zichichi, LAA project leader, and would like to thank my colleagues R. K. Bock, V. Buzuloiu, A. Gheorghe.

REFERENCES

[1] S.Lone,R.K.Bock,Y.Ermolin,W.Krischer,C.Ljuslin,K.Zografos : Fine-grain parallel computer architectures in future triggers - NIM A288 (1990) pp. 507-516
      R.Bock,W.Krischer,S.Lone : Benchmarking computer architectures for HEP real-time algorithms,CERN-LAA/RT/88-08

[2] M.Lea,ASP : A Cost-effective Parallel Microcomputer,IEEE Micro Oct 1988
      F.Rohrbach,THE MPPC PROJECT,CERN/DRDC/90-76

[3]   The SPACAL Collaboration : Scintillating fibre calorimetry at the LHC -
        CERN/DRDC/P1,1990
[4]   R.K.Bock,V.Buzuloiu,W.Krischer : Algorithms and an image processing architecture
        for on-line electron identification from lateral profiles in SPACAL
        CERN LAA RT/90-01
[5]   DATACUBE : "MAXVIDEO User's Manual",Peabody,MA,1988
[6]   A.K.Jain,P.M.Farrelle,V.R.Algazi : Image Data Compression in "Digital Image
        Processing Techniques "(editor M.P.Ekstrom)-ACADEMIC PRESS 1984
[7]   G.Strang,Wavelets and dilation equations: A Brief Introduction
        SIAM Review,Vol..31,No.4,pp. 614-627,December 1989
[8]   H.T.Kung , C.E. Leiserson : Systolic arrays for VLSI
        in Duff,Stewart (eds.) : "Sparse Matrix Proceedings" 1979,
        SIAM 1979 ,pp. 256-282
        H.T.Kung : Why systolic architectures ? IEEE Computer,vol.15, 1982 ,pp.37-46.
        K.Bromley (editor) : "Highly Parallel Signal Processing Architectures"
        SPIE Critical Review of Technology Series 19,vol. 614,1986
        K.Bromley,H.J.Whitehouse : Signal Processing Algorithms,Architectures and
        Implementations,SPIE : O-E/LASE 1986
        S.Y.Kung : "VLSI Array Processors "- Prentice Hall 1988
[9]   G.H.Golub,C.F.van Loan : "Matrix computations", John Hopkins Press,
        Baltimore, 1989
[10]   W.M.Gentleman and H.T.Kung : Matrix triangularization by systolic arrays,
        Proc.SPIE Vol. 298 Real-Time Signal Processing IV (1981), pp. 19-26
[11]   J.G.McWhirter : Recursive Least Squares Minimisation using a systolic array
        Proc. SPIE, Vol. 431, Real-Time Signal Processing VI (1983), pp. 105-112
[12]   U.Schmidt,K.Caesar,T.Himmel : Data driven single chip video processor
        IEEE Transactions on Consumer Electronics 8/90

SIGNAL PROCESSING, TRIGGERING AND DATA ACQUISITION

N. Ellis

University of Birmingham, Birmingham, United Kingdom
S. Cittolin and L. Mapelli
CERN, Geneva, Switzerland

ABSTRACT

We address issues of triggering and data acquisition for the LHC. The parameters of the accelerator together with predictions for cross sections and a detector model are used to evaluate the system requirements. We then review the available technology and discuss how a trigger and data acquisition system might be constructed. We comment on areas in which the design of detectors, and of trigger and data acquisition systems affect each other. We also indicate areas of rapid technological advance which are relevant to triggering and data acquisition. Finally, we highlight areas in which further R&D is required.

1. INTRODUCTION

Some of the general issues of signal processing, triggering and data acquisition for the Large Hadron Collider (LHC) were addressed at previous workshops [1]. Here we attempt to outline in more detail how a trigger and data acquisition system might be built. Obviously much needs to be done before such a system could be realized. However, our studies suggest that it is possible, in principle, to construct a system with the required performance using technologies which are either available now or which are in an advanced state of development.

In sect. 2 we try to set out the assumptions we have made regarding the performance which is required for the trigger and data acquisition system. Starting from a beam crossing interval of 15 ns, we estimate trigger rates and data volumes at various points in the system. The trigger rates are affected by the parameters of the machine, the predicted cross sections and the physics signatures which are deemed to be of interest; data volumes are, in addition, affected by the number of channels and occupancy of the detectors. We use estimates for a typical general-purpose detector operating at a luminosity of $10^{34}\,\mathrm{cm}^{-2}\mathrm{s}^{-1}$.

In sect. 3 we outline a possible architecture for the trigger and data acquisition system. This is based on a multilevel trigger in which the crude trigger decisions made quickly at the first level are progressively refined at higher trigger levels. Data are buffered in digital or analogue memories during each stage of trigger processing. We pay attention to the interaction between different trigger levels and the data acquisition system. We also emphasize areas in which the design of detectors affects the trigger and data-acquisition systems and *vice versa*.

The design of trigger systems is discussed in more detail in sect. 4. We consider triggers based on calorimeter information for high transverse momentum (p_T) electrons and jets, and for large missing transverse energy (p_T^{miss}). Next, we discuss triggers for high p_T muons, based on information

New Technologies for Supercolliders, Edited by L. Cifarelli and
T. Ypsilantis, Plenum Press, New York, 1991

from external muon detectors. Finally, we discuss more specialized triggers based on electron identification detectors which are unlikely to contribute at the first-trigger level, but could be important at the second-trigger level.

Section 5 describes in more detail developments for some of the key components of the front-end of the data-acquisition system. These include very fast analogue-to-digital converters (ADCs), analogue and digital pipeline memories, and digital signal processors (DSPs). We also discuss developments in microelectronics: it is worth noting that custom chips are now accessible to the electronics designer. Developments being made by industry for high-definition television (HDTV) are also mentioned.

In sect. 6 we review some components which will be important at higher levels in the data acquisition and trigger system. These include buses and data links, and mass-storage media. We also discuss possible scenarios for event building.

The complexity of trigger and data-acquisition systems for the LHC will require much more use of software tools than was previously the case. A brief discussion of software aspects can be found in sect. 7.

In our conclusions (sect. 8) we summarize and try to identify areas in which the design of the trigger and data-acquisition systems affect the overall design of experiments. We also indicate areas where we feel more R&D work is required in order to build trigger and data-acquisition systems to be ready for the LHC.

2. PARAMETER OVERVIEW

In order to establish the framework that signal processing, triggering and data acquisition will have to face at LHC, an analysis of the expected trigger and event rates, and of data volumes must be performed. Event rates are calculated in sect. 2.1 from the estimated total inelastic cross section at centre-of-mass energy $\sqrt{s}$ = 16 TeV. The average event size is extracted in sect. 2.2 from a model of a general-purpose detector. Results of Monte Carlo physics studies performed in this and in other working groups are used in sect. 2.3 to estimate trigger rates. Finally, in sect. 2.4 we combine these parameters to predict the overall data volumes and estimate the required bandwidths.

At this stage any evaluation is necessarily very rough leading to enormous uncertainties in the conclusions. This parameter overview should be considered only as an exercise to establish a working hypothesis.

2.1. Event rate

The cross section, σ_{inel}, visible to experiments for high-p_T physics is estimated by subtracting the contribution of elastic and single diffractive interactions from the total cross section, σ_{tot}. Extrapolations of existing p$\bar{\text{p}}$ collider data [2] suggest the value σ_{tot} = 110 ± 20 mb for the total cross section at $\sqrt{s}$ = 16 TeV, which corresponds to a visible cross section in the experiment of σ_{inel} ~ 60 mb. For luminosities in the range 10^{33} - 4×10^{34}cm^{-2}s^{-1}, the rate of visible interactions will be in the range 6×10^7 - 2×10^9 interactions/s. However, given the bunch structure of the beam with bunch crossings every 15 ns, the relevant rate will be 6.7×10^7 bunch crossings/s with between one and forty overlapping events per bunch crossing.

In reality, the bunch structure of the machine contains eleven holes of ~ 1 μs and one of ~ 3 μs [3] due to beam injection and dump, giving an average bunch crossing rate of 54 MHz. Independently of any effect that such beam structure might have on the performance of the various detectors, it could be used for the synchronization (local clocks, reset, etc.).

2.2. Event size

In order to stress the extent of the triggering and data acquisition problem at LHC, we will consider the case of a general-purpose detector, i.e. a detector capable of electron, muon and jet identification and of missing transverse-energy measurement. Such a detector would consist of inner tracking detectors with electron identification power, electromagnetic and hadronic calorimetry and muon chambers. The pseudorapidity coverage required for most of the physics signals studied in the physics working groups of Aachen workshop is $|\eta| < 3$. The measurement of the missing transverse energy requires a bigger coverage, possibly to $|\eta| < 5$. Detectors with good resistance to the high level of radiation will have to be used at such small angles to the beams. The granularity of these forward detectors will be coarse, providing a negligible contribution to the overall event size. A general-purpose detector of this kind will presumably be located in a "medium" luminosity area, where peak luminosities up to about $2 \times 10^{34} cm^{-2} s^{-1}$ are expected [3]. The following estimates are therefore based on a luminosity of $10^{34} cm^{-2} s^{-1}$.

From a parametrization of the energy dependence of the charged particle multiplicity, n_{ch}, using CDF results at 0.63 and 1.8 TeV [4], the expected number of charged particles per unit of rapidity at $\sqrt{s} = 16$ TeV is $dn_{ch}/d\eta = 6$ for $\eta = 1$. Consequently, the expected total number of particles in the calorimeter is 12 per unit of rapidity. Studies of the rate of particles reaching the muon chambers [5] indicate that for an absorber thickness of 12 absorption lengths, only one track will be seen in the muon chambers for $\sim 10^4$ incident charged hadrons. Hence, at a luminosity of $10^{34} cm^{-2} s^{-1}$ and for a detector coverage of $|\eta| < 3$ one expects:

- ~ 350 tracks/15 ns   in the inner tracking detectors,
- ~ 700 particles/15 ns   in the calorimeters,
- < 1 tracks/15 ns   in the muon detectors.

2.3. Trigger rate

Prompt triggers will be based on signatures of electrons, muons, jets and missing transverse energy, based on the rough particle identification achievable in the shortest possible time with information from calorimetry and fast muon detectors. The basic selection parameter at the first trigger level will be a threshold on the transverse energy or momentum. Table 1

Table 1.   Inclusive rates

Muons	
Full single-muon rate after 12 l of lead	$10^5 - 10^6$ Hz
Single-muon rate for $p_T > 20$ GeV	2×10^3 Hz
Two-muon rate for $p_T > 20$ GeV (both muons)	25 Hz
Electrons [a]	
Single-electron rate for $p_T > 20$ GeV	10^5 Hz
Two-electron rate for $p_T > 20$ GeV (both electrons)	10^3 Hz
Jets [b]	
Two-jet rate for $p_T > 180$ GeV (both jets)	10^4 Hz
Two-jet rate for $p_T > 300$ GeV (both jets)	10^3 Hz

(a)   Note that the lowest possible rate for a first-level trigger based on calorimeter cuts only is the one of the single π^0, which for $p_T > 20$ GeV is 2×10^4 Hz.

(b)   There is an uncertainty of almost an order of magnitude in the Monte Carlo prediction of jet production at $p_T \sim 20$ GeV due to the choice of structure functions, fragmentation parametrization and other factors.

summarizes the expected inclusive rates [6] for a luminosity of $10^{34} cm^{-2} s^{-1}$ and a coverage of $|\eta| < 3$.

The inclusive rates of table 1 give an indication of the thresholds necessary for a first-level trigger rate not exceeding 10^5 Hz. These have to be confronted with the thresholds required by the physics signals that are sought. In table 2, three important physics channels expected at LHC are used as examples to demonstrate the feasibility of first-level triggers. The thresholds listed correspond to acceptable efficiencies for the detection of the physics signals indicated, and are the result of studies performed in other working groups of the Aachen workshop.

Table 2. First-level trigger rates with thresholds required by physics signatures

Top [7]	
$t\bar{t} \to$ (e or μ) + 3 jets:	
Single muon with $p_T^\mu > 40$ GeV	2×10^2 Hz
Single electron with $p_T^e > 40$ GeV	5×10^3 Hz
$t\bar{t} \to e + \mu + X$	
Single muon with $p_T^\mu > 50$ GeV	10^2 Hz
Single electron with $p_T^e > 50$ GeV	4×10^2 Hz
Higgs [8]	
$H \to ZZ \to$ 4 leptons:	
Two muons with $p_T^\mu > 20$ GeV	25 Hz
Two electrons with $p_T^e > 20$ GeV	10^3 Hz
SUSY [9]	
$\tilde{g}\tilde{g} \to p_T^{miss}$ + multijets:	
Three jets with $p_T > 200$ GeV	5×10^2 Hz
$\tilde{g}\tilde{g} \to ZZ + 2$ jets + $\tilde{g}\tilde{g} \to$ 4 leptons + jets + p_T^{miss}	
Two muons with $p_T^\mu > 30$ GeV	10 Hz
Two electrons with $p_T^e > 30$ GeV	10^2 Hz

From table 2, it appears that the prompt trigger rate will be dominated by triggers requiring electrons. Assuming that thresholds can be kept reasonably sharp, it seems possible to provide prompt triggers with sufficient acceptance for physics at rates of the order of 10 kHz. Rates might be reduced further by demanding more complicated signatures. It should be noted that an inclusive missing transverse energy trigger, if implemented, might also give a high rate. We assume a total first-level trigger rate of 10^4-10^5 Hz. As described in more detail in sect. 3, after the fast (~ 1 μs) first-level trigger, higher levels of trigger are needed with increasing background rejection power. Tracking information combined with more refined calorimeter cuts for electron identification, precise p_T cuts on muon candidates, and kinematic and topological selections can be applied by second-level trigger processors. A rejection compared to the first-level trigger of at least 10^2 should be possible, leaving an overall rate after the second-level trigger of 10^2-10^3 Hz.

Further reduction has to be gained to meet data storage and off-line analysis capabilities. No detailed evaluation of third-level trigger algorithms has been done by our working group. However, full event reconstruction and on-line filtering probably performed by general-purpose processor farms must reduce the data rate to acceptable levels.

2.4. Data volumes and bandwidths

An accurate estimate of the data volumes that an LHC data-acquisition system will have to face could only be done if the detector configuration were known. At this stage we can only make an educated guess based on the number of electronic channels of typical subsystem examples. The expected granularity of detectors under study in other working groups, which are also the object of detector R&D projects, are summarized in table 3.

Table 3. Number of electronics channels

Inner Tracking:	
Silicon tracking and preshower [10]	2×10^7 channels
TRD straw tube [11]	4×10^5 channels
Scintillating fibres:	
Expected occupancy: 10^{-2}–10^{-3}	10^6 channels
Calorimetry [12,13] (dominated by electromagnetic compartment)	
for a granularity of $\Delta\eta \times \Delta\phi = 0.02 \times 0.02$ with at	
least two readout samplings and for $|\eta| < 3$	2×10^5 channels
Expected occupancy: 10^{-1}–10^{-2}	
Muon Tracking	
Resistive plate chambers [14] or drift chambers	$10^5 - 10^6$ channels
Expected occupancy: 10^{-5}–10^{-6}	

Combining the numbers of channels and the expected occupancies from table 3, we obtain estimated data volumes, assuming the use of zero suppression, which are shown in table 4.

Table 4. Data volumes

Inner Tracking: bytes/15 ns	10^4–10^5 hits/15 ns	or ~ 10^6
Calorimetry: bytes/15 ns	10^3–10^4 cells/15 ns	or ~ 10^5
Muon detector:	negligible	

Data rates after the different trigger levels are obtained by combining the data volumes from table 4 with the trigger rates evaluated in sect. 2.3. The required bandwidths are summarized in table 5, where we assume zero suppression but no further data compression.

Table 5. Bandwidths

First level: bytes/s	10^{10}–10^{11}
Second level:	10^8–10^9 bytes/s
Third level:	10^7–10^8 bytes/s

As described in more detailed later, the enormous amount of data corresponding to the first-level trigger accepts might not need to be moved out of local memories. A possible scheme, in which the first-level trigger is used to drive the second-level algorithms, would require that only sections of the detector which are flagged by the first-level trigger are readout by the second-level processors, leaving the bulk of the data in local storage until the second-level trigger decision is made.

The rate of second-level triggers is such that bandwidths of the order of GB/s are needed in order to empty the buffer memories, build the events and transfer the data to processor farms for higher-level filtering and data storage. Buses, networks and data links capable of such throughput are discussed in sect. 6.2.

The 10 - 100 MB/s rate for data storage has been estimated by arbitrarily assuming a further reduction of a factor of 10 at the highest-trigger level. No systematic study has so far been done on how to achieve such a rejection. However, it is common prejudice to believe that enough processing power will be available in a processor farm to perform quite advanced physics analysis in order to reach a manageable level of data storage.

3. AN ARCHITECTURE BASED ON MULTI-LEVEL TRIGGERS

Irrespective of the detailed form that a trigger and data-acquisition system might have in a LHC experiment, the boundary conditions described in sect. 2 impose a few basic features that must be satisfied to cope with LHC requirements:

- The system must be designed for the highest-possible performance.
- The first-level trigger processors, whether based on analogue or digital electronics, must be pipelined to handle a new event every 15 ns.
- Pipelined buffering and hierarchical data collection are essential.
- Where possible, the system should be characterized by a high degree of parallelism for easy scaling and adaptation to evolution of required performance.

In the following, we explain in more detail how these requirements come about.

Following experience from experi-ments at the Sp$\bar{\text{p}}$S [15,16] and Tevatron [17,18] proton-antiproton colliders, and from the preparation of the future e-p collider HERA [19,20], we envisage a trigger system for experiments at the LHC based on several levels. The same conclusion has been reached in studies for the SSC [21]. Relatively crude decisions made quickly at the first level can be refined at the second and third levels using more detailed information from the detectors and more complicated algorithms. The role of each trigger level is to reduce the trigger rate to the point where it can be accepted by the next higher level. In fig. 1 we illustrate the structure of such a trigger system, based on three levels, where we have indicated order-of-magnitude rates which might be achieved after each level of triggering.

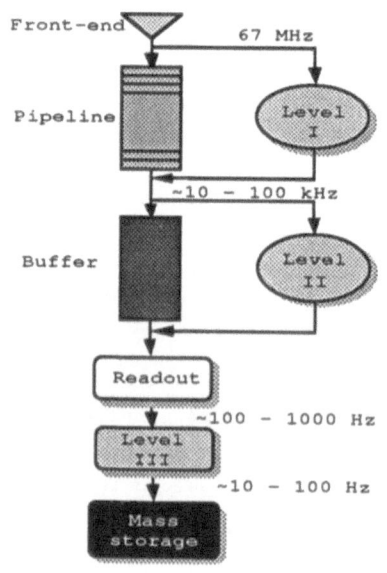

Fig. 1(a). Model of a multi-level trigger system (conceptual model).

Here we discuss the feasibility of such a model system, but it must be stressed that this architecture only reflects our current ideas for a powerful trigger and data-acquisition system. We feel that it is not appropriate to try to define a final solution for trigger and data acquisition now, since we do not yet have at our disposal the technology needed, especially in the front-end. Instead we review the state-of-the-art of the technologies involved in order to define a plan of work and provide recommendations to the high-energy physics community. The first-level of trigger processing must be very fast because information from all channels of the whole detector (perhaps 10^7 to 10^8 channels for a detector with central tracking) must be stored until the decision is available. It is likely that the first-level storage will be done in electronics on, or near, the detector.

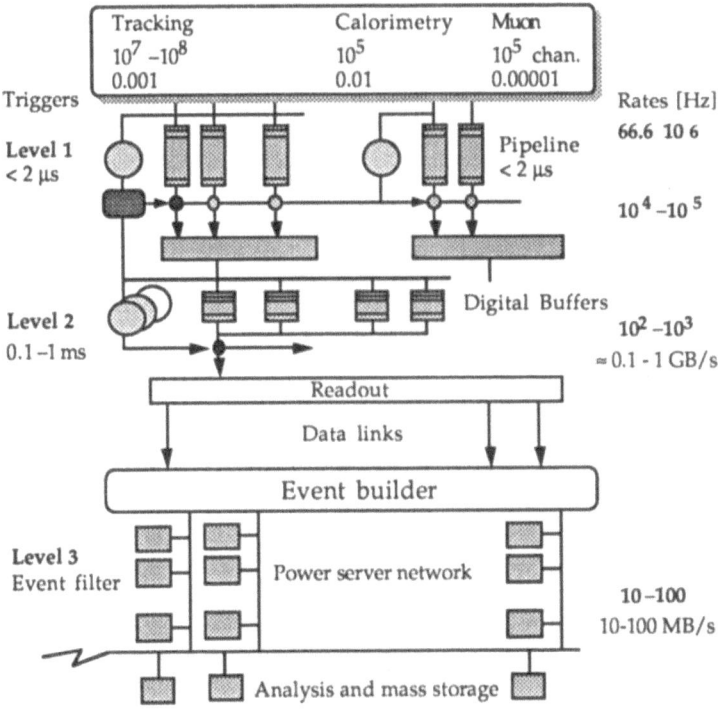

Fig. 1(b). Model of a multi-level trigger system (showing more details).

The size of a typical LHC detector is ~ 30 m long by 20 m across. For such a large detector, the time to form a first-level trigger decision is at least 400 ns, allowing only for cable delays to bring information from the whole detector to one central place and then to send the trigger decision back to the electronics mounted on the detector. From this it should be clear that the routing of cables carrying first-level trigger information will be critical in determining the decision time. This must be taken into account at a sufficiently early stage in detector design.

Given that the first-level decision time must be at least several hundred nanoseconds and since a bunch crossing occurs every 15 ns, information for many events must be stored for each channel during first-level processing; each event may contain several interactions. Storage devices known as pipelines are expected to fulfil this role. The pipeline memories, which may be analogue or digital-storage devices, accept new information every 15 ns. The data are stored until the first-level decision is available, after which they are either discarded (first-level reject) or transferred to the second-level (first-level accept). The number of storage elements in the pipeline for each detector channel is determined by the first-level decision time. For a first-level decision time of 1 μs, 67 storage elements are required. Design studies for such devices are discussed in sect. 5.1.2.

The first-level trigger may take up to ~ 2 μs to deliver its verdict to the detector electronics, this latency being made up of a combination of the response times of detectors and their associated electronics, cable delays and propagation delays through trigger electronics. However, the trigger must be able to start analyzing a new event every 15 ns. We must therefore think of a trigger system concurrently processing many events each separated in time from the next one by 15 ns. The idea of pipelined processing, commonly employed in high-performance computers, can be adopted. Possible architectures for first-level processors are discussed separately for calorimeter-based triggers and for muon triggers in sects 4.1 and 4.2 respectively.

We consider it important to combine information from different detectors in the first-level trigger. An obvious example is the need for a trigger based on electron-muon coincidences. To facilitate this, careful consideration must be given to the layout of the whole first-level trigger system. Thought must also be given to problems of synchronizing triggers from different sources. The need for a single first-level central-trigger processor must be foreseen in the overall detector design.

The trigger rate after the first-level trigger will still be very large. Rates studies, discussed in sect. 2.3, suggest that with thresholds for electron and muon triggers chosen to have good efficiency for interesting physics channels, the rate after the first-level trigger will still be in the range $10^4 - 10^5$ Hz. To achieve this we require a reduction of at least 10^3 compared to the beam crossing rate, and more than a factor of 10^4 compared to the interaction rate for luminosities in excess of $10^{34} \mathrm{cm}^{-2} \mathrm{s}^{-1}$.

When an event is accepted by the first-level trigger, the data from the whole detector are moved into a buffer memory where they are stored during the second-level trigger processing. In our model, we envisage using digital second-level buffers, so for detectors with analogue first-level pipelines an ADC must be included in the scheme as shown in fig. 2(a). The alternative of prompt digitization and a digital first-level pipeline is shown in fig. 2(b). These two alternatives can coexist, some subdetectors using analogue and others using digital pipelines. However, independently of whether or not the storage elements are analogue or digital, all first-level pipelines from all subdetectors must be synchronized. This essential consideration, which is already important in HERA detectors, means that a detector-wide view must be taken before detailed design of the readout electronics for individual subdetectors.

A particularly delicate area is the interface between the first-level pipeline and the second-level buffer. While some proposed systems involve deadtime following a first-level trigger (during which time data are

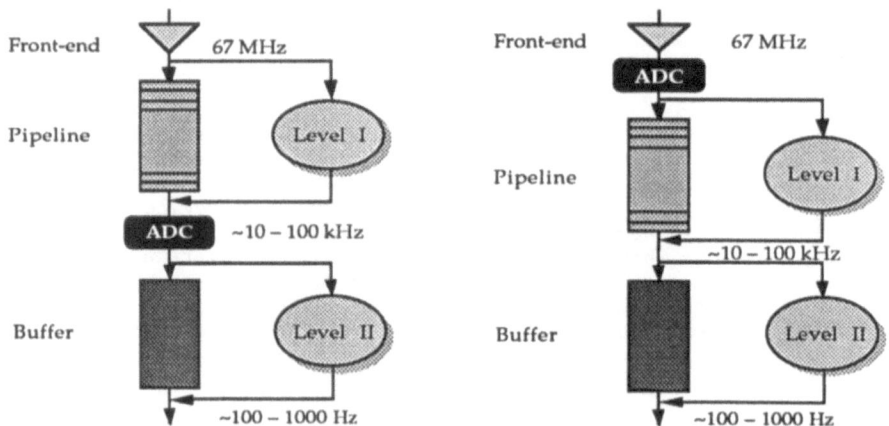

Fig. 2(a). Second-level buffer memory with analogue first-level pipeline.

Fig. 2(b). Second-level buffer memory with digital first-level pipeline.

converted and moved into a second-level buffer), other systems avoid this. Although deadtime is in general unwelcome, it may be at an acceptably low level. This is another issue which needs to be addressed detector wide.

The second-level trigger system must access information from the detector. There is therefore an intimate relationship between second-level trigger processing and second-level data storage. One extreme view is to consider the second-level buffer as an integral part of the second-level trigger and to move all the data from the detector (or subdetector) into the trigger processor system. At the other extreme, one could buffer the data on the detector, only moving those data which are needed by the second-level trigger processors.

For some subdetectors very large volumes of data are involved — perhaps 10^5 bytes per event for a calorimeter and even more for a central tracking detector, even after some data compression (or zero suppression) has been performed. Moving such large amounts of data after each first-level trigger will not be easy given trigger rates up to 10^5 Hz. Where possible, it may be preferable to buffer the data locally until after the second-level trigger has accepted an event. We note that external muon detectors are a special case because the data are expected to be very sparse.

We discuss the benefits of storing data locally during second-level processing in the context of a second-level electron trigger in sect. 4.1.2. Note that for triggers in which the processing is localized to a small part of the detector, such as electron or muon triggers, the first-level trigger can be used to flag regions of the detector containing candidates. This information can then be used by the second-level trigger, avoiding the need to access data from the whole detector.

While we think of second-level trigger systems loosely associated with specific detectors, it may be important to combine information from different detectors in the second-level trigger. This is particularly true in the case of electron identification where the second-level trigger may only be able to reduce the background rate to an acceptable level by combining information from calorimetry and another detector such as a preshower/tracker or a transition-radiation detector. We discuss this further in sect. 4.3.

Detailed rate calculations for second-level triggers are still at an early stage. However, we believe that the second-level trigger should be able to gain a factor of more than 10^2 in rejection by refining the first-level trigger decision using more precise information from the detectors, and by combining information from different detectors. Thus, the rate from the second-level trigger will be in the region 10^2-10^3 Hz. Note that the information available at this stage need not be crude: if up-to-date calibration is available, electrons, muons and jets could be measured using the full precision of the detector.

In our model shown in fig. 1, events selected by the second-level trigger are fully read out before third-level processing is performed. The second-level trigger will already have made sophisticated decisions based on information from several (maybe all) subdetectors. Third-level processing may have to do full event reconstruction and make physics analysis cuts: at a luminosity of $4 \cdot 10^{34} cm^{-2} s^{-1}$, the rate for $W \rightarrow e\nu$ and $W \rightarrow \mu\nu$ decays is predicted to be hundreds per second!

We have not considered in detail the structure of a third-level trigger system. However, we note that very powerful (and relatively inexpensive) commercial processor systems are already available, with yet more powerful products promised in the near future. Such processor systems, connected to make a processing farm, could form the basis of a third-level trigger. Each incoming event is allocated to a processor which performs the third-level processing for that event (i.e. the events are farmed out). Concurrently, other processors are working on different events. With such a system, the limiting factor is not so much computational power (one can always add more processors to the farm), but the problem of moving data from the readout electronics to the processors. We discuss data transmission and networking in sect. 6.2.

4. TRIGGER TECHNIQUES

In this section we discuss how first and second-level triggers based on calorimeter or muon-detector information could be implemented. Triggers based on other detectors, probably used to refine the electron signature at the second-trigger level, are also discussed.

4.1. Calorimeter triggers

As discussed in sect. 3, calorimeter triggers should allow one to select events on the basis of high p_T electrons and photons, jets, and also missing transverse energy. The isolation of electrons and photons can provide an additional handle for separating interesting physics from the dominant background of jets.

4.1.1. First-level calorimeter trigger

One can envisage implementing a first-level calorimeter trigger using either analogue [22] or digital electronics [23]. A trigger for high-p_T electrons (or photons) could be implemented by a system of discriminators attached to the front-end electronics of the electromagnetic calorimeter. One would have to make an analogue sum-over samplings in depth (if more than one), and might in addition perform lateral summation. Even if one wishes to retain the full granularity of the calorimeter, it is desirable to form overlapping windows so that showers which share their energy between calorimeter-cells trigger with good efficiency. Such a system was implemented for the UA2 experiment [24]. Summing over larger areas of the calorimeter has the advantage of reducing the number of channels to be handled by the first-level trigger. It should be noted that even for this "analogue" trigger, a considerable amount of digital electronics will be required — for example to count the number of clusters in the calorimeter — and that this will have to be pipelined to handle the 15 ns bunch-crossing period.

The identification of isolated electrons in the first-level trigger is potentially very useful, giving a substantial rejection against the background from jets. This is not easy to implement using analogue electronics, although resistive networks [25] could provide a solution.

A trigger for high-p_T jets might also be possible by making analogue sums over jet-sized areas of the calorimeter, and discriminating on the sum. For this, analogue information from the electromagnetic and hadronic calorimeters would have to be combined. In such a scheme it is important to control carefully the relative calibration of the different calorimeters which enter the analogue summing logic, otherwise the resolution will be degraded and the trigger threshold smeared out.

The problem of implementing a missing transverse energy trigger using analogue electronics looks very hard. Whereas electron and jet triggers can be implemented with the electronics mounted locally on, or near to, the detector, missing transverse energy requires a global sum. With such a large number of calorimeter cells involved, it is hard to see how such a system could be built.

An alternative approach to first-level calorimeter triggering is to use a digital trigger processor. This technique, which is illustrated in fig. 3, has already been used in a number of experiments. One generally makes an analogue sum of a small number of calorimeter channels, before digitizing, using a fast ADC. This sum combines the different depth samplings of the electromagnetic calorimeter and generally involves some lateral summation as well, typically over an area of $Dh \times Df = 0.2 \times 0.2$.

This reduces the number of channels to be digitized for the first-level trigger. It is desirable to make independent digitization of the hadronic calorimeter so that calibration differences between the different calorimeters can be corrected. This also makes possible the implementation, in the first-level trigger, of electron identification based on the longitudinal profile of the shower, as well as lateral isolation.

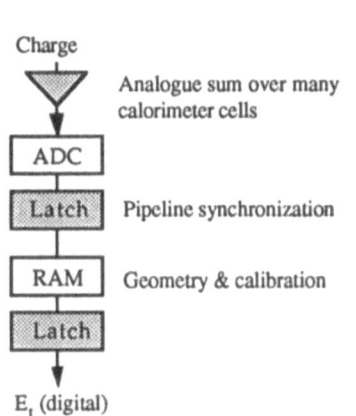

Fig. 3(a). ADC and look-up table for a pipelined digital calorimeter trigger.

Fig. 3(b). Illustration of pipelined trigger processing for a first-level calorimeter trigger.

Having digitized information from the detector, a look-up table is used to convert to energy units: probably transverse energy. Note that flash ADCs operating at 100 MHz (for 8-bit resolution) and RAMs with 15 ns access times are standard items. In fig. 3(a), latches are included after the ADC and after the RAM. These are memory registers which synchronize the pipelining of data: every 15 ns the data from a given beam crossing move downwards from one latch to the next. Thus, the digital transverse energy value is available 30 ns after the signal was presented to the ADC.

It is worth noting that a more ambitious approach [26] is to fully digitize all channels from the calorimeter before the first-level pipeline. If this were done, separate digitization for the first-level trigger would be unnecessary. This would allow the trigger to use information from the detector calibrated channel by channel.

The principle of operation of a pipelined digital processor is illustrated in fig. 3(b). This logic, which would be repeated for every electromagnetic trigger channel in the calorimeter, sums over two-by-two (overlapping) windows and compares the sum with a threshold. The latches again control the flow of data through the processor; three pipeline steps are included between the input to the first adder and the output of the logic, corresponding to 45 ns. Much more elaborate logic would be used in a practical processor.

A major advantage of the digital compared to the analogue scheme is that it is possible to make more complicated decisions at the first level. A digital processor can offer electron triggers with overlapping sliding windows, several thresholds with different multiplicity requirements, and (optional) isolation requirements. For jet triggers, the relative calibration of the electromagnetic and hadronic calorimeters can be adjusted. Most important of all, a full missing transverse-energy calculation can be implemented. In addition, a digital implementation gives easy control of the calibration, good monitoring and lots of flexibility.

Preliminary studies for a trigger processor offering all the features described above look encouraging. Simple calculations give a calculation time (excluding cable delays) of less than 400 ns. Recent technological advances make custom chip design much more accessible to us than a few years ago. Even very complicated processors could be made using only a few custom chip designs, each containing a large amount of logic. A single custom chip could in the future perform functions which in previous digital processors occupied almost a whole circuit board. This gives higher speed and better reliability; it can also be very cost effective. As in existing systems, the number of interconnections will be a serious problem. Thankfully, custom chip packaging now allows very large numbers of external connections.

Of course, there are many considerations which have to be taken into account when comparing analogue and digital solutions for first-level triggering [22]. The amount of power dissipated by digital electronics may be larger than that for an analogue system; for some calorimeters this may be an important consideration which is related to the question of where one should locate the electronics and also to problems of cabling. Also relevant is the issue of radiation hardness of any electronics which is installed inside the detector; it is worth noting that radiation-hard digital and analogue electronics are now available to us from industry as discussed in sect. 5.3. While a digital solution may require larger numbers of connections (several bits in parallel instead of a single analogue signal), there are not the same problems with noise.

Both for digital and for analogue first-level calorimeter triggers, the implementation of synchronized pipelined processing distributed over the area of the detector is not going to be easy. However, it is encouraging that both solutions look possible in principle, at least for electron triggers. Further work is required in order to master the techniques required to build a real system. It is important to develop both analogue and digital trigger designs further so that a detailed comparison can be made between them. The best choice may depend on the calorimeter technique selected; equally, the choice of calorimeter technique should be influenced by how well the calorimeter can be used in the trigger.

4.1.2. Second-level calorimeter trigger

The longer time available for the second-level trigger allows one to use programmable devices instead of hard-wired processors. There is a large variety of commercial processors from which to choose. High-performance general-purpose pro-cessors, including Reduced Instruction Set Computers (RISC) are the most flexible and easiest to program. Digital Signal Processors (DSPs) offer more computing power for certain applications [27]. Parallel computers (e.g. transputers [28], distributed array processors or associative string processors [29]) are an even more powerful alternative for problems which can be solved by a parallel algorithm. Image processors, used in the television industry, may also have a role to play; new developments for high-definition television (HDTV) may be particularly relevant here, as discussed in sect. 5.2. One should also mention that neural networks [30] could offer an alternative to traditional computing techniques.

The second-level trigger system is more than just processing power. Equally important are the data links, buses or networks which allow the processors to access the data. The overall architecture of the second-level system — buffer memories, data transmission systems and processors — must be considered as a whole. Different problems require different architectures, particularly depending on whether the problem is a local one (such as finding clusters in a calorimeter) or a global one (such as calculating the missing transverse energy).

The interaction between the first and second-level triggers is also relevant. If, for example, the first-level trigger has already identified the location of all candidate high-p_T electrons, this information can be used by the second-level trigger [31]. This is illustrated in fig. 4 which shows a system in which local processors, distributed over the detector,

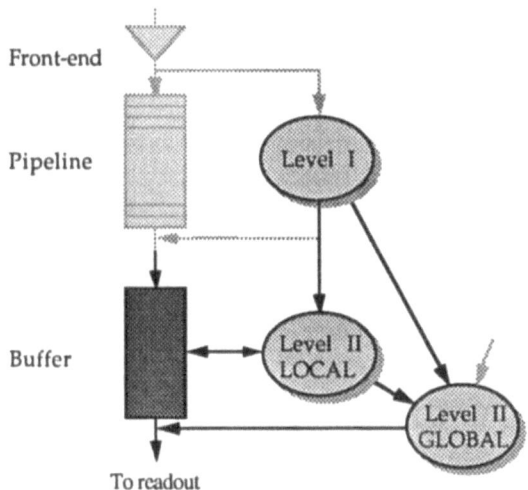

Fig. 4, Example of a second-level architecture for an electron trigger. The ovals represent trigger processors.

validate electron candidates identified by the first-level electron trigger. In this case, the need for extraordinary processing power and data-transfer rates is avoided, as described below.

Following a first-level trigger, the data are stored in the second-level buffer memory which may be located on the detector. The first-level trigger informs local processors if there is an electron candidate in the part of the calorimeter for which they are responsible. Assume 1000 local processors each responsible for 1/1000 of the detector; then if the global first-level trigger rate is 10^5 Hz, each processor only has to respond to 10^2 Hz! There must, in addition, be global processors which gather together information from the local processors before making an overall decision. These could be implemented as a processor farm, with ~ 100 processors taking turns to process events; each global processor must then respond to 1% of the first-level triggers, giving a rate per processor of 1000 Hz. Thus, the time scale for processing an event is ~ 1 ms instead of ~ 10 µs required if different events are not processed in parallel. Given the processing power of today's microprocessors, very sophisticated algorithms can be executed given an ~ 1 ms time scale.

The data rates in such a second-level architecture must also be considered. Suppose each local processor in the above example has to access an area of the detector corresponding to ~ 2000 calorimeter cells (~ 1% of the calorimeter) in order to validate the electron candidate. Then the data rate into each local processor is only ~ 0.4 MB/s. Very little data need to be sent from the local processors to the global processors — just a few words per candidate electron (E_T, η, ϕ, χ^2) — with a first-level trigger rate of 10^5 Hz and ~ 100 bytes per event, this corresponds to a total of only 10 MB/s into the farm of global processors.

Thus, by making use of information from the first-level trigger about the location of candidate electrons and by moving only those data which are required, one can envisage an architecture in which each processor has a long decision time and in which data-transfer rates are modest. The full data from the events must, of course, be stored somewhere during the second-level trigger processing: this can be done locally. The requirement that the second-level processors should selectively access the data, must be included in the design of the buffer memories; a system which includes this possibility is under study [26]. The communication between the buffer

373

memories, the local processors and the global processors is also not trivial despite the low data transfer rates.

The electron trigger architecture described above demonstrates how the different elements of a second-level trigger system — buffer memory, local and global processors, and data communications — need to be considered together. It is useful to model such architectures using simulation programs such as SIMSCRIPT or VERILOG [32]. We believe that detailed simulation is essential in order to evaluate and compare a variety of second-level architectures. Simulation allows the performance of a system to be studied as a function of parameters such as buffer depth or algorithm execution time. It also allows one to study the response under more realistic conditions by simulating error conditions: in some existing experiments error recovery is a dominant source of dead time.

It must be made clear that the architecture described above for an electron trigger is only one possibility. Alternative schemes using massively parallel processors and image processors have also been studied [29]. These studies have concentrated on the comparison of different computer architectures for cluster analysis.

The problem of making a second-level missing transverse energy trigger is very different from that of finding or validating electron candidates. One needs to access data from the whole detector, but the algorithm is very simple and well defined: essentially a weighted sum over calorimeter cells. This could already be done by a digital first-level trigger.

It must be remembered that the second-level trigger must be powerful enough to gain a factor of 10^2 in rate compared to the first-level trigger. Simulation studies for second-level algorithms should take account of the fact that the first-level trigger may already have used many of the easy signatures. Ideally, the first and second-level simulation should be integrated.

4.2. Muon triggers

The physics requirements described in sect. 2.3 demand triggers on high-p_T muons. Such triggers can be implemented using information from external muon detectors which are shielded from the interaction region by many interaction lengths of material [33]. These detectors provide several-position measurements along each track, with good precision in at least one coordinate. High-p_T muons are identified by their ability to penetrate the hadron absorber and by their small angular deflection by bending in a strong magnetic field and by multiple scattering. The detector must provide fast signals for first-level triggering; this requirement must be considered in the detector design. At the second level it will be necessary to measure accurately the momentum of very high-p_T muons, possibly requiring a detailed analysis of the track trajectory in the magnetic field.

4.2.1. First-level muon trigger

We envisage a first-level trigger which uses as input a pattern of hit elements from the muon detector. The logic compares the pattern on hits in the muon chambers with patterns which are valid for high-p_T muons that originate from the interaction region. Several techniques are possible [14] such as look-up tables stored in RAMs or programmable logic; both of these are very fast (~ 15 ns). The result is a flag indicating the presence of a muon candidate: the first-level trigger provides the position of candidates but no information on their transverse momenta.

Thresholds can be controlled by changing the list of valid hit patterns. Essentially, one defines roads from the interaction region through the muon chambers in which one requires hits. Narrow roads correspond to high-p_T muons, wider roads to lower-p_T muons. While the implementation of such a trigger system appears to be relatively straightforward, the details depend on the geometry chosen for the muon detector.

4.2.2. Second-level muon trigger

As for calorimeter triggers, there is a large choice of commercial processors which might form the basis of a second-level muon trigger system. Alternative solutions include neural networks, associative memories and data-driven processors. The possibility of using information from the first-level trigger about the location of muon candidates should be considered. However, the external muon chambers are expected to have very low occupancy, so there is much less to be gained from selective movement of data than in the case of calorimetry. Various architectures are possible for second-level muon triggering such as processor farms with each processor handling a different event, or architectures using a massively parallel computer system. The latter possibility has been studied [34]; using a system of associative string processors execution times of ~ 10 μs should be possible without making any reference to information from the first-level trigger.

4.3. Triggers based on other detectors

Triggers based on the calorimeters and external muon chambers alone should give sufficient rejection against background at the first level. A number of other detectors, particularly those associated with electron identification, can help to provide additional rejection at the second level. We illustrate this with two examples: a preshower/tracker detector [35] and a transition-radiation detector [36].

4.3.1. Preshower/tracker

A preshower/tracker detector described in ref. [35] can be used at the trigger level to select electrons. The detector consists of several layers of silicon tracking, then a converter followed by several more layers of silicon tracking. Electrons are signalled by a single ionizing track before the converter which starts to shower before the second-tracking stage. Charged hadrons are unlikely to interact in the converter, while electron pairs from converted photons (in a non-magnetic detector) will give a twice-ionizing signal in the tracking before the converter. Good rejection against background is obtained by exploiting the excellent granularity of the detector when forming a coincidence between the silicon layers before and after the converter, and with a high-energy cluster in the calorimeter.

It is apparent that the preshower/tracker trigger only makes sense when combined with information from the calorimeter. This would probably be done in the second rather than the first-level trigger, although the fast response of the silicon detectors is compatible with the first-level time scale. Combining information from different detectors will require a common effort on trigger design. It is worth noting that there are several similarities between architectures presented for calorimeter [31] and preshower/tracker [35] second-level triggers.

4.3.2. Transition radiation detector (TRD)

A TRD has the capability to separate electron and hadron tracks on the basis of pulse height. The detector and trigger which were described at Aachen workshop [36] produce transition radiation X-rays in a foam radiator and detects them using straw tubes. Digitization is performed using only three thresholds, yielding 2-bit values for each channel. Roughly speaking, the first threshold is set to be sensitive to a single-pion track, the second one to be sensitive to two overlapping pion tracks, and the third one to be sensitive to an electron track. Good separation of electrons is obtained by comparing the number of straws above the third threshold with the number between the first two thresholds. A possible second-level trigger architecture was presented which makes use of a massively parallel computer system to find the tracks.

5. FRONT-END ELECTRONICS

Undoubtedly, one of the most difficult technological challenges for LHC experi-mentation is in the area of front-end electronics. Although good progress is being made in many areas, the state-of-the-art of electronics is still inadequate and it is widely recognized that a big R&D effort is necessary in order to meet the LHC time scale. In particular, industrial developments must be watched closely in order to take full advantage of the fast progress in this domain.

In this section, we briefly mention a few examples of the features required of some basic components (sect. 5.1) and we discuss how technological developments might influence our preparations for the LHC (sect. 5.2). A few comments on micro-electronics are given in sect. 5.4.

5.1. Examples of components

A number of basic components will be essential elements of any trigger and data acquisition system for the LHC. Analogue-to-Digital Converters (ADCs), analogue and digital memories for pipelined readout structures, first-level trigger processors, data-flow processor systems and Digital Signal Processors (DSPs) will all have to be developed. The demands of the LHC include:

- very high speed with clocks at multiples of 67 MHz for front-end applications,
- low-power dissipation, and compact packaging and cabling where required,
- excellent reliability for detector electronics in closed areas,
- radiation hardness for electronics mounted on the detectors (up to hundreds of MRad in some cases).

Prototypes exist which satisfy one or the other of the above requirements, but no one satisfies all of them.

5.1.1. Analogue-to-digital conversion

Critical requirements come from calorimetry, for which a dynamic range equivalent to 15-16 bits and a precision equivalent to 9-10 bits are needed. Low-power consumption and radiation hardness will be critical issues for inner detectors, including an electromagnetic calorimeter, if digitization is done locally. Speed is critical for every detector in which digitization is done before the first-level trigger.

Table 6. Some current developments of A/D converters [40]

Author	Type	Rate [MHz]	No. of bits	Tech-nology	Power [Watt]
Tsumoto, NTT	Flash	2000	6	26 GBip	2
Peetz, HP	Flash	250	9	7 GBip	12
Yoshi, Sony	Flash	350	8	10 GBip	1.5
Fukishima, Sony	2 steps	40	8	1.4 CMOS	0.1
Akazawa, NTT	Flash	400	8	18 GBip	2.7
Tsukada, Hitachi	Flash	25	8	2 CMOS	0.3
Matsuura, Hitachi	2 steps	20	8	2 CMOS	0.2
Zojer, Siemens	2 steps	74	10	7 GBip	2

Many techniques of conversion are available today. Alternatives to the standard flash ADC are the multistage flash ADC [37], ADCs based on the S-D technique [38] and pipelined ADCs [39]. These have some advantages and disadvantages, either in reality or in the promises of their development trends. Table 6 shows a comparative analysis of ADCs presently under development [40]. In any case, progress in this field is very fast based on new technology. The solution adopted will depend on the detector type and the front-end electronics (e.g. at which stage in the trigger hierarchy the digitization takes place).

5.1.2. Pipeline memories

As discussed in sect. 3, the first-level trigger decision time will be much longer than the bunch crossing period of 15 ns. The information from all the detector channels must therefore be buffered until the first-level decision is available, probably for ~ 2 ms.

Given the high-data flow, the simplest organisation of the buffering at this level is by means of pipelines with steps of (submultiples of) 15 ns and a fixed length equivalent to the decision time of the first-level trigger.

Several architectures could be envisaged depending on the pulse characteristics and on the occupancy of the detector channel. Figure 5 shows schemes of three different pipelines with somewhat complementary fields of application. Analogue pipelines may be preferred where high-packaging and low-power dissipation are critical issues, but they might be restricted to limited dynamic range and their timing and calibration are likely to be complex. Digital pipelines are simpler to control and offer large dynamic range, but their use might be limited to areas where high-power dissipation is acceptable, allowing fast digitization. A third type consists of a mixture of analogue and digital electronics, namely a one-stage analogue memory to store the pulse, coupled to a one bit shift register to identify the bunch crossing to which the content of the analogue memory belongs. The latter scheme is applicable to detectors for which the occupancy is such that the probability of having two hits in the same cell during the trigger decision time is negligible. A prototype of an analogue pipeline [41] has already been operated at 66 MHz and details were presented at Aachen workshop [42]. A scheme for a chip for calorimetry readout, containing fast digitizers and digital pipelines was also presented at this workshop [26].

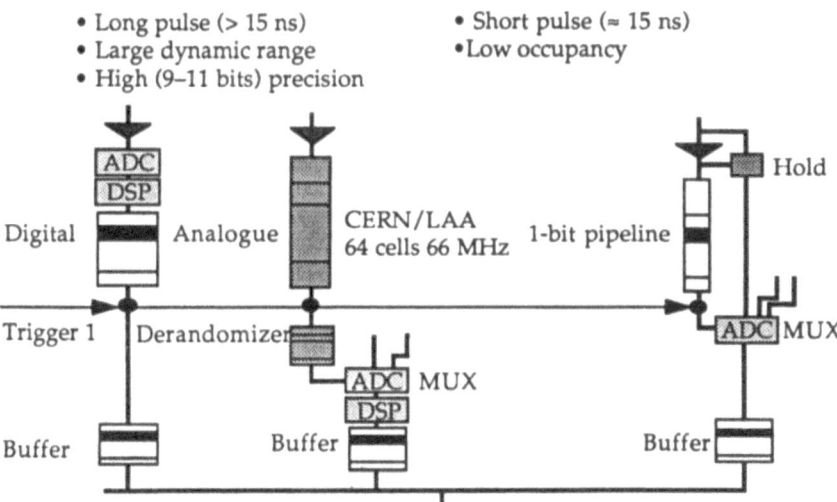

Fig. 5. Pipeline architectures.

5.1.3. Digital signal processing

The signals generated by a LHC calorimeter (or other detector) may extend over more than one bunch crossing. In such cases, a digital channel with programmable filter capability can be used to extract the physical information and to associate events to a given bunch crossing. For detectors where particle identification can be done by signal shape analysis in single channels, this can be a part of the trigger process.

Ignoring considerations of power consumption, density of packaging and radiation hardness, it is very attractive to go digital at an early stage, complementing the analogue shaping with a pipelined digital-signal processor; this is included in the design of ref. [26].

If prompt digitization is not possible, digital-signal processing can still be done after the first-level trigger, probably forming part of the second-level trigger, and fulfilling a data compression role. A review of the state-of-the-art in DSPs was given at Aachen workshop [27].

5.2. Technology trends

The technology of CMOS and BiCMOS is quickly moving towards its physical limit. Compared with what was available ten years ago there is an increase of at least two orders of magnitude in memory density (fig. 6(a)) and in microprocessor performance (fig. 6(b)).

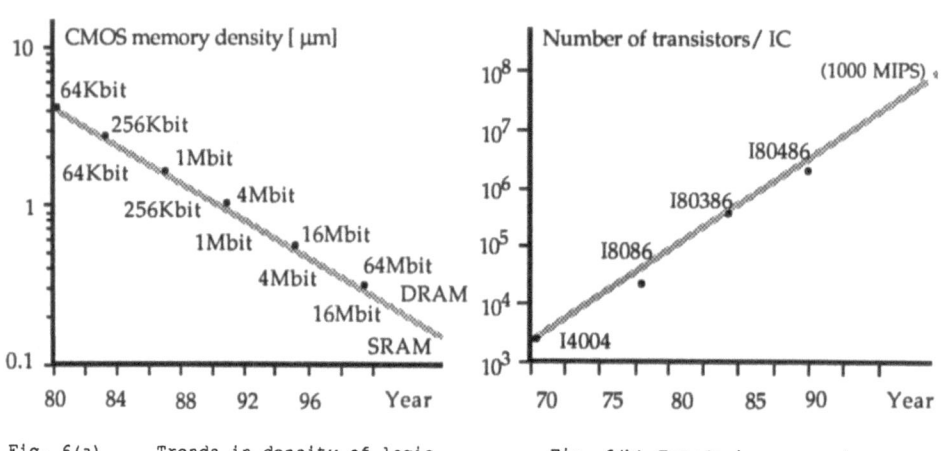

Fig. 6(a). Trends in density of logic
in micro-electronics.

Fig. 6(b).Trends in processing
power per chip.

Provided such trends continue, the projected compactness seems to match the requirements for the readout of highly granular detectors, and the projected speed matches the needs of fast front-end electronics and trigger systems. These projections might even be overtaken by technology changes, such as the use of GaAs components or the introduction of new processor architectures (e.g. massive parallelism). Obviously, in order to take full advantage of the technology the high-energy physics community must develop the capability of exploiting this progress.

5.3. High-definition television

One of the main driving forces for the rapid innovation and the fast appearance on the market of technological developments is the programme for the production of a new television standard. In the last 20 years the needs related to telecommunications and, particularly, to television have strongly contributed to the large scale industrial exploitation and development of standard technologies (CMOS, BiCMOS). The popular video market provides the justification for the enormous financial investment needed. Together with people in other fields, researchers in fundamental physics have made extensive use of this "spin-off". Flash ADCs, analogue memories, personal computers, helical-scan recording, data-compression techniques and image-processing systems are all examples of television industry spin-off from which we can profit.

In the last few years the extensive intellectual and technological resources of the European studio and consumer television industry, broadcasters, PTTs and universities have been brought together in the EUREKA project, EU95, to meet a new challenge: the development of a High Definition Television (HDTV) [43]. Designed to progress in an evolutionary way, the HDTV project represents a tremendous effort of R&D in the field of the standard technology.

The HDTV specifications are in a similar range of the ones needed for LHC experiments. Transmission encoding of 144 MHz will require fast ADCs and high speed, high-density mass storage devices; high-performance imaging will need powerful image processors, digital filters, etc.; data communications of 1.3 Gbits/s are also planned. Independently of the signal transmission technique, the signal treatment at reception is fully performed digitally. Therefore, a TV channel might be seen as an LHC detector channel both from the performance and from the functionality points of view: fast digitization; enormous throughputs; inherent pipelining (one pixel after another instead of one bunch crossing after the other).

In addition, the need to adapt multi-standard formats in a digital framework imposes the development of powerful programmable video-signal processors [44]. Their announced performance figures are very promising and the modular and programmable structure makes them ideal candidates for application in the first levels of LHC triggers. In conclusion, given the time scale defined for the EU95 project, analogue bandwidths of ~ 100 MHz and digital throughputs of ~ 1 Gbit/s can be expected in the middle of the 1990's as a spin-off from the European HDTV development. This does not mean that HDTV components could be used directly in LHC experiments. Other requirements do not match our needs: power consumption, density of packaging and radiation hardness are additional problems for the LHC. However, it is important to be aware that an extensive R&D program is under way in Europe addressing problems similar to ours. The use of HDTV prototypes to exercise architectures in realistic conditions and the combination of our expertise and competence with their methods and means would be beneficial for the solution of LHC problems.

5.4. Micro-electronics

An evaluation performed in the context of the Superconducting Super Collider (SSC) has concluded that the investment necessary for the electronics in a typical experiment would amount to as much as 40% of the total cost. A good fraction of the electronics will need to be micro-electronics: in the front end where it will have to be mounted directly on detectors, in the trigger and data acquisition systems in order to meet the requirements of complexity and power con-sumption, and at the interconnection level to provide high density, reliability and low-unit cost. Micro-electronics is, therefore, a crucial issue for LHC experiments. Digital micro-electronics developments will be increasingly more re-usable and easily adapted to the changes of technology. In the near future increasingly big libraries of macrocells will be available, development systems will become more standard, and they should be used more easily by the designer. In order to avoid duplication of effort, a coordination of European activities is desirable. European initiatives, such as EUROCHIP [45], will help this by providing universities and laboratories with the proper tools for the training of a new generation of engineers. By taking advantage of such programmes, high-energy physics laboratories and institutions should coordinate their efforts in order to pursue the development of this cost-effective, advanced technology necessary for LHC experiments. Discussions in this direction have been initiated in this working group and will be continued in the immediate future with the aim of establishing a suitable European micro-electronics environment.

A very important issue for electronics developments is the need of radiation hardness, at least in some areas of the detectors. Although no detailed investigation has been performed in our working group we do not underestimate the importance of this problem for the LHC. A few contacts with representatives of some major industries [46] have shown us encouraging signs of industrial interest. Processes for the production of radiation-hard components seem to be available today. The task of the high-energy physics community is to develop test systems to ascertain their usefulness.

6. DATA FILTERING AND ACQUISITION

The last stage of the architecture introduced in sect. 3 has to deal with the events accepted by the second-level trigger. The data from individual detector subsystems have to be merged into full events at an estimated rate of 100 to 1000 events/s and, after an extra level of data reduction, have to be stored on suitable recording media for final off-line data analysis. In this section we present the results of studies of event-building techniques (sect. 6.1) and of a detailed compilation of all the buses which are used today in high-energy physics experiments or are likely to become available in a near future (sect. 6.2). Section 6.3 deals with mass storage of the finally accepted data.

6.1. Event building

Event building is the process of connecting the data sources (i.e. the front-end electronics and local memories) to the data destinations (i.e. higher-trigger level processor farms and data-storage media). Various approaches are considered at present. Among the more traditional ones are shared bus architectures and tree structures. Less traditional alternatives have a greater degree of parallelism based on dual-port memory inter-connection networks, processor networks, or switching networks (e.g. cross-bar or barrel switches). A review of these techniques was presented at Aachen workshop [47], including examples of existing systems and giving relevant references.

The more traditional techniques mentioned above imply the movement of the totality of the data accepted by the first and second-level trigger from local memories to full-event buffers. Consequently, the data communication system requires very high performance. The cache memory techniques available in many recent microprocessor architectures and extended to data-link protocols, such as SCI (Scalable Coherent Interconnect) (sect. 6.2), may allow a logically simpler interface between the front and the back ends. Often the highest-trigger level needs to process only a small fraction of the data. In a virtual and cache memory environment the trigger processing unit can see the full-event data as a logical structure in its local virtual memory. An SCI link between the processing unit and the front-end local buffers would provide, in a transparent way, the physical access to the data; only those data needed by the selection algorithm would actually be moved. Consequently, the readout of the full-event data would be done only for the finally select-ed events, giving a big saving in required bandwidth (e.g. 10 to 100 MB/s on several parallel paths).

A cache-memory scheme of this kind would be advantageous only if the fraction of data required by the final-trigger algorithms did not exceed a few percent of the total. In any case, the choice of the optimal event building technique is also closely coupled to the overall data acquisition architecture, and will depend on the type of buses and data links used, the data control paths, and the location and amount of processing power. It is, therefore, too early to concentrate on any event building architecture. Instead, as in many other areas, the activity should concentrate on simulating different ideas and initiating demonstrator projects in order to be able to make a choice suitable for a given architecture, and to build prototypes and study integration into the full-data acquisition architecture.

6.2. Buses and links

The basic advantage of using standard buses is the homogeneous environment with guaranteed, well-defined properties. Industry provides standard components for use in complex systems. Standards provide a safe environment for the chip and board-level designer as well as for the system architect. Standard buses fulfil the following requirements:

- uniform transmission properties of data paths;
- no cross talk, error detection and recovery;
- unique addressing schemes for single and multiple access;
- minimal, well-defined read/write protocols and access protocols (arbitration);
- live insertion and removal of modules;
- modular mechanical framework for replacement, renewal and maintenance;
- a reliable power supply and cooling environment.

A fundamental question for LHC is whether buses are adequate for the requirements of a data-acquisition system and whether they are needed at all. In this section we summarize a detailed study, prepared in the context of Aachen workshop [48], of existing standard buses, such as Fastbus, VMEbus, VXI and VME64, and other buses under development in industry, such as Futurebus+ and SCI.

Good performance is a question of implementation. Even conventional buses can be pushed to 40-100 MB/s. New, powerful bus generations are evolving, driven by interest from the high-end computing industry, to provide performance and interconnectivity far beyond the limits of conventional buses. For example:

- VME64 practically doubles the performance of VMEbus;
- Futurebus+ may achieve 3.2 GB/s on 256-bit backplanes;
- SCI, a point-to-point cable bus, will have a performance of 1 GB/s/node.

Note that Fastbus could also increase significantly its real performance if necessary investments were made.

A loss of performance can be due to several factors:
- embedded "de-luxe" software,
- lack of parallelism for the sake of economy,
- failure to make use of data driven concepts where applicable,
- curing of "flakey" implementations by slow-speed operation,
- general-purpose designs which have inferior performance compared to purpose-built ones,
- module design with functionality specifications instead of performance.

Although careful implementations taking into account the above aspects might make use of existing buses (e.g. Fastbus and VMEbus) in areas where independent units with local buses do not require high bandwidths, it is clear that for the main data streams new bus standards will have to be used. The global second-level trigger system, the third-level processing farm and the data logger all require bandwidths in the order of 10^8-10^9 bytes/s. New bus standards give bandwidths which extend well into the 1 GB/s range. Futurebus+ could be used in the area of bandwidth beyond 100 MB/s. The event-builder stage and the interconnection between local and global second-level triggers may use this choice if a link-type interconnection between Futurebus+ backplanes becomes available. This might possibly be the SCI; both standards are being developed in parallel and have common parts of specification concerning crates and control registers. Point-to-point buses like SCI provide a novel way of interconnecting processor and memory nodes at very high speed, largely independent of the number of nodes. Any node may send or receive data or commands to or from other nodes in the network. Topologies of interconnected ringlets can easily be changed and optimized by changing the cabling. Interconnections between high-end (commercial) processors with back-plane bus units are expected to become available in addition to a node chip for memory interface design. The availability of new features like caching, split transaction and virtual memory access will allow for better performance, more throughput and new data-acquisition concepts (for example, event building in sect. 6.1). Further studies in connection with test set-ups and system simulation are, however, required. The SCI seems to be the most promising candidate for the global second and third-level triggers and for event building. The link-type nature of SCI presents a considerable advantage since all three areas can be covered by a three-layer topology of 50 to 100 SCI ringlets, interconnecting (commercial) high-end processor farms with uniform backplane-bus units. Features like caching, virtual address and very high bandwidth are available to simplify architectures both in hardware and software.

Buses will also be suitable backbones for slow control systems. A VME-based system with VXIbus for interconnection seems to be a good choice for this role.

High-speed links will certainly complement buses in data acquisition systems for massive data transfers. Gigabit optical links are starting to appear on the market. Major computer manufacturers including IBM, DEC and HP all have working prototypes. Components to build such links are now also available from companies like Gigabit Logic, Vitesse and BT&D.

Examples of applications of optical links in high-energy physics are:

- A 160 Mbits/s VMEbus inter-crate module (VMExi) using the ADM TAXI chip set.
- A 16.6 Mbits/s system used on UA1, based on the PArallel TRansmission Optical Link (PATROL).
- A prototype of a 100 Mbits/s version of PATROL.
- A project for a 10 km, 1 Gbit/s optical link between the L3 on-line and off-line computers, which should be completed within two years. A detailed compilation of several projects, currently under way, has been prepared in the framework of Aachen workshop and can be found in ref. [49].

Finally, one should not exclude the possibility of the development of specialized systems designed to achieve maximum speed. An example of a readout controller capable of data transfers of 80 MB/s has been presented at Aachen workshop [50].

6.3. Mass storage

According to the evaluation described in sect. 2.4, data bandwidth for mass storage at LHC could be of the order of 100 MB/s (i.e. the equivalent of one 3480 cartridge every 2 s). No device commercially available today is capable of such rates, but developments are under way, pushed by the commercial broadcast industry.

Although the best performances are expected from optical devices, mass storage for LHC will probably be based on the products derived from studio high-definition Video Tape Recording (VTR) systems which are being developed with the helical-scan tech-nique.

Helical scan is the most cost-effective technique today, providing higher recording capacity in a smaller physical package than any other computer storage technique at unbeatable prices. Sony has recently announced a Digital Instrumentation Recorder (DIR 1000) with a recording/playing rate of 1 to 30 MB/s based on a 19 mm video-cassette tape with a total capacity of up to 96 GB. Moreover, one should not forget that the HDTV program requires tape recording at 200 MB/s.

7. SOFTWARE

Although it is too early to start the development of data acquisition systems for LHC experiments, one cannot over-emphasize the important role of a well-designed on-line system. The control and the monitoring of the experiment, the management of thousands of CPUs distributed at the various levels of the trigger and data-acquisition system and the organisation of the multiplicity of tasks in the several-thousand processor environment will require a software system of such a complexity that modern software engineering must be used to handle it.

The high-energy physics community is not used to modern programming tech-niques, but it should now become acquainted with them in order to develop the necessary expertise in time. At this stage, software activities should proceed in two directions.

(a) Develop tools for program design and code generation with the use of new techniques such as:
- specification and description languages (graphical and textual description),
- object-oriented programming,
- platform-independent operating systems (UNIX),
- CASE techniques.
(b) Exercise architecture modelling by means of simulation techniques, animation, timing and statistics analysis. Modelling tools such as Verilog, normally used for electronics chip layouts, have been or are being used for full-architecture simulations, but languages like SIMSCRIPT and MODSIM are considered more suited for full-trigger and data-acquisition system modelling [32].

8. CONCLUSIONS

Our investigations show that although building front-end electronics, triggers and data-acquisition systems for the LHC will be very difficult, but it is not impossible provided the necessary R&D commences immediately with sufficient investment of manpower and funding. The areas which we feel are most in need of further study are those closest to the detectors where high speed is needed, in some cases in conjunction with low-power dissipation and radiation hardness. In particular, front-end analogue electronics, pipeline memories, fast ADCs, and first-level trigger processors need to be developed. In some of these areas, we may be able to benefit from developments in industry which must be followed closely.

Further away from the detector (beyond the first-level trigger) we feel that more of the components of the trigger and data-acquisition system can come from industry. High-performance processors, data links, networks and buses are already available and the technology is moving fast. Here we see a need to follow developments from industry by testing and evaluating the latest products. The challenge will be to combine the available technology to build the trigger and data-acquisition system. The complexity of such systems will require the use of modern software methods and software tools for design and modelling.

We would also like to emphasize the need to consider requirements of the trigger and data-acquisition system in the overall detector design. Clearly, all detectors used in the first-level trigger must provide fast signals. In addition, physically large detectors will necessarily have a long first-level trigger decision time due to cable delays; it is important to make provision for short-cable runs from the outset. The choice of detectors may also affect the location of the front-end and first-level trigger electronics; problems of cabling must be weighed against problems of power dissipation and radiation hardness. Synchronization is a major issue that must not be neglected given the short-bunch crossing period. For every detector channel there will be pipeline and buffer memories, distributed over the detector, which must be synchronized to much better than 15 ns (the synchronization of movement of data from first-level pipelines to second-level buffers is particularly delicate). The message here is that the first-level readout electronics of all detectors and all the first-level trigger systems must be designed in a coherent way.

Considering the large event size and high first-level trigger rate, one is faced with very large data transfer rates. It is not desirable to move these data over large distances before second-level trigger processing unless there are constraints from the overall detector design. A related issue is that each trigger level can benefit from calculations performed at lower levels. For example, the first-level electron and muon trigger processors could be used to flag areas of the detector containing candidates to be validated by the second level. There is a general need for a coherent design of the trigger system both between different detectors and between trigger levels.

ACKNOWLEDGMENTS

We would like to thank all members of the working group on signal processing, triggering and data acquisition who helped to make the Aachen workshop a success.

REFERENCES

[1] Proc. ECFA Study Week on Instrumentation Technology for High-luminosity Hadron Colliders, eds E. Fernandez and G. Jarlskog, Yellow Report CERN 89-10 (1989); Proc. Workshop on Physics at Future Accelerators, ed. J. Mulvey, Yellow Report CERN 87-07 (1987); J.R. Hansen, Triggering with high luminosity in the LHC, in the Feasibility of Experiments at High Luminosity at the Large Hadron Collider, ed. J. Mulvey, Yellow Report CERN 88-02 (1988) 41.

[2] Z. Kunszt et al., Large cross section processes, Proc. Workshop on Physics at Future Accelerators, ed. J. Mulvey, Yellow Report CERN 87-07 (1987) 123.

[3] G. Brianti, The LHC project, Plenary talk given at Aachen workshop (1990).

[4] F. Abe et al., Phys. Rev. Lett. 61 (1988) 1819.

[5] M. Della Negra et al., Study of muon triggers and momentum reconstruction in a strong magnetic field for a muon detector at LHC, Proposal CERN/DRDC 90-36, P7 (1990); G. Carboni, private communication.

[6] S. Hellman, L. Mapelli and G. Polesello, Trigger rates at the LHC, presented by S. Hellman at the ECFA Large Hadron Collider Workshop (1990), and references therein.

[7] G. Unal, Communication to the study group on experimentation at the LHC.

[8] D. Froidevaux, Communication to the study group on experimentation at the LHC.

[9] C. Albajar et al., Experimental aspects on SUSY searches at the LHC, Aachen workshop (1990).

[10] D.J. Munday et al., A proposal to study a tracking/preshower detector for the LHC, Proposal CERN/DRDC 90-27, P3 (1990).

[11] V.A. Polychronakos et al., Integrated high-rate transition radiation detector and tracking chamber for the LHC, Proposal CERN/DRDC 90-38, P8 (1990).

[12] D. Acosta et al., Scintillating fibre calorimetry for the LHC, Proposal CERN/DRDC 90-23, P1 (1990).

[13] B. Aubert et al., Liquid argon calorimetry with LHC-performance specifications, Proposal CERN/DRDC 90-31, P5 (1990).

[14] F. Ceradini et al., A fast tracking level-1 muon trigger for high-luminosity colliders using resistive plate chambers, presented by E. Petrolo at Aachen workshop (1990).

[15] N. Bains et al., Nucl. Instr. and Meth. A292 (1990) 401.

[16] G. Blaylock et al., The UA2 data acquisition system, Proc. of the Int. Conf. on the Impact of Digital Micro-electronics and Microprocessors on Particle Physics, eds M. Budich, E. Castelli and A. Colavita, Trieste, Italy (1988) 247.

[17] D. Amidei et al., Nucl. Instr. and Meth. A269 (1988) 51.

[18] A.P. White, Status of the D^0 detector at Fermilab, Proc. of the 8th Topical Workshop on $p\bar{p}$ Collider Physics, 1-5 September 1989, Castiglione della Pescaia, Italy, eds G. Bellettini and A. Scribano, World Scientific (1990).

[19] R.J. Ellison and U. Straumann, The H1 trigger, H1 Internal Note H1-05 90-137 (unpublished).

[20] The ZEUS detector, Status Report 1989, DESY (1989).

[21] Proc. of the Workshop on Triggering and Data Acquisition for Experiments at the Supercollider, ed. R. Donaldson, SSC-SR-1039 (1989); A.J. Lankford, E. Barsotti and I. Gaines, Nucl. Instr. and Meth. A289 (1990) 597.

[22] G. Grayer, The role of analogue circuitry in LHC/SSC triggering, Aachen workshop (1990).

[23] N. Ellis and J. Garvey, A digital first-level calorimeter trigger, presented by N. Ellis, Aachen workshop (1990).

[24] V. Hungerbühler, Yellow Report CERN 81-07 (1981) 46;

A. Beer et al., The central calorimeter of the UA2 experiment at the CERN p̄p collider, Nucl. Instr. and Meth. 224 (1984) 360.

[25] R. Bonino, A cluster finding analogue network, Aachen workshop (1990).

[26] G. Goggi and B. Lofstedt, Digital front-end for calorimeters at LHC, presented by L. Mapelli at Aachen workshop (1990).

[27] D. Crosetto, DSP review and applications, Aachen workshop (1990).

[28] J.C. Vermeulen, Data acquisition and triggering with transputers, Aachen workshop (1990).

[29] W. Krischer, Image processing in LHC detectors, Aachen workshop (1990).

[30] S. Amendolia and B. Denby, Ongoing approaches to the trigger problem using neural networks, presented by S. Amendolia at Aachen workshop (1990).

[31] D. Crosetto, N. Ellis, G. Mornacchi and J. Strong, A level-II architecture for calorimeter triggers, presented by J. Strong at Aachen workshop (1990).

[32] J.P. Porte et al., DAQ simulation for LHC, Aachen workshop (1990).

[33] K. Eggert, Muon detection and trigger, Plenary talk, Aachen workshop (1990).

[34] G. Vesztergombi, Second-level muon trigger concept for the Large Hadron Collider, Aachen workshop (1990).

[35] A. Poppleton, Pad detector trigger, Aachen workshop (1990).

[36] R.K. Bock and J. Pfennig, The TRD second-level trigger, presented by J. Pfennig at Aachen workshop (1990).

[37] S. Hosotani et al., An 8-bit 20 MS/s CMOS A/D converter with 50 mW power consumption, IEEE Journal of Solid-state Circuits, Vol. 25, No 1 (1990).

[38] R. Koch et al., A 12-bit sigma-delta analog-to-digital converter with a 15 MHz clock rate, IEEE Journal of Solid-state Circuits, Vol. SC-21, No 6 (1986).

[39] F. Anghinolfi et al., One megahertz sampling rate low-power analog-to-digital converter for data processing in particle detectors, submitted to Nucl. Phys. B.

[40] P. Gray, Recent developments in high-speed CMOS A/D conversion, CERN Seminar (28 June 1990).

[41] F. Anghinolfi et al., HARP: Hierarchical analog readout processor with analog pipelining in CMOS, to appear as CERN/ECP preprint.

[42] P. Jarron, Trends in analog front-end electronics, Aachen workshop (1990).

[43] EUREKA HDTV EU95, Bosch, Philips, Thomson et al. (19..).

[44] U. Schmidt et al., ITT Intermetall, Data-driven array processor for video signal processing, to be published in IEEE Transactions on Consumer Electronics 8/90 (1990).

[45] P. Sharp, Micro-electronics, Aachen workshop (1990).

[46] C. Terrier, TMS, Radhard, ASIC methodology and high-speed FADC, Aachen workshop (1990);

See also contributions to the parallel session on radiation studies at Aachen workshop.

[47] S. Quinton, Event-building review, Aachen workshop (1990).

[48] H. Muller et al., Buses and standards for LHC, Aachen workshop (1990);

F. Renardy et al., SCI at LHC, Aachen workshop (1990);

D. Linnhofer et al., Use of FASTBUS at LHC, Aachen workshop (1990);

G. Heyes et al., The "V"-bus family, Futurebus+ and SCI, Aachen workshop (1990).

[49] R. McLaren et al., Summary of the ECFA working subgroup on fast data links, Aachen workshop (1990).

[50] R. Belusevic and G. Nixon, A pedestrian approach to fast DAQ or how to outbus the buses, presented by R. Belusevic at Aachen workshop (1990).

PARTICIPANTS

J. ALBERTY	ECP Division, CERN, Geneva, CH
F. ANSELMO	ECP Division, CERN, Geneva, CH
R. ARNOLD	C.R.N., Strasbourg, France
F. BERGSMA	PPE Division, CERN, Geneva, CH
G. BRUNI	I.N.F.N.-Bologna, Italy
V. BUZULOIU	PPE Division, CERN, Geneva, CH
G. CHARPAK	PPE Division, CERN, Geneva, CH
S. CHATURVEDI	EP Division, CERN,Geneva, CH
E. CHESI	ECP Division, CERN, Geneva, CH
M. CHIARINI	PPE Division, CERN, Geneva, CH
L. CIFARELLI	University of Naples, Napoli, Italy and
	PPE Division , CERN, Geneva, CH
S. CITTOLIN	ECP Division, CERN, Geneva,CH
A. CONTIN	PPE Division, CERN, Geneva, CH
C. D'AMBROSIO	PPE Division, CERN, Geneva, CH
S. DE PASQUALE	PPE Division, CERN, Geneva, CH
R. DE SALVO	PPE Division, CERN, Geneva, CH
T. EKELOF	PPE Division , CERN, Geneva, CH and
	Uppsala University, Uppsala, Sweden
Q. FAN	EP Division, CERN, Geneva, CH
Y. GIOMATARIS	EP Division, CERN, Geneva, CH
J.L. GUYONNET	C.R.N.,Strasbourg, France
T. GYS	PPE Division, CERN, Geneva, CH
W. HAO	EP Division, CERN, Geneva, CH
D. HATZIFOTIADOU	EP Division, CERN, Geneva, CH
P. JARRON	ECP Division, CERN, Geneva, CH
S. KHOKHAR	EP Division, CERN, Geneva, CH
V. A.KHOZE	TH Division, CERN, Geneva, CH and
	Leningrad Institute for Nuclear Physics, Gatchina, U.S.S.R.
V. KHOZE	TH Division, CERN, Geneva, CH
W. KRISCHER	ECP Division, CERN, Geneva, CH
V. KUMAR	EP Division, CERN, Geneva, CH
G. LAURENTI	PPE Division, CERN, Geneva, CH
H. LEUTZ	PPE Division, CERN, Geneva, CH
M. MARINO	PPE Division, CERN, Geneva, CH
G. MARCHESINI	University of Parma, Parma, Italy
J.P. MENDIBURU	PPE Division, CERN, Geneva CH and Collège de France, Paris

R. MENG	Argonne National Laboratory, Argonne, IL, USA
P. NEVSKI	PPE Division, CERN, Geneva, CH and
	Moskow Physical Engineering Institute, U.S.S.R.
T. NIINIKOSKI	PPE Division, CERN, Geneva, CH
V. PESKOV	EP Division, CERN, Geneva , CH and
	World Laboratory, Lausanne, CH
L. QITANG	EP Division, CERN, Geneva, CH
A. RACZ	EP Division, CERN, Geneva, CH
F. SAULI	PPE Division, CERN, Geneva, CH
G. SCHULER	DESY, Hamburg, Germany
J. SEGUINOT	PPE Division, CERN, Geneva , CH and Collège de France, Paris
R. SEHGAL	EP Division, CERN, Geneva, CH
P. SHARP	Rutherford Appleton Laboratory, Chilton, Didcot, Oxon , GB
K. SMITH	University of Glasgow, Glasgow , Great Britain
G. SUSINNO	PPE Division, CERN, Geneva , CH and
	University of Calabria, Cosenza, Italy
Y.Y. WANG	EP Division, CERN, Geneva, CH
C. WILLIAMS	PPE Division, CERN, Geneva, CH
C. YE	EP Division, CERN, Geneva, CH
T. YPSILANTIS	PPE Division, CERN, Geneva, CH and Collège de France, Paris
L. XIAGUANG	CCAST/World Lab, Peking, CHINA
M. ZEYREK	PPE Division, CERN, Geneva , CH and
	Middle East Technical University, Ankara, Turkey
Q. ZHAOMING	CCAST/World Lab, Peking, CHINA
A. ZICHICHI	PPE Division, CERN, Geneva, CH

INDEX

ADAMO, see Aleph Data Model
Aleph Data Model, 317, 335, 344
Altarelli-Parisi
   equation, 293, 295, 307
   splitting function, 292
AME, see Analog Memory Elements
Analog Memory Elements, 109
Angular ordering, 294
Antenna patterns, 280-281
Argon, 20, 24, 76, 164
Auger
   effect, 80
   spectrometry, 75
Avalanches chambers, 38-39

Barium fluoride, 24-47
Baryon number violation, 301-305
Boron Nitride, 17, 20
Branching algorithm
   coherent, 287
Bunching principle, 187-188

CAB, see COSMOS Application Builder
CAESAR program, 317
Calcium fluoride, 51
Calorimeter, 117
   cluster analysis, 348-349
   electromagnetic, 9-28, 29-46
   hadronic, 40, 42
   lead-glass, 79
   spaghetti, 57-63, 85
   trigger, 370-374
   warm-liquid, 65-76
CCD, see Charged Coupled Devices
Central tracking, 173
Čerenkov counters, 79
Charge collection efficiency, 25
Charge trapping, 202
Charged Coupled Devices, 87-88, 93-95
   107, 191,194-195, 238
Coated Cathode Conductive Layer
   Chamber, 216-222
COCA COLA see Coated Cathode
   Conductive Layer Chamber

Coherence,
   intrajet, 264
   interjet, 289,293
Computing, 323-333
Contrast Transfer Function, 190
COSMOS Project, 307-321
COSMOS Application Builder, 317
Cesium iodide, 31, 54, 95
   photocatode, 15, 16, 36-40, 54
   photodiodes, 11, 21, 57
CTF, see Contrast Transfer Function

Data acquisition, 85-105, 361-384
Deep inelastic scattering, 288, 292
Delay tube, 185
δ ray, 26
Drag phenomena
   collective, 280
Drell-Yan process, 288, 292, 317
Drift chambers, 65
Drift velocity
   of electrons, 9-10

Electron identification, 77-84
ELOISATRON, see Eurasiatic Long
           Intersecting Storage Accelerator
EF, see Ethylferrocene
Energy-multiplicity correlation, 291
Entity-relationship model, 336-345
E_6 Grand Unification Theory, 168
Ethylferrocene, 31-33, 35
EURODEC program, 317
EUROJET program, 307
Eurasiatic Long Intersecting Storage
      Accelerator, 1, 3, 5-7, 9-10, 15,
      27, 29, 40, 42, 77, 153, 165,
      267, 287

Faraday cage, 17
Förster transitions, 176-177
Forward-backward asymmetry, 168
Frisch grid, 10-11
Front-end electronics, 105-123, 376-380

Rapidity gaps, 298
RD6 collaboration, 79

Scintillation, 9-28
Scintillators, 44, 46, 49-55, 202
Scintillating fibres, 173-183, 185-195
SCSI, see Small Computer System
    Interface
Set Up Descriptor, 317, 335-345
Signal processing, 361-384
Silicon, 11, 30, 57
   detectors, 174, 195, 199
   microstrip detectors, 219, 252
   photocatode, 15
   strip photodiode, 15-17
Small Computer System Interface, 325-
    326
Small-x region, 293, 295
SPACAL see Spaghetti calorimeter
Spaghetti calorimeter, 57-63, 85
SSC, see Superconducting Supercollider
Standard Model, 2
Station farms, 328
Straw chambers, 79, 84
Straw tubes, 79
Stokes shifts, 173, 176-177
Structure functions, 297, 308
SUD, see Set Up Descriptor
Sudakov form factor, 288, 293, 295
Superconducting Strips detectors, 223
Superconducting Supercollider, 1, 2, 10,
    15, 27, 29, 40, 42, 77, 270,
    301-302, 374

Synchrotrons, 3

TEA, 31-32
Tetrakis (dimethylamine) ethylene, 24-47
Tetramethyl germanium, 65
Tetramethyl silane, 65-76
Tetramethyl pentane, 65
TMAE, see Tetrakis (dimethylamine)
    ethylene
TMG, see Tetramethyl germanium
TMP, see Tetramethyl pentane
TMS, see Tetramethyl silane
Top quark, 167, 173, 181-182
Transition Radiation Detectors, 77-84
Track finder, 348
TRD, see Transition Radiation Detectors
Trigger, 12, 77, 85-105, 137, 166, 173-
    174, 185, 202, 361-385

Vapour phase epitaxy, 198
VEGAS program, 317
Vertices
   primary, 173
   secondary, 173
Viton O rings, 17, 20

WALIC collaboration, 65

Xenon, 9-28
X rays, 49, 54, 78, 211, 239